HIGHLIGHTS OF ASTRONOMY

INTERNATIONAL ASTRONOMICAL UNION
UNION ASTRONOMIQUE INTERNATIONALE

HIGHLIGHTS OF ASTRONOMY

VOLUME 4

PART II

AS PRESENTED AT THE XVIth GENERAL ASSEMBLY

1976

EDITED BY

EDITH A. MÜLLER

(General Secretary of the Union)

D. REIDEL PUBLISHING COMPANY

DORDRECHT-HOLLAND/BOSTON-U.S.A.

1977

Library of Congress Cataloging in Publication Data

Highlights of Astronomy as presented at the General Assembly
    of the IAU.
International Astronomical Union.
    Highlights of Astronomy as presented at the XVIth General
    Assembly of the IAU v. 4 (2 parts)
    1977
    Dordrecht, D. Reidel
    illus.
    Editor: 1967, L. Perek; 1970, C. de Jager; 1973,
        G. Contopoulos; 1977, E. Müller.
    1. Astronomy—Congresses. I. Perek, Lubos, ed.
    II. Jager, Cornelis de, ed. III. Title.
QB51.I 57   520   71-159657
ISBN 90-277-0849-5 (part I); 90-277-0830-4 (pbk.)
     90-277-0850-9 (part II); 90-277-0832-0 (pbk.)

Published on behalf of
the International Astronomical Union
by
D. Reidel Publishing Company, P.O. Box 17, Dordrecht, Holland

All Rights Reserved
Copyright © 1977 by the International Astronomical Union

Sold and distributed in the U.S.A., Canada, and Mexico
by D. Reidel Publishing Company, Inc.
Lincoln Building, 160 Old Derby Street, Hingham,
Mass. 02043, U.S.A.

No part of the material protected by this copyright notice may be reproduced or
utilized in any form or by any means, electronic or mechanical, including
photocopying, recording or by any informational storage and retrieval system,
without written permission from the publisher

Printed in The Netherlands

TABLE OF CONTENTS

PREFACE VII

JOINT DISCUSSIONS

1. Galactic Structure in the Direction of the Polar Caps
   (edited by M.F. McCarthy and A.G.D. Philip) 3

   CONTENTS 5

5. Stellar Atmospheres as Indicator and Factor of Stellar Evolution
   (edited by R. Cayrel) 99

   CONTENTS 101

6. The Small Scale Structure of Solar Magnetic Fields
   (edited by F.L. Deubner) 219

   CONTENTS 221

7. The Impact of Ultraviolet Observations on Spectral Classification
   (edited by L. Houziaux) 277

   CONTENTS 279

JOINT MEETINGS

Observational Evidence of the Heterogeneities of the Stellar Surfaces
   (edited by M. Hack and J.P. Swings) 373

   CONTENTS 375

CONTENTS OF VOL. 4 - PART I

INVITED DISCOURSES

J.-C. PECKER / L'Astronomie Infrarouge et les Poussières Galactiques

P. MORRISON / Astronomy and the Laws of Physics

C. SAGAN / Exploration of the Planets

JOINT DISCUSSIONS

2. X-Ray Binaries and Compact Objects
   (edited by E.P.J. van den Heuvel)

3. Space Missions to the Moon and Planets
   (edited by S.K. Runcorn)

4. Clusters of Galaxies, Cosmology and Intergalactic Matter
   (edited by M.S. Longair and J.M. Riley)

JOINT MEETINGS

Prospects in Space Astrometry (edited by P. Lacroute)

PREFACE

It has become a tradition in the Union to publish the Invited Discourses and the Proceedings of the Joint Discussions held at a General Assembly in a separate volume entitled HIGHLIGHTS OF ASTRONOMY. This is the fourth volume of its kind and it contains some of the scientific highlights of the Grenoble General Assembly.

In order to reduce its size it was decided to publish its content in two separate parts. The part (I) contains the full texts of the Invited Discourses given by Prof. J.-C. Pecker and by Dr. C. Sagan, and an abstract of Dr. P. Morrison's paper, thus complying with his wish to forego publication of the full text of his Discourse. Furthermore it collects the proceedings of three Joint Discussions and one Joint Meeting all of which are related essentially to observations from space, to external galaxies and to cosmology. Part (II) contains the proceedings of the four Joint Discussions and one Joint Meeting related essentially to stars and the structure of our Galaxy.

Clearly Volume 4 (parts I and II) of the Highlights reflects only a part of the scientific activities which took place at the Grenoble General Assembly. Many more important papers and discussions were held during Commission meetings and joint meetings. They may be found in the Commissions' reports published in the Transactions Vol. XVI B, 1977. Furthermore, the proceedings of three joint Commission meetings on the topics "CNO Isotopes", Supernovae", and "Topics in Interstellar Matter" are being published in separate volumes by the D. Reidel Publishing Company, and the discussion of Commission 10 on "How can solar flares be understood?" will be published in an issue of Solar Physics.

The new publication policy of the IAU to use for its publications the method of offset printing from camera-ready manuscripts requires a special effort from all authors and from the editors of the individual Joint Discussions and Joint Meetings. My sincere thanks are due to all those who made this special effort. I also wish to thank Mme. R. Bertschi and Mme. R. Läubli for their fine and most helpful cooperation in editing this volume.

Edith A. Müller
General Secretary

JOINT DISCUSSIONS

JOINT DISCUSSION NO.1

GALACTIC STRUCTURE IN THE DIRECTION OF THE POLAR CAPS

(Edited by M. F. McCarthy and A.G.D. Philip)

Organizing Committee

M. F. McCarthy (Chairman), R. Bartaya, J. Delhaye, T. Elvius,
C. Jaschek, D. H. P. Jones, I. King, L. Martinet, T. Walraven.

C O N T E N T S

1. GALACTIC STRUCTURE IN THE DIRECTION OF THE POLAR CAPS

M.F. McCARTHY, S.J. / Introductory Remarks   9

PART ONE

    Galactic Structure at Lower z Distances - Space Densities and Motions - The Solar Neighbourhood - M Dwarfs - Missing Mass
        <u>Chairman</u>: K.Aa. Strand

W. GLIESE / M Dwarfs at Lower z Distances   11

C. JASCHEK and B. HAUCK / Available Stellar Data in Galactic Polar Areas   21

W. GLIESE, C. JASCHEK, and M.F. McCARTHY / Bibliography on Galactic Structure in the Direction of Polar Caps   23

P.S. THÉ / M Dwarfs and the Missing Mass   25

D.H.P. JONES / Some Research Programmes into Galactic Structure at the Galactic Caps Under Way at the Royal Greenwich Observatory   27

P. PESCH / The Space Density of M Dwarfs - An Observational Program   29

D. WEISTROP / The Luminosity Function of Late-Type Main-Sequence Stars in the Direction of the North Galactic Cap   31

M. JOEVEER and J. EINASTO / Galactic Mass Density in the Vicinity of the Sun   33

N. SANDULEAK / The Frequency of Faint M Giant Stars at High Galactic Latitudes   35

GENERAL DISCUSSION OF TOPICS IN PART ONE   37

PART TWO

Galactic Structure at Higher z Distances - Space Densities and Motions - Different Stellar Populations and Chemical Abundances

Chairman: T. Elvius

I.R. KING / The Stellar Distribution Above the Galactic Plane: An Introduction — 41

E.K. KHARADZE and R.A. BARTAYA / Some Results of Classification of Stars in Kapteyn Areas Applied to Galactic Structure in NGP — 49

A. BLAAUW / Mean Chemical Abundance of the F Stars as a Function of Distance from the Galactic Plane — 51

R.P. FENKART and U.W. STEINLIN / RGU Three-Colour-Photometry Towards the NGP: Halo-to-Disk Mass Ratio

M. GRENON / Density Law, Vertical Distribution and Vertical Gradient of Metal Abundances for G and K Giants — 55

T. MARKKANEN / Polarization Measurements and Extinction Near the NGP — 57

K. Aa. STRAND, C.C. DAHN, and R.S. HARRINGTON / Parallaxes of Selected Stars Near the NGP — 59

C.A. MURRAY / Progress Report on a Search for Parallax Stars in the Region of the South Galactic Cap — 61

A. FLORSCH / Search for Large Radial Velocities in the Direction of NGP Using the Fehrenbach Techniques — 63

R.F. GRIFFIN / The Gravitational Field of the Galaxy in the z Direction — 65

R.G. KRON, LIANG-TAI GEORGE CHIU, and K.O. BROOKS / Berkeley Studies of Faint Stars at High Latitudes — 67

T. OJA / Swedish Programmes Concerning the z Distribution of Stars — 69

R.W. HILDITCH and G. HILL / Studies of A and F Stars in the Region of the NGP — 71

A.G.D. PHILIP / The Distribution of Field Horizontal-Branch Stars in the Galactic Halo — 73

A.R. UPGREN / A New Program to Determine K (z) from Main Sequence Stars — 75

CONTENTS

J. STOCK, W. OSBORN, and A.R. UPGREN / Objective-Prism Radial Velocities at High Latitudes — 77

C. TURON-LACARRIEU / Discussion of the Calculation of the Density Law (Dz) — 79

A.N. BALAKIREV / Motions of Near-Polar K-Giants Along the z Coordinate — 83

GENERAL DISCUSSION OF TOPICS IN PART TWO — 87

PART THREE

Summary and Conclusions - Present Results and Future Problems for Study

Chairman: B.J. Bok

W.J. LUYTEN / The Role of Proper Motions in Determining the Luminosity and Density Functions — 89

A. BLAAUW / Concluding Remarks Part Two — 95

M. McCARTHY, A.G.D. PHILIP, I. KING, and U. STEINLIN / Current Results and Suggestions for Future Work — 97

INTRODUCTORY REMARKS

           Martin F. McCarthy S.J.
           Vatican Observatory, Vatican City

After urging you all to briefness, I should set a good example. I welcome you all: participating colleagues and all IAU members and guests, especially those from our host commissions 25, 33 and 45. I thank you all: the contributors to today's events, to the members of the Organizing Committee, to the Presidents of the sponsoring Commissions, IAU officials and all those who are not here today but have helped to make the arrangements for today's discussion.

To assist with the direction of our work I have asked Prof. K. Aa Strand, Prof. T. Elvius and Prof. B. Bok to preside over Parts One, Two and Three. Questions and comments may be made to the invited speakers and to the discussion contributors at the end of each Part of our program.

We shall be considering the different stellar populations encountered as one moves away from the galactic plane towards the polar caps, namely the disc and the halo populations. We know that at our eccentric position in the galaxy this disc has a thickness of some 1000 parsecs while the galactic halo extends several thousands of parsecs. The research work discussed today will help us to discern and distinguish these stellar populations.

Basically there are two problems under discussion. One involves the "missing mass" especially in the solar neighborhood. And while we are occupied with this problem so close to home we shall recall that this problem is not limited to our own galaxy, as the rotational curves of other galactic systems have demonstrated. This problem of "missing mass" even occurs in clusters of galaxies, as studies of the Coma group have indicated. So we are not alone with our problem.

The second problem concerns the variation in the luminosity function and the different stellar populations and different chemical abundances encountered at higher z distances from the plane of the galaxy.

We should not be surprised that the luminosity function may not be the same as that found at $z = 10$ pcs. The photographs of nearby galaxies, the pioneering studies of Oort and later by Bok and MacRae and later still by McCuskey and by Luyten have indicated this for our own galactic system. Evidence of how and why all this occurred is still to be explored and we are still in the early stages of studies of galactic structure and evolution. Now, *"Ad laborem"* and again my thanks.

M DWARFS AT LOWER z DISTANCES

W. Gliese
Astronomisches Rechen-Institut, Heidelberg, F.R.G.

Twelve years ago, during the International Summer Course at Lagonissi, Bart Bok (1966) presented a paper "Desiderata for Future Galactic Research". He pointed to the importance of studies in the galactic polar caps which would allow us, as he hoped, to determine some basic data for galactic structure.

The significance of his statements was demonstrated already in the following years when the gradual increase of our knowledge of the objects in the solar neighbourhood was suddenly interrupted by new observational data from which diverging conclusions have been drawn. Never before have such controversial opinions on problems in the vicinity of the sun been so eagerly discussed: A formerly unknown numerous population of low-velocity red dwarfs, the mass density resulting from stars three times that formerly thought, a remarkably increased luminosity function - yes or no? Obviously, the controversies have smoothed down again during the last months. But what is the situation after this troubled period? Which problems are still unsolved? Which promising programmes will be proposed? I think this Joint Discussion is justified in 1976!

From studies of stars at high galactic latitudes, the stellar density distribution perpendicular to the galactic plane can be determined. Furthermore, such investigations yield the possibility of deriving the luminosity function for intrinsically faint stars near the sun. In these areas, the number of contaminating distant giants is at a minimum.

Bart Bok (1966) put special significance on the value of the average density of matter in the galactic plane near the sun. At that time Oort's value (1965) of 0.15 solar masses per cubic parsec was accepted. As the total mass of known stars came out to about 0.06 solar masses and the density of interstellar matter was estimated to about 0.02 or 0.03 solar masses, at least 0.06 solar masses per cubic parsec remained unaccounted for. That is the well-known "problem of the missing mass" in the solar neighbourhood!

TABLE I

Observations of a large space density of M dwarfs in the directions of the galactic poles

Author(s)		NGP N SGP S	Squ. degr.	Nos. of stars	Objects	Observed quantities	x	Results
Sanduleak	1964	N	120	1261	M stars, $12<V<17$	Colours V-I	3	
Pesch, Sanduleak	1970	N			Sanduleak stars	Proper motions		Low velocities
Weistrop	1972	N	13.5	13820	$12<V<18$	B,V photogr.	5-10	$10<M_v<13$ (15)
Murray, Sanduleak	1972	N		21	Sanduleak stars $m_v = 13-16.5$	Proper motions	5	Low velocities
Pesch	1972	N		21	Sanduleak stars small μ, $9<V<15$	dM, spect. types U,B,V photoel.	2	Low velocities
Sanduleak	1976	N	190	273	Sanduleak stars M3 and later	crude sp. types crude m_R		
Gliese	1972	S	550	75	very red stars $m_{pg} = 15.0-17.3$	μ, colour class (Giclas et al.)	>1	"normal" velocity dispersion
D.H.P. Jones	1973	S	165	22	McCarthy stars (1964); $11<m_v<15$	Colours	~1	"normal" velocity dispersion
Thé, Staller	1974	S	238	96	Proper motion stars, M2-M4	Spectral types μ, m_{pg}	1-1.5	"normal" velocity dispersion
Dolan	1975	S			Thé-Staller stars		?	Space density uncertain

Does this mass come from stars of very low luminosity, from a large percentage of faint red degenerates, from dark stars, from faint companions or unseen companions, from molecular hydrogen? Have we to make a marked correction to the luminosity function?

Different possibilities have been studied without discovering the "missing mass". However, a very new development was initiated in 1964 by Sanduleak who detected over 1200 faint red stars in a 120 square degree area in the north galactic polar region. The low dispersion plates did not permit a luminosity class separation. But, assuming that the great majority of these objects are dwarfs, the total space density of M dwarfs was estimated to be nearly three times the number previously thought to exist.

Luyten's luminosity function (1968), derived from the frequency of stars with large proper motions, contradicted this assumption. However, Weistrop (1972a, 1972b) also detected large numbers of faint red stars near the NGP and, from star counts as a function of colour and apparent magnitude, she derived a luminosity function of M dwarfs several times larger than Luyten's function, at least between $+10 < M_V < +13$ or even $M_V < +15$.

Further investigations seemed to confirm the existence of a numerous population of low-velocity M dwarfs in the solar neighbourhood hitherto unknown. Table I shows the essential data of these publications. Special attention should be given to column "x" which states whether or not the resulting space density of red dwarfs is higher than previously thought: the resulting space density is about x times that given by Luyten's luminosity function.

From the observations cited in Table I, the following conclusions were drawn: In the luminosity range $+10 < M_V < +15$ a population of red low-velocity stars exists in the solar neighbourhood which is derived to be two to ten times as numerous as given by Luyten's luminosity function. Such objects with small proper motions could not be detected by Luyten and by Giclas in the extensive searches for faint stars with annual proper motions exceeding 0.''2. They seemed to be highly concentrated to the galactic plane in a thin layer similar to the interstellar gas and the B stars. Therefore, the total mass density in the vicinity of the sun would not be $0.15 \, M_\odot \, pc^{-3}$ but even $0.21 \, M_\odot \, pc^{-3}$ (Oort, 1965). Nevertheless, Weistrop (1972b, 1974) and Veeder (1974) showed that the existence of this M-dwarf population seemed to yield the "missing mass". About two thirds of the total mass appeared to be due to the M dwarfs.

Near the SGP, since Luyten (1960) no programme comparable to Sanduleak's and to Weistrop's investigations has been carried out. The mass density derived from the few data seemed to be somewhat lower than in the NGP regions.

The new results implied either that most of the mass in the disk component formed into stars only recently or that stellar velocity dispersions are not closely correlated with age (Biermann, 1974; Biermann and Tinsley, 1974). The assumption that the existence of this population had solved the problem of the missing mass created new problems related to the stability of the galactic disk (M. Schmidt, 1975).

From the beginning of these discussions the new results were called in question by Luyten. The discrepancy between his luminosity function and the frequency of red dwarfs assumed from the observations in the vicinity of the NGP was too amazing. Weistrop's investigations (1972a, 1972b) were the most convincing evidence for the overabundance of M dwarfs. These studies were virtually based on colour measurements. But even if red stars are classified decidedly as dwarfs, moderate errors in spectral type or colours will produce fairly large errors in the derived absolute magnitudes and distances. Among red dwarfs an error $\Delta(B-V) = +0^m.1$ corresponds roughly to $\Delta M_V = +1.3$ which reduces the derived distance to 55 per cent. Therefore, systematic errors in the observed quantities can be responsible for wrong values of the supposed star density.

The colour system used by Weistrop (1972a, 1972b) has been challenged by Luyten (1974) and by Jôeveer (1975). Finally, Weistrop herself (1975, 1976a) and Faber et al.(1976) could show that the original (B-V) values actually had been determined too red. It appears now that Weistrop's observations are compatible with the previously assumed data.

In further investigations the results obtained by Sanduleak, Murray/Sanduleak, Pesch and Gliese also have been questioned or corrected (Gliese, 1974; Jôeveer, 1974; Koo and Kron, 1975; Weistrop 1976b, 1976c).

Have we now returned to the status of 1964 before these discussions started? Have the papers cited in Table I proved to be useless? I do not think so - numerous observational programmes have been initiated by these controversies.

We have today a fairly complete knowledge of the objects with large proper motions in the galactic polar caps. On plate pairs taken with the Forty-eight Inch Schmidt-Telescope, Luyten detected many thousands of red stars with $\mu > 0".179$ annually between $m_{pg} = 13$ and 21. The SGP catalogue (Luyten and La Bonte, 1973) and the NGP catalogue (Luyten, 1976) also give crude apparent magnitudes m_{pg} and m_R and estimated colour classes (k, k-m, m). Near the SGP Giclas et al. (1972, 1973, 1975) have listed several hundred "very red objects" (mostly colour class "+4"), the vast majority with proper motions exceeding $0".06$ annually.

Proper motions, magnitudes and colours of so many red stars - could these data be a basis for solving our problems? The large number of objects in these proper motion surveys did not permit more than crude estimates of magnitudes and colours which may vary systematically among different plate pairs and among different parts of the sky.

TABLE II

Mean differences between the m_{pg} estimated by Luyten or by Giclas and the photoelectric B measured by Eggen near the SGP

Luyten and La Bonte (1973)			Giclas et al. (1972)		
m_{pg}(LP)	N	m_{pg}-B	m_{pg}(G)	N	m_{pg}-B
10.0 - 12.9	42	-0.01	< 13	82	+0.82
13.0 - 14.9	61	-0.20	13.0 - 15.9	79	+0.39
15.0 - 16.5	34	-0.54	16.0 - 17	18	+0.11

Within $10°$ of the SGP, these estimates have been compared with B and (B-V) data from Eggen (1976a, 1976b) who had measured the (UBVRI) photometry of the known M dwarfs brighter than V = 15 and with proper motions exceeding 0".096 annually. Table II does not show only systematic differences up to $0^m.8$ among the three sets of magnitudes but also significant scale differences. The (B-V, colour class) diagram of Fig. 1 gives evidence of the dispersions among the B-V of the same colour class

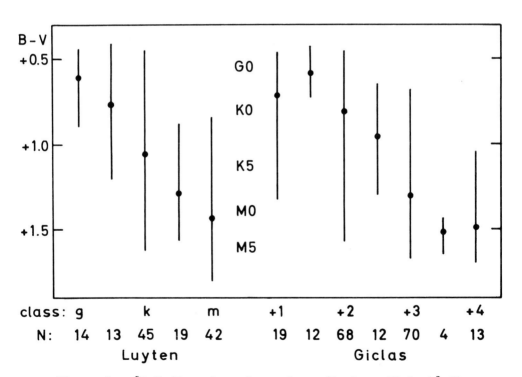

Figure 1. [B-V (Eggen), colour class (Luyten, Giclas)] diagram. Maximum, minimum, mean (B-V), and numbers N of stars for each colour class.

TABLE III

Observed (O) and computed (LM) numbers of red proper motion stars within 10° of the South Galactic Pole

B = 12.26 - 14.25

Parallax	0".10 ≤ μ < 0".20		0".20 ≤ μ < 0".30		0".30 ≤ μ	
	O	LM	O	LM	O	LM
> 0".095					1	1
0".051-0".095	1		1			3
.039 .050	1	1		1		3
.029 .038	1	3	1	2	3	4
.022 .028	2	8	3	5		4
.016 .021	4	29	7	10	1	6
.010 .015	13	54	15	10	1	3
.005 .009	5	18		0.5		
Sum	27	113	27	28.	6	24

B = 13.96 - 15.95

Parallax	O	LM	O	LM	O	LM
> 0".095						1
0".051-0".095	1	1	1		1	4
.039 .050		1		1	1	4
.029 .038	1	4	1	3	8	5
.022 .028	5	10	7	6	5	6
.016 .021	18	39	23	13	5	8
.010 .015	28	119	13	21	7	6
< 0".010	26	236	8	2	1	
Sum	79	410	52	47	28	34

and of the remarkable overlap of the adjacent classes (see also Gliese, 1976; Table II and Table III). Unfortunately, there are still no photoelectric data available for comparisons with the estimated magnitudes and colours of stars fainter than m_{pg} = 17.

We have made several attempts to compare the observed numbers of red stars with large proper motions with numbers computed on the basis of assumed luminosity functions and velocity distributions. The results are not very encouraging since we do not know the real velocity distribution of the low-luminosity stars and since the observed magnitude scales and colours may have systematic errors (Gliese, 1976). Certainly, it has been shown that Luyten's numbers of very faint proper motion stars detected in both galactic polar caps are in contradiction to a dense population of intrinsically faint low-velocity stars. But, on the other hand, it has not been possible to confirm definitely Luyten's luminosity function in this way.

Many of our difficulties will become evident when, finally, I show you in Table III some results from Eggen's data (1976a, 1976b) which he had measured in a 315 square degree area (= 0.76 per cent of the whole sphere) at the SGP. In two magnitude ranges $12.26 \leq B \leq 14.25$ and $13.96 \leq B \leq 15.95$ the numbers of red objects with large proper motions are given in different distances. The observed numbers in the columns "O" are based on the photometric parallaxes published by Eggen (1976b). Columns "LM" have been computed with Luyten's luminosity function (1968) and with a velocity distribution like that of the early M dwarfs ("McCormick stars"). The data are subdivided into three proper motion ranges from 0''10 to 0''20, 0''20 to 0''30 and proper motions exceeding 0''30 annually.

There is no agreement between computed and observed numbers of the objects with smaller proper motions. We may suppose that this discrepancy is due to incomplete knowledge of the stars with $\mu < 0''20$, but we cannot yet prove it. Significant numbers of members of a low-velocity population would appear only in distances nearer than 25 pc. Eggen stated that no support is found for the recently suggested superabundance of low-velocity M dwarfs near the sun.

But we are looking for the true luminosity function in the solar neighbourhood. Doe these numbers confirm the Luyten-McCormick model? In the immediate vicinity of the sun they do not! Only a few proper motion stars have been found nearer than 26 pc which means in a volume of about 500 pc^3. In larger distances, the differences "O-LM" fluctuate and it appears to be impossible to verify the observed data exactly by a simple model of one luminosity function with one uniform velocity distribution for stars of all luminosities and for different distances from the galactic plane.

Most of the stars are nearer than 100 pc to the galactic plane, but a uniform density distribution should not be assumed for all distances. On the other hand, with increasing z, an increasing velocity dispersion can be expected. Furthermore, we know of the correlation between velocities and ages (Wielen, 1974), but what do we know of the intensities of Ca II emission as indicators of the ages of these dwarfs? Last, not least, uncertainties of the photometric parallaxes can introduce serious biases into the data in Table III. Random and systematic errors must be considered. All these effects need a careful analysis. But the statistically low numbers of objects in our samples leave some doubt whether or not such a detailed investigation can solve the problems.

Nevertheless, Eggen's set of data very clearly shows which observations are required additionally.

Much progress would be made by the measurement of reliable trigonometric parallaxes of an unbiased sample of faint red stars in both polar caps. With eager expectation we are looking forward to the papers by Murray and by Strand. Sanduleak's new catalogue of 273 faint stars of spectral types M3 and later near the NGP (1976) presents such a sample of objects recommendable for further investigations. Luyten (1976b), already has shown that there are stars among them without noticeable proper motions.

It is desirable to determine spectroscopic and photometric data for a large number of faint red objects in both galactic polar regions to supplement the knowledge of their proper motions. I think that D.H.P. Jones and Donna Weistrop will report on further studies of objects at the faint end of the luminosity function near the NGP. Pesch will report on a survey for M dwarfs in a region of the SGP to parallel Sanduleak's work in the north. Furthermore, I have been informed that Warren is measuring the U, B, V data of the very red Giclas stars in the south.

The non-existence of the numerous population of low-velocity red dwarfs seems to be confirmed. But what do we know about the real velocity distribution of the low-luminosity red stars? What is their space density? What is the luminosity function of M dwarfs intrinsically fainter than $M_v = 13$? What do we know about the frequency of late degenerates? How shall we solve the problem of the missing mass now? By assuming a lower value for the total mass density, as our colleagues from Tartu do? Let us begin our Joint Discussion!

Acknowledgement: The English version of the text was graciously corrected by Dr. R. Scholl.

References

Biermann, P.: 1974, *Astron. Astrophys.* 30, 31.
Biermann, P. and Tinsley, B.M.: 1974, *Astron. Astrophys.* 30, 1.
Bok, B.J.: 1965, in A. Blaauw and L.N. Mavridis (eds.), *Observational Aspects of Galactic Structure, Int. Summer Course (Lagonissi 1964)*, Athens, Paper XXIX.
Dolan, J.F.: 1975, *Astron. Astrophys.* 39, 463.
Eggen, O.J.: 1976a, *Astrophys. J. Suppl.* 30, 351.
Eggen, O.J.: 1976b, *Astrophys. J.* 204, 101.
Faber, S.M., Burstein, D., Tinsley, B.M. and King, I.R.: 1976, *Astron. J.* 81, 45.

Giclas, H.L., Burnham, Jr., R. and Thomas, N.G.: 1972, *Lowell Obs. Bull.* (No. 158) 7, 217.
Giclas, H.L., Burnham, Jr., R. and Thomas, N.G.: 1973, *Lowell Obs. Bull.* (No. 160) 7, 273.
Giclas, H.L., Burnham, Jr., R. and Thomas, N.G.: 1975, *Lowell Obs. Bull.* (No. 162) 8, 9.
Gliese, W.: 1972, *Astron. Astrophys.* 21, 431.
Gliese, W.: 1974, *Astron. Astrophys.* 34, 147.
Gliese, W.: 1976, in E.K. Kharadze (ed.), *Stars and Galaxies from Observational Points of View, Proc. of the 3rd European Astron. Meeting (Tbilisi 1975)*, Abastumani Astrophys. Obs., p. 463.
Jõeveer, M.: 1974, *Tartu Astrofiz. Obs., Teated* 46, 3.
Jõeveer, M.: 1975, *Publ. Tartu Astrofiz. Obs.* 43, 89.
Jones, D.H.P.: 1973, *Monthly Notices Roy. Astron. Soc.* 161, 19P.
Koo, D.C. and Kron, R.G.: 1975, *Publ. Astron. Soc. Pacific* 87, 885.
Luyten, W.J.: 1960, *A Search for Faint Blue Stars*, No. 22. Obs. Univ. Minnesota, Minneapolis, Minn.
Luyten, W.J.: 1968, *Monthly Notices Roy. Astron. Soc.* 139, 221.
Luyten, W.J.: 1974, *The Observatory* 94, 136.
Luyten, W.J.: 1976a *The North Galactic Pole*, Univ. Minnesota, Minneapolis, Minn.
Luyten, W.J.: 1976b, *Proper Motion Survey with the 48-Inch Schmidt Telescope*, No. 46. Univ. Minnesota, Minneapolis, Minn.
Luyten, W.J. and La Bonte, A.E.: 1973, *The South Galactic Pole*, Univ. Minnesota, Minneapolis, Minn.
McCarthy, M.F., Bertiau, F.C. and Treanor, P.J.: 1964, *Ric. Astron. Specola Astron. Vaticana* 6, 571.
Murray, C.A. and Sanduleak, N.: 1972, *Monthly Notices Roy. Astron. Soc.* 157, 273.
Oort, J.H.: 1965, *Stars and Stellar Systems* 5, 455.
Pesch, P.: 1972, *Astrophys. J.* 177, 519.
Pesch, P. and Sanduleak, N.: 1970, in S.W. McCuskey, *Warner and Swasey Observatory Report 1968-1969, Bull. AAS* 2, 151.
Sanduleak, N.: 1964, in S.W. McCuskey, *Warner and Swasey Observatory Report 1963-1964, Astron. J.* 69, 720.
Sanduleak, N.: 1976, 'A Catalogue of Probable Dwarf Stars of Type M3 and Later in the Direction of the North Galactic Pole', *Astron. J.* 81, 350.
Schmidt, M.: 1975, in A. Hayli (ed.), *Dynamics of Stellar Systems, IAU Symp.* 69 (Besançon 1974), p. 325.
Thé, P.S. and Staller, R.F.A.: 1974, *Astron. Astrophys.* 36, 155.
Veeder, G.J.: 1974, *Astrophys. J.* 191, L57.
Weistrop, D.: 1972a, *Astron. J.* 77, 366.
Weistrop, D.: 1972b, *Astron. J.* 77, 849.
Weistrop, D.: 1974, *Astron. J.* 79, 954.
Weistrop, D.: 1975, *Bull. AAS* 7, 411.
Weistrop, D.: 1976a, *Astrophys. J.* 204, 313.
Weistrop, D.: 1976b, *Astron. J.* 81, 427.
Weistrop, D.: 1976c, 'Further Comments on the Evidence for a High Density of red Dwarf Stars' (preprint).

AVAILABLE STELLAR DATA IN GALACTIC POLAR AREAS

C. Jaschek
Observatoire de Strasbourg, France   and
B. Hauck
Institut d'Astronomie de Lausanne, Switzerland

One of the aims of the Strasbourg Stellar Data Centre is to collect and distribute the different stellar data available, so that they can be used for statistical work. As an example of the statistics which can be made, we present in the table some results for the galactic pole areas. The photometric data were obtained from a general file (Magnenat 1976) which lists the stars observed photoelectrically and the system in which observations were made. The results show clearly how fragmentary our knowledge still is; completeness exists at most up to m = 9.

North Galactic pole area (b > $75°$)

	<6	6-7	7-8	8-9	9-10	10-11	11-12	>12	Total
HD	54	97	227	535	301	56	1	-	1 284
IDS	4	10	8	19	19	34	9	12	117
Var.	2	1	3	3	3	3	4	1	23
UBV	52	60	108	167	259	172	55	40	915
RV	53	76	105	112	62	18	7	3	441
MK	49	72	108	220	369	372	101	91	1 496
ST	28	49	49	59	46	23	12	29	296
AG	49	96	229	625	1 020	871	219	15	3 126
Total	49	96	226	622	1 535	3 950			

South Galactic pole area ($|b| > 75°$)

	<6	6-7	7-8	8-9	9-10	10-11	11-12	>12	Total
HD	40	95	219	641	957	210	14	1	2 185
IDS	1	10	21	36	66	29	14	17	197
Var.	1	2	-	3	2	-	1	-	23
UBV	39	53	48	100	89	40	7	15	391
RV	36	39	34	50	27	2	1	1	195
MK	31	42	39	63	61	18	3	-	268
ST	15	12	8	24	34	21	1	10	187
AG C	11	30	85	226	517	283	7	-	1 164

Notes: HD   = stars classified in the HD system
       IDS   =   "   contained in the Index Catalog of Double Stars
       Var. =   "   listed as Variables in Kukarkin et al.
       UBV   =   "   measured in the UBV system (Blanco + Mermilliod)
       RV   =   "   with measured radial velocity (Wilson + Abt)
       MK   =   "   having MK classifications (La Plata + Kennedy)
       ST   =   "   measured in Strömgren's system
       AG,C =   "   with position and proper motion
       Total = number of stars to be expected (Allen)

Reference

Magnenat, P.: 1976, *Information Bulletin*, CDS Strasbourg, No. 11, p. 17.

BIBLIOGRAPHY ON GALACTIC STRUCTURE IN THE DIRECTION OF POLAR CAPS

W. Gliese
Astronomisches Rechen-Institut, Heidelberg, F.R.G.
C. Jaschek
Observatoire de Strasbourg, France   and
M.F. McCarthy
Specola Vaticana, Vatican City State

The notion of preparing a bibliography on the topic of today's Joint Discussion originated at about the same time in Heidelberg, Strasbourg and Castel Gandolfo. Letters and lists were exchanged as we prepared for our work here today; you must judge the results of this draft and help us with your suggestions. New IAU policy prefers that bibliographical materials be limited to *Reports on Astronomy* of the various commissions; hence there is no place in *Highlights of Astronomy* for the full Bibliography. Both the Director of the Vatican Observatory and the Director of the Strasbourg Centre de Données Stellaires have offered to publish the final listing. We are grateful to both and shall accept the Strasbourg offer; it is planned that it will appear in the January 1977 issue of the *Information Bulletin*.

After listing the author, year of publication, an abbreviated publication reference together with Volume and page references and the title of the paper, we have listed, wherever possible, three additional sets of numbers. The first gives the approximate number of stars discussed in the paper; the second, the approximate magnitude range; and the third, the lower limit of the (absolute) latitude of stars contained in the paper. It should be noted that it was not always possible to furnish all these data for each article listed. Still to be added to the present compilation are the Reference Numbers to the Astronomy and Astrophysics Numeration in preparation by Gliese, a full list of bibliographical abbreviations used prepared by Jaschek and an Author Index and a preface written by McCarthy.

Two points of interest may be noted here. One observation is that two independent lists (one by WG and MFM, the other by CJ) were combined; each contained about 120 references; only 20 were common to both lists. Gliese began his original listing with the year 1964; Jaschek started his with 1955; McCarthy began his list with Oort's 1932 paper and there are several lacunae; no list or combination of lists can pretend to completion. For the past 10 years the present bibliography can be considered reasonably complete and references to other and to most earlier published works will be found therein.

Thanks to Dr. A. Florsch and colleagues, about 100 copies of the first version of this bibliographical listing were xeroxed at Strasbourg Observatory immediately before this meeting. We welcome all corrections, additions and suggestions for deletions. There are now 233 entries on 21 pages. We await your comments and suggestions.

M DWARFS AND THE MISSING MASS.

P.S. Thé.
Astronomical Institute, University of Amsterdam.

This is a short report on our study of the M dwarfs in the South Galactic Pole region in connection with the problem of the missing mass and the luminosity function of intrinsically faint stars. The M-type stars found in our objective prism survey with the Lembang Schmidt telescope have been studied in collaboration with Dr. D.H.P. Jones for the discrimination between dwarfs and giants. Furthermore, these stars are all observed photo-electrically in Kron's R,I-system at La Silla, Chile. Results of the work will be published soon. A blink survey has been initiated at Nymegen using copies of red and blue Mt. Palomar Sky survey plates to find faint red stars in the S.G.P.-region.

Kumar has put forward the idea that Oort's missing mass is to be found as a large number of very low mass stars (0.01-0.07 $M_\odot$) in the solar neighbourhood. His idea is backed up by the fact that stars within this mass range have been found by double star observers. Staller suggested that the low velocity M dwarfs found in the directions of the galactic poles represent Kumar's very low mass stars. This proposal is ment not only for giving an answer to the problem of the missing mass but also for those problem connected with the observed small velocity dispersion (10-15 km/sec) of the low velocity M dwarfs. Staller estimated the contracting and cooling times of these stars to be about 5×10^8 to 10^9 years. In the lifetime of our galaxy (about 10^{10} years) 10 to 20 generations of these low mass stars have been created and evolve to very faint degenerated black dwarfs. The mass density of these stars, in the solar neighbourhood, will be large enough to solve the problem of the missing mass. The stars we observe at present as low velocity M stars are genuinely young red stars, and their small velocity dispersion will therefore not contradict Spitzer and Schwarzschild's mechanism for the creation of velocity dispersion by encounters with interstellar clouds. The age of the older very faint black dwarfs is comparable to that of our galaxy. These stars should therefore have a large velocity dispersion. Staller has shown that on the average the low mass stars have a velocity dispersion larger than 21 km/sec, and therefore the whole group of low mass stars is satisfying Toomre's stability criterion for the galactic disk, such as newly derived by Biermann.

SOME RESEARCH PROGRAMMES INTO GALACTIC STRUCTURE AT THE GALACTIC CAPS
UNDER WAY AT THE ROYAL GREENWICH OBSERVATORY

D.H.P. Jones
Royal Greenwich Observatory

The R.G.O. has a continuing programme to investigate both the
density and motions of the stellar content of the solar neighbourhood.
Gaps in current knowledge are wider for the fainter stars which are
found in proportionally greater numbers near the galactic poles. Our
approach divides naturally into two questions: (i) What are the
proportions of red dwarfs to giants at each apparent magnitude? and
(ii) What are the corresponding total numbers of red stars?

Question (i) was first tackled by a narrow-band photometric
search for M dwarfs (Jones 1973). This approach is continuing and a
progress report has been delivered (Jones 1976). To push the search
deeper 41 stars in the South Galactic Cap (mostly fainter than V = 15)
selected by P.S. Thé from a survey with the Lembang Schmidt as being
M2 or later were observed with the Robinson-Wampler Scanner on the 150-
inch A.A.T. The Na D lines, CaH and the band at $\lambda 5500$ tentatively
identified by Pesch with CaOH have been measured. Column density
calculations of CaOH for giants and dwarfs of different temperatures
support Pesch's identification. Nearly all the stars are dwarfs but
only seven appear in proper motion catalogues, (Thé and Staller 1974).
At the galactic caps most stars brighter than V = 12 are giants and
fainter dwarfs. The transition zone is only 10-14.

Question (ii) is being tackled in an observational programme
begun in 1976 March. V and I plates of SA 57 and the North Galactic
Pole field have been taken with the Palomar Schmidt. To date 2 V and
2 I plates of SA 57 have been measured on GALAXY, after a search on an
I plate. Sequences in V, R & I have been set up in both fields with
the 60-inch Infrared Flux Collector on Tenerife, running from 7 to 17
in V and 7 to 16 in I. There were 60 standards available to calibrate
the GALAXY measures of SA 57 and they fitted with r.m.s. errors of
$\pm 0\overset{m}{.}08$ (both V plates) and $\pm 0\overset{m}{.}11$ (both I plates). In 21 square degrees
we found 1590 stars brighter than V = 17 and I = 16 and redder than
V-I = 1.0. (about K4 on the main sequence). Estimating their distances
photometrically (Upgren 1975, Sanduleak 1965) we find 0.055 stars per
cubic parsec. The figure would be materially increased by the

inclusion of one further red star beyond the limit of the V sequence. Of course, this one star has a Poissonian uncertainty of one.

The Oxford Astrographic Zone crosses the North Galactic Cap. Plates covering 64 square degrees and extending over 70 years are being measured for proper motion. In addition repeat plates of the POSS were taken on the same run as the photometric plates and proper motions will be measured for all faint red stars.

All these programmes are team efforts and their results will be published either by the RAS or the RGO.

REFERENCES

Jones, D.H.P.:1973, Monthly Notices Roy. Astron. Soc. 161, 19P.
Jones, D.H.P.:1976, in R.J. Dickens and J. Perry (eds.), "RGO Tercentenary Symposium", R.O. Bull. (in press).
Sanduleak, N.:1965, Ph.D. Thesis, Case Institute of Technology, p.60.
Thé, P.S. and Staller, R.F.A.:1974, Astron. Astrophys. 36, 155.
Upgren, A.R.:1975, in A.G. Davis Philip and D.S. Hayes (eds.), "Multicolor Photometry and the Theoretical HR Diagram". p.441.

THE SPACE DENSITY OF M DWARFS - AN OBSERVATIONAL PROGRAM

Peter Pesch
Case Western Reserve University-Warner and Swasey Observatory

In spite of extensive proper motion surveys, there remains some uncertainty about the space density and the kinematics of late-type dwarf stars. This uncertainty is primarily due to the unknown kinematic bias introduced by the proper motion discovery technique.

A spectrographic search, especially one using an objective prism on a Schmidt telescope, can identify significant numbers of intrinsically (and actually) faint stars, free from any kinematic bias. To find faint and cool stars, one chooses an objective prism with low dispersion and emulsions sensitive to the red and near photographic infrared. At low dispersions, in this wavelength region, there are no reliable luminosity criteria. Thus Sanduleak conducted his survey in the direction of the north galactic pole, with the expectation that very few remote giants would contaminate his results. A catalog of 273 probable dwarf stars of type M3 and later (Sanduleak, N. 1976, A.J., 81, 350) based on his objective prism survey in the direction of the north galactic pole is now available. Thanks to W.J. Luyten (1976, Proper Motion Survey with the Forty-Eight Inch Schmidt Telescope XLVI (Univ. Minnesota, Minneapolis)) proper motions have been measured for all of these stars.

Two independent programs to obtain photoelectric photometry have begun. Thanks to the kindness of the Hale Observatories, Pesch was able to observe 41 (51% of the catalog) stars in V,R,I (Kron System). Most of the 41 stars were of Sanduleak's spectral category b (types M3-M4) and in general, represent the brighter stars (in apparent magnitude) in the catalog. It is to be emphasized that no proper motion criterion was used in selecting these 41 stars. This is in distinction to a previous paper (Pesch, P. 1972, Ap.J., 177, 519) where an even smaller sample of 27 stars from Sanduleak's unpublished thesis was selected for observation on the basis of low proper motion. Absolute visual magnitudes were determined using the observed R-I colors and the calibration of Gliese (Gliese, W. 1969, Low Luminosity stars, edited by S. S. Kumar (Gordon and Breach, New York), page 41). The 41 stars are characterized by median values of 1.1 mag., +11 mag., 40 pc and 33 km sec^{-1} respec-

respectively, for R-I, M_v, distance and tangential velocity (uncorrected for solar motion). The standard deviation about the mean tangential velocity is 30 km sec^{-1}.

Image-tube slit spectra were obtained for 24 of the 41 stars. The spectra cover the wavelength region $\lambda\lambda 3800-7200$ Å at a dispersion of 285 Å mm^{-1}, and were used for classification purposes. With 3 exceptions, the types were dM3-dM5; 4 showed hydrogen emission lines.

Conard C. Dahn at the U. S. Naval Observatory, Flagstaff station, has observed 62 stars from Sanduleak's catalog on the B,V,I system. Dahn selected the later and fainter stars to observe, so his 62 are primarily of spectral groups c and c+ (later than M3-M4), with only 10 stars in common with Pesch's sample of 41. Dahn's 62 stars are characterized by mean values of 0.022 sec of arc, 1.60 mag., 2.61 mag. and 11.8 mag. respectively, for photometric parallax, B-V, V-I and M_v. The dispersions in U and V are 33.8 km sec^{-1} and 24.8 km sec^{-1} respectively.

Although Sanduleak's catalog contains only 1/3 as many stars as the McCormick spectrographic survey (Vyssotsky, A.N. 1963, **Basic Astronomical** Data, edited by K. Aa. Strand Univ. Chicago P., Chicago , page 192) it refers to significantly less luminous stars. Because of the importance of this catalog, we plan to continue the observations until photoelectric photometry is available for all the stars in it. We feel that this is the most satisfactory way of learning what - if any - has been the effect of the kinematic bias on our understanding of the M dwarf population. Until that time we feel that discussions of fragmentary observations are premature.

To augment Sanduleak's sample and to investigate reports of a low observed space density of M stars in the south galactic pole, Pesch undertook a deep objective-prism survey in that direction at CTIO. Because the Curtis Schmidt telescope at CTIO is similar to the Burrell Schmidt of the Warner and Swasey Observatory, these two surveys are nearly identical. A catalog based on the CTIO data is now in preparation. To date it appears that the surface density is comparable and the catalog will contain approximately the same number of faint, late M stars. We also plan to observe these stars photoelectrically so as to obtain accurate magnitude and colors so that distances, etc. can be determined. It should be noted that although there is some overlap with the extensive photometry of Eggen (1976, Ap.J. Suppl., 30, 351), most of the stars in the forthcoming catalog have not been observed photometrically.

THE LUMINOSITY FUNCTION OF LATE-TYPE MAIN-SEQUENCE STARS IN THE DIRECTION OF THE NORTH GALACTIC CAP

Donna Weistrop
Kitt Peak National Observatory

As a result of the recent discussion concerning the luminosity function of late-type main-sequence stars (Weistrop 1976 and references therein), a program of photoelectric photometry of all red stars in a field near the North Galactic Pole was undertaken. The sample is complete for stars redder than (B - V) = 1.40 magnitude for the following apparent magnitude and area limits: V = 12.0-14.0 magnitudes, 13.5 square degrees; V = 14.0-15.0 magnitudes, 3.0 square degrees; V = 15.0-17.5 magnitudes, 1.0 square degree. Observations in BVRI have been obtained for the 44 stars in the sample. Giants and dwarfs are distinguished by their location in the (B - V)-(V - I) diagram or from published proper motion data, where available. The absolute magnitudes of the dwarfs are determined from the M_R-(R - I) relation.

The density distribution perpendicular to the galactic plane of the dwarfs is consistent with the distribution for K giants found by Oort (1960). The derived luminosity function does not differ significantly from that determined by Wielen (1974) for stars close to the Sun. The local space density for stars in the interval M_V = 8.5-14.0 magnitudes is 0.099 stars pc^{-3}. The corresponding stellar density derived from Wielen's luminosity function is >0.066 stars pc^{-3}. Sixty-six percent of the density derived here is contributed by two stars with absolute magnitude in the range M_V = 13.0-14.0 magnitudes.

These conclusions are not significantly altered if all the stars redder than (B - V) = 1.40 magnitude are assumed to be dwarfs.

A complete treatment of the observations and results summarized here will be submitted to THE ASTROPHYSICAL JOURNAL.

REFERENCES

Oort, J.: 1960, *Bull. Astron. Inst. Neth.* 15, 45.
Weistrop, D.: 1976, *Astron. J.* 81, 427.
Wielen, R.: 1974, in G. Contopoulos (ed.), *Highlights of Astronomy*, D. Reidel Publishing Co., Dordrecht, p. 395.

GALACTIC MASS DENSITY IN THE VICINITY OF THE SUN

M. Joeveer and J. Einasto
W. Struve Tartu Astrophysical Observatory

It is possible to estimate the galactic mass density in the solar neighbourhood either directly by summing up the mass densities of individual subsystems of stars and interstellar matter or indirectly from dynamical considerations.

Observational data on the number density of visible stars lead to mutually consistent results on the stellar component of the mass density. The mean of different estimates is $\rho_{stars}=0.052 \pm 0.010\ M_\odot pc^{-3}$. By adding the probable contributions of intrinsically faint undetected objects and of interstellar matter the value $\rho = 0.09 \pm 0.02\ M_\odot pc^{-3}$ has been obtained for the total mass density.

Dynamical estimations have given rather discrepant results ($\rho_{dyn} = 0.05 \div 0.30\ M_\odot pc^{-3}$). The authors who obtained $\rho_{dyn} \gtrsim 0.15\ M_\odot pc^{-3}$ have often interpreted their results as an indication of the presence of a population of unknown objects (missing mass) in the solar vicinity.

More probable explanation of the discrepancy between the direct and dynamical density determinations lies in the fact that the dynamical determinations are subject to large systematical and accidental errors. In the first approximation $\rho_{dyn} \propto (\sigma_z/\sigma_z)^2$, where σ_z is the mean dispersion of the galactovertical velocities and σ_z is the dispersion of the distances z from the galactic plane in a flat steady-state subsystem. The values of ρ_{dyn} are very sensitive to any errors in the accepted values of both dispersions, σ_z and σ_z. The inclusion of few halo or old disk population stars into statistical sample of young stars may significantly increase the velocity dispersion in the sample whereas the space distribution remains practically the same. Hence inhomogeneous statistical samples based on HD or MK spectral classifications lead to exaggerated values of ρ_{dyn}. Among the available determinations of ρ_{dyn} the lowest values are the most trustworthy. These values are in agreement with direct density estimates. Thus so far there is no real dynamical evidence for the presence of significant amount of missing mass in the solar vicinity.

The full text of the contribution is published in Tartu Astron. Obs. Teated NO 54, 1976, 77.

THE FREQUENCY OF FAINT M GIANT STARS AT HIGH GALACTIC LATITUDES

N. Sanduleak
Case Western Reserve University-Warner and Swasey Observatory

Based on the observations of M giant stars in the north galactic polar objective-prism survey of Upgren (1960) and the data summarized by Blanco (1965) the overall space density of all M-type giants as a function of distance from the galactic plane at the position of the sun can be approximated by,

$$\rho(z) = 9.0 \; e^{-3.0z}, \qquad (1)$$

where z is in kpc and $\rho(z)$ is the number of stars per $10^6 pc^3$. This relationship is derived from the observed fall-off in space densities up to a distance of about 2 kpc.

The question arises as to the validity of extrapolation equation (1) to larger z distances so as to predict the number of faint M giants expected per unit area near the galactic poles. Adopting for the M giants a mean visual absolute magnitude of -1.0 (Blanco 1965), one finds that equation (1) predicts that less than one giant fainter than V$\sim$12 should be expected in a region of 200 square degrees. This expectation formed the hypothesis of a thesis study (Sanduleak 1965) in which it was assumed that the very faint M stars detected in a deep, infrared objective-prism survey at the NGP were main-sequence stars, since this could not be ascertained spectroscopically on the very low-dispersion plates used.

The luminosity function for the assumed M dwarfs found in that thesis study showed an excess number of stars compared with other determinations such as that by Luyten (1968). It has been suggested that part of this discrepancy might result from the presence of an appreciable number of giants amongst the fainter stars. Conceivably, this might result if the densities beyond two kiloparsecs declined at a much slower rate than given by equation (1).

Various observers have obtained spectroscopic and photometric data for small samples drawn from this NGP survey. To date these studies have uncovered only three difinite giant stars having V>12.0. Two of these are the Mira-type variables T Com and FQ Com. The variablilty of the latter star was discovered by Kinman et al. (1966) during the course of a blink survey for RR Lyrae stars near the NGP. It was the only faint, new long-period variable detected in that survey which covered an area of 74 square degrees and reached a limiting magnitude of B$\sim$18.5. Although the Kinman et al. survey was not intended nor ideally suited for the detection of long period variables, this discovery rate suggests that the halo giants (at least of the Mira-type) are about as infrequent as would be inferred from equation (1).

The Sanduleak (1965) thesis was never published in detail. However, we recently (Sanduleak 1976) made available a catalog and finding charts for 273 probable dwarf stars of type M3 and later near the NGP. Nearly one-half of these stars, which have apparent magnitudes in the range 12<V<16, were found to have published proper motions sufficiently large to indicate that they are nearby dwarfs. Luyten (1976) has now measured the proper motions of the remaining stars and finds that 14 of them (Nos. 46, 63, 64, 97, 159, 172, 200, 205, 213, 219, 226, 239, 262, and 274) have total proper motions smaller than 0.020. Given the observational errors involved, Luyten notes that the true motions might be close to zero for these stars and suggests that they might all be giants or subgiants. It would, of course, be of great interest if observers with access to sufficiently large telescopes would undertake to investigate this possibility.

However, even if, surprisingly, all of these small proper motion stars proved to be giants, it would set an upper limit of about 5% for the frequency of faint giants in our thesis study. This would be insufficient to account for the excess in our luminosity function which therefore must either be real or the result of observational errors or quite possibly a combination of both factors.

REFERENCES

Blanco, V.M.: 1965, Stars and Stellar Systems, **5**, 241.
Kinman, T.D., Wirtanen, C.A. and Janes, K.A.: 1966, Ap. J. Supp. **13**, 379
Luyten, W.J.: 1968, Monthly Notices Roy. Astron. Soc. **139**, 221.
Luyten, W.J.: 1976, Proper Motion Survey with the 48-Inch Schmidt Telescope XLVI, Univ. of Minnesota Press, Minneapolis.
Sanduleak, N.: 1965, "The Luminosity Function of the M Dwarf Stars", Case Institute of Technology, Cleveland (Ph.D. Thesis).
Sanduleak, N.: 1976, Astron. J. **81**, 350.
Upgren, A.R.: 1960, Astron. J. **65**, 644.

DISCUSSION

Weidemann: The local mass density of white dwarfs as estimated by me in 1967, 0.02 or 0.04 $M_\odot/pc^3$ depended on observation of high latitude blue stars by Sandage and Luyten made that same year; they found one blue white dwarf per square degree down to the eighteenth magnitude. Are there any observational programs in progress or planned which might extend these studies to fainter magnitudes?

King: In a Berkeley program, George Chiu is determining proper motions of all stars in three fields of 1/10 square degree each. The accuracy is high down to the 19th magnitude and not quite as good from there to the 21st magnitude. A preliminary examination shows that some of the stars with large motions are yellow and white in color; there are about a dozen of these per field. Some of them may be halo main sequence stars, farther away but moving faster; but it is likely that something of the order of a dozen stars per field will turn out to be white dwarfs.

Gliese: We know two values for the total mass density near the sun: 0.09 $M_\odot/pc^3$ from Einasto and 0.15 $M_\odot/pc^3$ from Oort's paper in 1965. Can anyone comment on this discrepancy?

McCarthy: Both Oort and Einasto are here at Grenoble; perhaps we can ask them directly.

King: It appears to me that we no longer have large numbers of M dwarfs with low velocities; the space density and the velocity dispersions seem to be back to what we had expected. Yet one outstanding discrepancy remains: the statistical parallax of the Murray-Sanduleak stars, which gives a distance that now appears to be twice too small. Can anyone explain how this happened?

D. Jones: We realized three years ago that the ratio of the reverse solar motion to the dispersion in transverse velocity was erroneous for the Murray-Sanduleak stars. The statistical parallax was found by equating the mean proper motion to the basic solar motion. With the new photometric parallaxes the dispersion in transverse motion is nearly the same as for stars near the sun; the mean proper motion which should equal the reverse solar motion is now anomalous. We do not understand why.

Murray: Concerning the paper of Gliese I should like to speak in defense of so called "small" proper motions. We should remember that at 100 parsecs a velocity of 20 km/sec corresponds to a proper motion of 0".04/yr and any discussion which is limited to proper motions of 0".1 or 0".2 is bound to be very incomplete.

Gliese: The 21 Murray-Sanduleak stars proved to be very important for the problem of the space density of the M dwarfs. Therefore I am asking if photoelectric photometry exists for these stars. Furthermore: will it be possible to determine CaII emission intensities as age indicators?

D. Jones: Photoelectric colours $BVRI$ have now been measured for all these stars by Weistrop and in VRI most of them have been observed by Bingham, Wallis and myself. The two series are in good agreement. Regarding the second question, the CaII emission should be measurable, but I do not believe such observations have been made.

King: I would like to emphasize the importance of correct photometry on a system that has no color-equation errors. It appears that the apparent density of M dwarfs is closely correlated with the reliability of the photometry. When the magnitudes and colors are correct, the excessive density of M dwarfs goes away. I should like to ask Dr. Weistrop if it is correct that the large reduction in her density determination for M dwarfs is due completely to a change in her photometry of these stars?

Weistrop: Yes, the photographic sample was the group for which photoelectric data were obtained. Some of the stars in the photographic sample were eliminated since they turned out to be bluer than $B-V = 1.40$ but you are essentially correct.

Strand: In reference to a remark by King that the photometry of the red dwarfs was poor for the purpose of determining photometric parallax he might have referred to the $B-V$ photometry which is not suitable for these stars, whereas $V-R$ or $V-I$ photometry is suitable.

Bok: We should use the PDS equipment for establishing sequences to $V = 21$ and to similar faint limits in R and I. I nominate SA 57 for northern work and SA 141 for southern work. I hope that in each of these areas for a field of 16 square degrees, more or less, good standard sequences can be established by combined photoelectric to $V = 16$ and photographic PDS techniques to $V = 21$ techniques. I hope too that machine color-magnitude counts can be made in these fields. We simply lack good faint counts for the polar caps. This must be remedied promptly.

King: In response to Bok's recommendation, I wish to point out that the presentation this afternoon by Kron will refer to some detailed work of the type that you advocate.

McCarthy: The fine illustrations of the effective use by Jones of luminosity criteria for M dwarfs by measuring the strengths of sodium and calcium hydride features reminds us that the first to use sodium as a luminosity criterium was Luyten and the one who first used calcium hydride was Ohman. Both of them are with us today and must be happy to see their findings so splendidly employed. I have two questions for D. Jones: how late in spectral type does your M dwarfs survey extend? and second, what has been the limiting V magnitude reached in your work?

D. Jones: To answer your first question: about dM5. Regarding the second question, I note that our list was supplied by Thè who believes that these stars are all fainter than $V = 15$. Staller is currently observing these stars photoelectrically in R and I. I should add that only 20% can be found in the best proper motion survey of Luyten and Giclas.

McCarthy: I wish Dr. Wing could tell us of the additional spectral criteria which he and Ford discovered near 9900Å in the spectra of dwarf M stars. Have subsequent observations confirmed this and is anything more known of its identification?

Wing: There is a feature at 9900Å in the spectra of cool M dwarfs found by Ford and myself in 1969. It has the property that it is quite strong in the latest M dwarfs, much weaker in the earlier M dwarfs and not at all visible in the M giants. Although it has never been applied to the problem of separating individual dwarfs from giants it has been used by Whitford to place upper limits on the contribution of M dwarfs to the integrated light of galaxies. Recently an identification of this band with FeH has been proposed by Nordh, Lindgren and myself; our paper summarizing the evidence is now being prepared. Thus the 9900Å band appears to be another band of a metallic hydride of low dissociation energy which gives a clear separation of giants from dwarfs. It is not a very easy feature to observe but it should have useful applications in classifying the reddest stars.

THE STELLAR DISTRIBUTION ABOVE THE GALACTIC PLANE: AN INTRODUCTION

Ivan R. King
Astronomy Department, University of California, Berkeley

This paper will truly be an introduction — a presentation and a discussion of the basic problems, to set the stage for detailed research results that will be reported by others.

The problem of star distribution perpendicular to the galactic plane has one central theme, which can be played in two modes. The theme is the drop-off of star-density with z, and the two modes are the general density profile and the stratification of populations. (This last phrase is a convenient one, but I should warn against taking it too literally. By "stratification" I mean a gradual change, rather than a sharp separation; and I want to emphasize even more strongly that the word "population" should be used in a general and abstract sense, rather than as a sharp separation into two, or five, or any other discrete number of components.)

The density drop-off is a simple problem in principle; it is controlled by the velocity distribution of the stars and the force field that holds them to the plane. The interrelationships are conveniently described by the familiar "hydrodynamical" equation

$$\frac{\partial (N \langle Z^2 \rangle)}{\partial z} \; \boxed{ + \frac{\partial (N \langle \Pi Z \rangle)}{\partial \tilde{\omega}} + \frac{N}{\tilde{\omega}} \langle \Pi Z \rangle } \; = -N \frac{\partial V}{\partial z} \; . \quad (1)$$

(The terms in the box have often been omitted. Although studies of the third integral have shown that these terms are not quite equal to zero, they can safely be ignored in the present discussion, because even at z = 2000 pc they produce a correction of only 20 per cent or so.)

If the $\langle \Pi Z \rangle$ terms are omitted, the hydrodynamical equation takes the much simpler form

$$\frac{\partial (N \langle Z^2 \rangle)}{\partial z} = -N \frac{\partial V}{\partial z} \; . \quad (2)$$

Even so, this equation still has no straightforward solution, because $\langle Z^2 \rangle$ depends on z in a way that is determined by the form of the Z-velocity distribution.

In one case, however, the solution of Eq. (2) becomes very simple. If the velocity distribution is Gaussian, then it can easily be shown that $\langle Z^2 \rangle$ is independent of z, and along a line of constant $\tilde{\omega}$ the solution is then

$$N(z) = N_o \exp(-\Delta V / \langle Z^2 \rangle) , \qquad (3)$$

where ΔV is the increase of the potential over its minimum value, in the plane. Note the role of $\langle Z^2 \rangle$ in scaling ΔV; it determines how strongly the change in potential will affect the density. Note also the simple power-law relationship between the densities of different stellar groups; if group B has a value of $\langle Z^2 \rangle$ that is n times as large as that of group A, its density changes are only the $1/n$ power of those of group A. It is this sensitive dependence of $N(z)$ on $\langle Z^2 \rangle$ that produces the stratification of stellar types.

At the same time, this very behavior of Gaussian velocity distributions should remind us how dangerous it can be to represent stellar density distributions directly by Equation (3). Few distributions in nature are truly Gaussian, and real stellar velocity distributions tend to have a tail that is much more extended than the meager tail of a Gaussian. This tail can be represented by adding a small percentage of stars that have one or more Gaussian velocity distributions with higher values of $\langle Z^2 \rangle$. Although these contribute very little to the density at $z = 0$, at higher z they become dominant, and the overall $N(z)$ looks quite different from one simple term of the form given by Eq. (3).

This way of representing the relationship between velocities and densities was introduced by Oort (1932), in his pioneering discussion of motions perpendicular to the plane. It is an effective method of representation, because even though a family of Gaussians is not a mathematically complete basis set, a superposition of them does in fact represent the local distribution of Z velocities rather well. However, one cannot overemphasize the sensitivity of $N(z)$ to small admixtures of high-velocity stars. A graphic example is the difference in the treatment of the same observational data by Oort (1960) and by Hill (1960). What Oort concluded was that the observed numbers of faint stars demanded the addition of a higher-velocity group that contributed only 1% of the star density at $z = 0$; at higher z, however, this group made a crucial difference.

In the face of such sensitivity to the shape of the tail, it might seem hopeless to choose a reliable velocity distribution. The task is helped very much, however, by a few observations of velocity dispersions at appreciable distances from the galactic plane, where a great deal of segregation has already taken place and the high-velocity-dispersion component shows itself much more clearly.

Given the caution about the shape of the velocity distribution, we can proceed to examine the behavior of Gaussian groups in a more quantitative way. Since ΔV is simply $\int_0^z K_z \, dz$, where K_z is the acceleration in the z-direction, we can calculate ΔV directly from the values of K_z given by Oort (1960). Furthermore, noting that 10^{-9} cm sec^{-2} pc = 0.3084 (km/sec)^2, we can express the results in units of km/sec, in order to go quickly from velocity dispersions to density ratios. The numbers are given in Table 1. The last column can be interpreted as the velocity dispersion of a group whose density has dropped by a factor of e at the height z given in column 1. (Note, however, that the values in the last two lines of the table are at a z-level where Eq. (3) has lost some accuracy, because of the $\langle \Pi Z \rangle$ correction alluded to above, which would tend to raise the densities somewhat.)

Table 1. Force Components and Potential Differences

z (pc)	K_z (10^{-9} cm sec^{-2})	ΔV (10^{-9} cm sec^{-2})	$(\Delta V)^{1/2}$ (km sec^{-1})
0	0.00	0	0.0
100	2.50	133	6.4
200	4.28	476	12
400	6.17	1544	22
1000	8.05	5932	43
2000	8.93	14536	67
3000	9.09	23552	85

Thus a young population, with a root-mean-square Z-velocity dispersion of 10 km/sec, should have a scale height of about 150 pc, whereas an older population with $\langle Z^2 \rangle^{1/2}$ = 20 km/sec should have a scale height of nearly 400 pc. Furthermore, at that level the density of the lower-velocity population should have dropped by e^4, or a factor of 70.

These simple dynamical arguments have another interesting consequence that has not, as far as I know, been noted before: an excess hump of halo-star density in and around the disc. Since the Z-velocity dispersion of halo stars is only about 100 km/sec, the numbers in Table 1 suggest that halo stars are 2 or 3 times as numerous in the galactic plane as they would be if the gravitational force of the disc were absent. This excess should be taken into account when we use local estimates of halo density to scale the densities in an overall halo model. Thus it appears that Schmidt's (1975) recent estimate of the total mass of the halo should be lowered even further.

Since the question may occur to some of you, I should mention that "massive halos" are not an issue at all in the galactic-polar-cap problem that we are discussing here. The halo population on which Schmidt sets such a firm upper limit has no relation to the extended density distribution that has been suggested by Ostriker and Peebles

(1973) and by others. The conventional halo, which is the one that concerns us here, is reasonably well represented by an inverse-cube density distribution, whereas the halo that has been postulated to explain the disturbingly flat rotation curves in spiral galaxies must have a much gentler drop-off — somewhere around inverse-square. Or, to look at the problem from the other direction, if our Galaxy has a "massive halo," its predominant mass must lie much father out, and we should expect vanishingly few of its objects to be found in the solar neighborhood.

This whole dynamical discussion depends on our knowing the function $K_z(z)$. But in fact, this function is of even greater interest for its own sake, and its determination is one of the prime problems of high-latitude studies. The slope of its initial rise tells us the spatial mass density in the solar neighborhood, and the value at which K_z levels off tells us the surface density. In principle the determination of $K_z(z)$ is simple. We need only compare the spatial stellar density $N(z)$ with the local distribution of velocity components $\phi(Z)_{z=0}$. Then with the aid of Eq. (1), or with the use that has been described for Eq. (3), or with some other way of applying the Liouville equation from which these equations are derived, $K_z(z)$ can be found directly from these two functions.

In practice, however, the determination of K_z has been an unending headache. Different studies have found quite different values of the local density, ρ_0; and, worse, attempts to derive the form of $K_z(z)$ have consistently produced a dip that implies the absurdity of a negative density somewhere. Obviously the problem is a lack of adequate observational data. There are three ways in which this trouble may arise. First, the stellar group used may not be one to which the dynamical hypotheses apply. An example is the A stars, which might not be distributed as smoothly as the theory assumes. Second, different parts of the data may refer to different subgroups within the sample studied. This is a particular danger among the K giants, which occupy the most inhomogeneous region of the whole HR diagram. The relative proportion of M11-type red giants to M67-type red giants varies by a large factor in the very z-range in which we need to do our study. Furthermore, these stars have different absolute magnitudes even though they are spectroscopically very similar. The third source of possible trouble is systematic errors in the data. Photometric errors are distressingly common, and spectral classes have their problems too. Note, for example, the suggestion by Oort (1960) that the spectral classifications with which he is dealing shift systematically at fainter magnitudes.

As a result of these difficulties, we still have no reliable determination of $K_z(z)$, and even the local volume density and surface density are rather uncertain. It has been a long journey through the wilderness; Professor Oort was given a glimpse of the promised land in 1932, but we still have not yet entered it.

Before leaving the area of dynamics I should mention one more consequence of the vertical separation of groups of different velocity dispersion. This is a progressive change of mean transverse velocity with height above the plane. It results from the correlation between mean rotational velocity and velocity dispersion, which is described by the equation (cf. Oort 1965, Eq. [38])

$$\Theta^2_{circ} - \Theta^2_m = \langle \Pi^2 \rangle \left\{ -\frac{\partial \ln \nu}{\partial \ln \tilde{\omega}} - \left[1 - \frac{\langle \Theta^2_{pec} \rangle}{\langle \Pi^2 \rangle}\right] - \left[1 - \frac{\langle Z^2 \rangle}{\langle \Pi^2 \rangle}\right] \right\} \quad (4)$$

If we note that for nearly all stellar types $\langle \Pi^2 \rangle \approx 4 \langle Z^2 \rangle$, then with reasonable values for various of its quantities Eq. (4) can be approximated as

$$\Theta \approx \Theta_{circ} - \langle Z^2 \rangle / 7^2 \quad (5)$$

Thus at large $\bar{z}$, as $\langle Z^2 \rangle$ increases the solar motion should shift systematically. Whereas this effect is unlikely to be significant for nearby stars such as those studied by Murray and Sanduleak (1972), for more distant stars it can become important.

Having talked about the behavior of stellar groups in general, I should now like to discuss a more specific question. When we look at the stars in a high-latitude field, what should we expect to see? More specifically, if we distinguish by magnitude and color, what stars dominate each part of the color-magnitude array? In answering this question I am fortunate to be able to draw on the calculations of a Berkeley student, Kate Brooks, who is modeling the star distribution in high latitudes. She does this by representing the distribution as a mixture of disc and halo populations, with a luminosity function and a density distribution for each, and by adjusting the parameters so that the computed numbers at each magnitude and color agree with the observed numbers. The model then gives a breakdown at each point in the color-magnitude array, showing which types of stars are the major contributors. The procedure is similar to that followed by Luyten (1960), but the data are accurately measured colors and magnitudes, and the stellar population is broken down in detail.

Using results kindly provided by Mrs. Brooks, I have prepared Figure 1, which is an attempt to show schematically what kinds of stars we see at the galactic pole. The figure does not extend brighter than 12th magnitude, because the small fields that we are studying have too few brighter stars to be significant. At the faint end, however, I have extended it beyond the range of our own observations, in order to show what stars are registered by the new 4-meter telescopes on fine-grain emulsions. Note the diagonality of the diagram. At the top right are some subgiants (and there would be some red giants too, if the diagram went brighter). The main-sequence stars of the disc population dominate a broad strip across the middle of the figure. Note that in general at fainter apparent magnitudes we tend to see stars that are both less luminous and farther away. Then among the faint white stars the halo

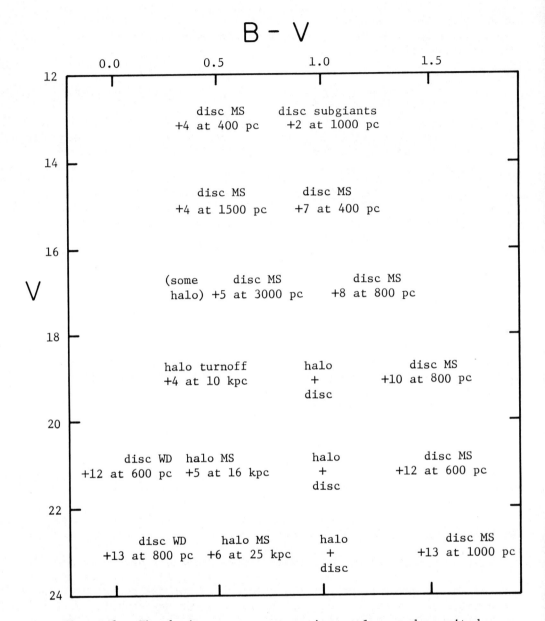

Figure 1. The dominant stars at various colors and magnitudes, at the galactic poles. The numbers are only rough approximations; at each point there is a mixture of absolute magnitudes and distances. Also, the diagram should be thought of as more continuously filled.

makes its appearance, first at the main-sequence turnoff and then on the main-sequence itself. Finally, among the bluer colors the white dwarfs of the disc take over.

Some stellar types do not appear in the figure. The red giants of the halo are not dominant in any region; apparently they can be distinguished only by methods more sophisticated than the simple sorting of colors and magnitudes. The halo main sequence barely shows up; the number of stars on its lower part appears to be a study reserved for the days of the Space Telescope, which should reach several magnitudes beyond the lower bound of this table.

A striking characteristic of the data in the figure is how far away the stars are that we see. For most stellar types, counting and classifying such faint stars is not a way of determining densities in our immediate neighborhood; the densities refer to far-away points. Take the halo population, for instance; we find out nothing about its local density but can only count up stars that are tens of kiloparsecs away. Even for the "local" M dwarfs, a survey such as this one relates predominantly to stars that are several hundred parsecs away. To count up the M dwarfs at smaller distances we need surveys at the brighter magnitudes, over larger areas of the sky, such as the color-magnitude survey of Weistrop (1972a, 1976a) or the spectral survey of Sanduleak (1964, 1976).

Consideration of these surveys brings us right back to the problem of systematic errors, however. The surveys of both Weistrop and Sanduleak claimed to find high space densities of M-dwarf stars, and the literature was then filled with a spate of less consequential papers that echoed this claim. It now seems clear, however, that there is no appreciable excess of M dwarfs over the numbers in Luyten's (1968) luminosity function and that both Weistrop and Sanduleak had been led astray by errors in their photometric systems or scales (Faber et al. 1976; Weistrop 1976a, 1976b). With wrong colors they had deduced wrong distances, which led to large errors in density.

If there is a generalization to be drawn from this whole discussion of high-latitude problems, it is that the principles are easy but the practice is difficult. And if there is a moral in the whole story, it is that we will never build a sound understanding until we have sound and reliable observational data.

REFERENCES

Faber, S. M., Burstein, D., Tinsley, B., and King, I. R.: 1976, Astron. J. $\underline{81}$, 45.
Hill, E. R.: 1960, Bull. Astron. Inst. Neth. $\underline{15}$, 1 (No. 494).
Luyten, W. J.: 1960, 'A Search for Faint Blue Stars. XXII. The Star Density in High Galactic Latitude,' University of Minnesota.
Luyten, W. J.: 1968, Monthly Notices Roy. Astron. Soc. $\underline{139}$, 221.
Murray, C. A., and Sanduleak, N.: 1972, Monthly Notices Roy. Astron.

Soc. 157, 273.
Oort, J. H.: 1932, Bull. Astron. Inst. Neth. 6, 249 (No. 238).
Oort, J. H.: 1960, Bull. Astron. Inst. Neth. 15, 45 (No. 494).
Oort, J. H.: 1965, in A. Blaauw and M. Schmidt (eds.), Galactic Structure, University of Chicago Press, Chicago, Ch. 21.
Ostriker, J. P., and Peebles, P. J. E.: 1973, Astrophys. J. 186, 467.
Sanduleak, N.: 1965, 'The Luminosity Function of the M Dwarf Stars,' Case Institute of Technology (Ph.D. thesis).
Sanduleak, N.: 1976, Astron. J. 81, 350.
Schmidt, M.: 1975, Astrophys. J. 202, 22.
Weistrop, D.: 1972a, Astron. J. 77, 366.
Weistrop, D.: 1972b, Astron. J. 77, 849.
Weistrop, D.: 1976a, Astrophys. J. 204, 113.
Weistrop, D.: 1976b, Astron. J. 81 (in press).

SOME RESULTS OF CLASSIFICATION OF STARS IN KAPTEYN AREAS APPLIED TO GALACTIC STRUCTURE IN NGP

E. K. Kharadze, R. A. Bartaya
Abastumani Astrophysical Observatory,
Georgian SSR, USSR.

The observational base of our investigation consists of a two-dimensional MK classification of about 11.000 stars in the 42 Kapteyn Areas situated along the gal.lat-s from $-17°$ up to $+72°$ and 200 Ap, Am stars discovered in the same KA. The dispersion of the applied objective-prism spectra is 160 Å per mm. The data are of high accuracy, close to the Michigan level, and uniformity, which make them reliable. The limit is close to the 12-th ph.mg.

The general conclusion of the undoubted importance is stated: the galactic concentration of dwarfs is closer than it had been assumed until now; on the other hand - the giants are not so closely concentrated to the galactic plane as it has been accepted.

Figure 1 is plotted for the F type stars of the III and V classes. Ordinates are apparent percentages of stars; abscissae - the increasing values of gal.latitudes ($1-0°$; $5-60°$). The numbers of the III class stars are nearly equal at all latitudes. Meanwhile those of the V class stars, unchangeable at the beginning, fall sharply at high latitudes.

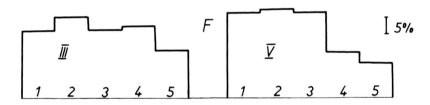

Figure 1.

The highly luminous stars of types F, G, K and almost M too are not so strongly concentrated to the galactic plane as it has been thought hitherto (Fig. 2). The space distribution of K supergiants is characterised by descrete grouping at different distances from the galactic plane.

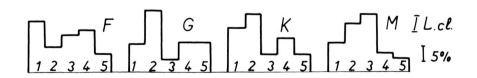

Figure 2.

The luminosity functions referring to the low latitudes differ insignificantly from each other, if one neglects the regular decrease of highly luminous stars here. And there is an evident difference at the high latitudes especially for the stars in the interval of photographic absolute magnitudes between -1 and +2 (Fig. 3).

Possibly this phenomenon is the result of a tendency of the Population II stars of the galactic halo to mix with the disc population stars. Dr. Upgren has pointed out a similar effect some years ago.

The distribution of Ap, Am stars is not uniform; Am stars significantly prevail in amount over the Ap stars; Ap stars occur mainly at the gal.lat-des less than $20°$; but Am stars up to $60°$; galactic concentration is not so sharply characteristic for Ap and Am stars as it is for the common B8-A5 stars (Fig. 4).

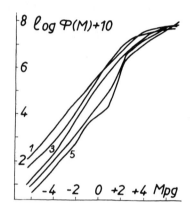

Figure 3.

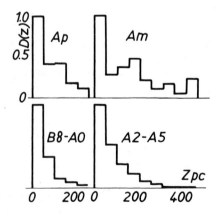

Figure 4.

MEAN CHEMICAL ABUNDANCE OF THE F STARS AS A FUNCTION OF DISTANCE FROM THE GALACTIC PLANE

A. Blaauw
Leiden Observatory

At the Tbilisi European Regional Meeting 1975, I reported on results of uvby, Hβ photometry of F-type stars in the north and south galactic polar caps (Blaauw and Garmany, 1976), based on stars in the McCormick proper motion fields between latitudes $60°$ and the poles. A relation was shown to exist between the quantity Δm_1 as determined in this photometric system, and the distance, z, from the galactic plane; Δm_1 being a measure of the metal abundance in these stars. The spectral range we deal with is defined by b-y = 0.25 to 0.40, corresponding with F0 to G2. It was found that, from the solar neighbourhood near z = 0, to z = 700 pc, the mean relative metal abundance M/H decreases by a factor of about one third.

This relation is understood in terms of the local chemical abundance distribution of the F stars and the distribution of their velocity components, W, perpendicular to the galactic plane, together with the field of force, K(z), in this direction. For K(z) I adopted the run with z as given by Oort (1965) for distances up to 1500 pc. For the distribution of the w velocity components, I made use of data for the bright stars generously made available to me by Dr. M. Mayor of the Geneva Observatory, forming the basis for his research on the local chemical evolution in the galactic disk and the radial metallicity gradient (Mayor, 1976). The distribution of the w velocity components was determined for six intervals of Δm_1. Two features of these distributions are relevant here: (a) The velocity dispersion, i.e. the r.m.s. value, σ_w, increases continuously with increasing Δm_1, from 10 to 31 km/sec. See the upper part of Table 1 where also the numbers of stars in these six groups are given; (b) It appears that these velocity distributions are gaussian to a sufficient degree of approximation to render it unnecessary for the present exploratory calculations to introduce more than one gaussian component.

We can then predict, for any subgroup of Δm_1, the density distribution in z according to the relation

$$D(z_1)/D(0) = e^{-\sigma_w^{-2} \int_0^{z_1} K(z)dz}$$

and subsequently, starting from the numbers of stars in the Δm_1 groups at $z = 0$, calculate the relative numbers at various distances z. From this, we obtain the average Δm_1 as a function of z, which we may compare with the observed relation. For the actual calculations the first three intervals of Δm_1 were combined so that we actually use the four subdivisions given in the lower part of Table 1.

Within $z = 200$ pc, the sample mainly consists of stars of subgroup I, whereas beyond 600 pc group IV dominates. At 1 kpc this latter represents about 90 percent of all stars. The resulting average values of Δm_1 give a good representation of the observed ones (except for a small zero point correction to be checked by further observations). This implies that the observed run with z can be entirely understood in terms of local dynamical considerations.

Table 1

Velocity Dispersion of Subgroups according to Δm_1 for Bright F0 to G2 Stars (b-y = 0.25 to 0.40)

Δm_1 (unit .001)	< 1	1-10	11-20	21-30	31-40	41-80	all
n	113	165	181	134	72	40	705
σ_w (km/sec)	9.8	10.4	11.2	15.4	20.0	30.6	
Subgroup	I		II	III	IV		
%	65		19	10	6		100
adopted σ_w (km/sec)	11		15	20	31		
<Δm_1>	+.006		+.026	+.035	+.050		

A more detailed report on this programme will be submitted in due course to Astronomy and Astrophysics.

REFERENCES

Blaauw, A. and Garmany, C.D.: 1976, in E.K.Kharadze (ed), Proceedings of the Third European Astronomical Meeting, Tbilisi, 1 - 5 July 1975, p. 351.

Mayor, M.: 1976, Astron. Astrophysics 48, 301.

Oort, J.H.: 1965, in A. Blaauw and M. Schmidt (ed), Galactic Structure, The University of Chicago Press, Chicago, London, Ch.21.

RGU THREE-COLOUR-PHOTOMETRY TOWARDS THE NORTH GALACTIC POLE: HALO-TO-DISK MASS RATIO

R. P. Fenkart and U. W. Steinlin
Astronomisches Institut der Universität Basel

The halo program of the Basel observatory, initiated by Becker in 1965, is based on a three colour photometry in test fields along the circle through the galactic centre and the galactic poles. The more favourable direction of the blanketing vector relative to the main sequence in the two colour diagram for RGU makes it possible to separate at least statistically the disk population and the halo population within the interval of absolute magnitudes $+3 \leq M_G \leq +8$. It is therefore possible to derive density functions for both populations and for different intervals in absolute magnitude for each test direction within the test plane defined above. This allows one to draw isodensity curves in the test plane and, assuming rotational symmetry of the halo, also isodensity surfaces. The last assumption is tested at least locally by test fields with different inclinations towards the test plane (Fenkart, R. P. and Wagner, R., 1975).

The investigations in Basel have led to two results which will be used here:

a) the combination of results in all eleven directions treated so far gives a determination of the local halo mass density in the solar neighbourhood;
b) the halo star density function in the direction of the galactic poles allows to calculate the halo-to-disk mass ratio (Fenkart, R. P., 1976).

1. The local halo mass density ρ_0. Since the classification with three colour photometry is only possible for stars with $3 \leq M \leq 8$ we get with those stars only a lower limit for ρ_0. But the luminosity function $\varphi(M)$ for the halo stars from our observation shows that

a) stars with $M < 3$ are very scarce and their contribution to the total mass can be neglected;
b) for stars with $M > 8$ we can extrapolate $\varphi(M)$ to faint magnitudes. It seems to extend at best horizontally rather far towards faint stars (a

similar result was recently obtained by Eggen (1976) using a very different approach), and we can include these stars for an upper limit for ρ_0.

For the masses of halo stars the mass-luminosity relation for stars with low metal content by M. Schmidt (1975) is used. With this we obtain the local mass density of halo stars ρ_0:

lower limit: $3.0 \cdot 10^{-4} \mathfrak{M}_\odot \text{ pc}^{-3}$; upper limit: $7.7 \cdot 10^{-4} \mathfrak{M}_\odot \text{ pc}^{-3}$.

2. The halo-to-disk mass ratio. The program in three colour photometry gives the stellar density distribution D(r) for the halo and for the disk for different absolute magnitudes as a function of the distance r from the sun. In the case of the polar fields SA 57 and SA 141, r is equal to the distance z from the galactic plane. We can, therefore, obtain the total mass within a column piercing vertically the galactic plane at the position of the sun, separately for the disk population and the halo population. The densities for corresponding luminosity groups in the direction of the north galactic pole (actually SA 57) and the south galactic pole (actually SA 141) agree well enough to assume a north-south symmetry of both disk and halo structure and to restrict the calculations to the northern polar direction (Fenkart, R., 1969).

Assuming furthermore that the luminosity function does not change with z, we have to integrate over D(z) up to 2 kpc for the disk and up to 8 kpc for the halo to reach a density of about 10^{-3} relative to that in the solar neighbourhood. The assumption of a uniform luminosity function for all z can to a large extent be checked by the numbers of stars for different z and M, except for the faintest stars at large z (where they are beyond the limiting magnitude) and for the brightest stars at small z (where their number is too small to be statistically significant). An integration only to 1.5 kpc and 6 kpc, respectively, would reduce the masses of the two populations by little more than 1 %.

As a result we obtain a halo to disk mass ratio of 0.10 and of 0.25, for the lower and for the upper limit of ρ_0, respectively.

References

Becker, W. 1965, Z. Astrophys. 62, 54.
Eggen, O. J. 1976, Astrophys. J. Suppl. 30, 351.
Fenkart, R. P. 1969, Astron. and Astrophys. 3, 228.
Fenkart, R. P., Wagner, R. 1975, Astron. and Astrophys. 41, 315.
Fenkart, R. P. 1976, Astron. and Astrophys. submitted.
Schmidt, M. 1975, Astrophys. J. 202, 22.

DENSITY LAW, VERTICAL DISTRIBUTION AND VERTICAL GRADIENT OF METAL ABUNDANCES FOR G AND K GIANTS

M. Grenon
Observatoire de Genève

A systematic photometric investigation on stars of spectral type G4 to K4 is performed in both polar caps in symmetric coaxial cones. According to a calibration of the Geneva system, Mv, Teff and [M/H] ratio have been now derived for about 400 stars. For the computation of density law, a special care has been taken to avoid selection biaises related to variations of mean age or metal abundance. A change of luminosity function for evolved stars in function of z has been detected. Using the geometrical properties of the cones and direct counts in the solar vicinity, we derive the following density law for stars with Mv $\in$ [0.0, 1.3] and (B-V) $\in$ [0.7, 1.3]

z(pc)	0	100	200	400	600	800
log $\rho(z)/\rho(o)$	.00	-.17	-.39	-.75	-1.04	-1.29

with a central density of $1.7 \cdot 10^{-3}$ star.pc^{-3}. This result confirms the old determination of Oort (1932), but deviates sensibly from the more recent ones of Upgren (1962) and Elvius (1951). The distribution of metal abundances at various heights is deduced from G and K stars of all luminosities and compared with that obtained from nearby G and K dwarfs. The increasing concentration to the galactic plane in function of the metal richness is a proeminent feature. The vertical gradient of [M/H] is estimated to -0.35.kpc^{-1} for z between 0 and 700 pc. This value is more than twice smaller than the recent ones of Blaauw (1975) or Jennens (1975). With a radial gradient of -0.05, (Grenon (1972), Mayor (1976)) we have a ratio of vertical on radial-gradient of about 7, which provides a new constraint for galaxy modelling.

References :
Oort, J.H.   1932, B.A.N., $\underline{6}$, 249
Elvius, T.   1951, Stock. Obs. Ann. $\underline{16}$, 1951
Upgren, A.R.   1962, A.J. $\underline{67}$, 37
Grenon, M.   1972, U.A.I. Coll. $\underline{17}$, LV
Blaauw, A.   1975, Proc. 3d Europ. Astr. Meet., Tbilisi, 351
Jennens, P.A. 1975, MNRAS $\underline{172}$, 301
Mayor, M.   1976, Astron. & Astrophys. $\underline{48}$, 301

POLARIZATION MEASUREMENTS AND EXTINCTION NEAR THE NGP

T. Markkanen
University of Helsinki, Observatory and
Astrophysics Laboratory

In order to study interstellar extinction near the NGP a polarization observation programme has been undertaken at the Metsähovi Observatory of the University of Helsinki. 32 stars ($b > 80°$) have been measured with an accuracy of a few hundredths of a percent. To this list observations of 42 stars made by Appenzeller (1968) were added. The stars measured have distances up to 400 pc (Figure 1).

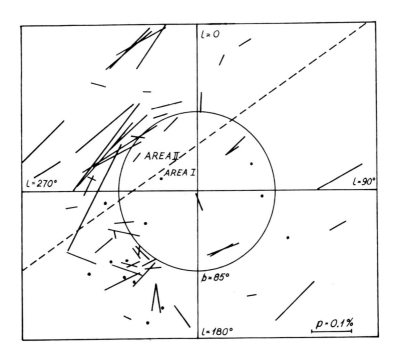

Fig.1. Polarization vectors near the NGP. Dots are stars with zero polarization.

There seems to be an area of higher polarization values at $l \approx 270°-360°$. In this area (II) the polarizations are low at small distances but between about 100 and 200 pc they increase to about 0.3 % and remain constant up to 400 pc. Elsewhere (I) the polarizations increase slowly with distance and reach a value of about 0.1 % at 200 pc (Figure 2).

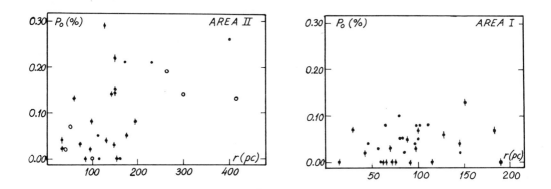

Fig.2. Amounts of polarization versus distance in areas II and I. Polarizations of the ticked dots are from Appenzeller (1968).

It can be concluded that there is a dust cloud at $l \approx 270°-360°$, at a distance of about 100-200 pc.

There is a correlation between the polarizations and the colour excesses E(b-y) measured by Feltz (1972). The excesses seem to be systematically too low by 0^m01. Similar results were obtained photometrically by Hilditch et al. (1976). By using the polarizations the lower limits of extinction can be estimated ($A_V \gtrsim 0.33p$). $A_V \gtrsim 0^m03$ is found generally at the NGP, $A_V \gtrsim 0^m1$ in the direction of the cloud in the fourth quadrant.

REFERENCES

Appenzeller,I.: 1968, Astrophys.J. <u>151</u>, 907.
Feltz, Jr.,K.A.: 1972, Publ. Astron. Soc. Pacific <u>84</u>, 497.
Hilditch,R.W., Hill,G. and Barnes,J.V.: 1976, Monthly
    Notices Roy. Astron. Soc. <u>176</u>, 175.

PARALLAXES OF SELECTED STARS NEAR THE NORTH GALACTIC POLE

K. Aa. Strand, C. C. Dahn, and R. S. Harrington
U. S. Naval Observatory, Washington, D. C.

As a contribution toward settling the question of the distances of Murray-Sanduleak (M-S) stars (Murray and Sanduleak, 1972), two of these stars were added to the U. S. Naval Observatory (USNO) program in 1973. After the initial plates had been taken, it appeared that parallax for an additional star could be measured on one series, while parallaxes for two additional stars were possible from the second. As a result provisional parallaxes are now available, Table I.

In terms of USNO parallaxes, these are considered provisional because they are based on half the number of plates on which the final parallaxes are ordinarily determined; only small changes are expected when additional material is added beyond the present.

In addition to the trigonometric parallaxes, photometric parallaxes have been obtained from V-I photometry by Dahn, based upon the M_V, (V-I) relation previously published (Strand et al, 1974). The mean trigonometric parallax obtained is $0\overset{''}{.}012 \pm 0\overset{''}{.}006$ s.d.; the photometric parallax for these five stars is $0\overset{''}{.}011 \pm 0\overset{''}{.}006$ s.d.; and the value for 15 of the 21 stars is $0\overset{''}{.}009 \pm 0\overset{''}{.}004$ s.d.

The data obtained here indicate that the M-S stars are more distant by a factor of two than originally expected, which is in agreement with recent results obtained elsewhere (Koo and Kron, 1975; Weistrop, 1976).

Table I

M-S	V	V-I	π_t	s.d.	μ/PA	v_T	π_p
10	13.18	1.77	+$\overset{''}{.}$023	+$\overset{''}{.}$004	$\overset{''}{.}$098/219°	20	$\overset{''}{.}$015
11	14.32	2.30	+$\overset{''}{.}$010	$\overset{''}{.}$007	$\overset{''}{.}$096/258	46	$\overset{''}{.}$018
16	15.47	1.54	+$\overset{''}{.}$013	$\overset{''}{.}$012	$\overset{''}{.}$088/270	32	$\overset{''}{.}$004
17	16.54	2.49	+$\overset{''}{.}$005	$\overset{''}{.}$013	$\overset{''}{.}$020/211	19	$\overset{''}{.}$009
18	15.25	2.03	+$\overset{''}{.}$007	$\overset{''}{.}$005	$\overset{''}{.}$121/283	82	$\overset{''}{.}$008
Mean			+$\overset{''}{.}$012	$\overset{''}{.}$006			$\overset{''}{.}$011

REFERENCES

1. Koo, D. C., and Kron, R. G. (1975). Publ. Astron. Soc. Pac., $\underline{87}$, 885.

2. Murray, C. A., and Sanduleak, N. (1972). Monthly Notices, Roy. Astron. Soc., $\underline{157}$, 273.

3. Strand, K. Aa., Harrington, R. S., and Dahn, C. C. (1974). I.A.U. Symp. No. 61, New Problems in Astrometry.

4. Weistrop, D. (1976). Astron. J., $\underline{81}$, 427.

PROGRESS REPORT ON A SEARCH FOR PARALLAX STARS IN THE REGION OF THE SOUTH GALACTIC CAP

C. A. Murray
Royal Greenwich Observatory

A series of plates is being taken with the U.K. 48-inch Schmidt Telescope at Siding Spring on a field centred on the star GC 1110 near the South Galactic Pole, in order to detect any stars brighter than about $m_{pg} = 18$, which may have significant trigonometric parallaxes.

A search list of more than 16,300 objects in an area $4\frac{1}{4}^\circ \times 4\frac{1}{4}^\circ$ has been formed from a preliminary scan on the GALAXY machine at R.G.O. From 1974 January to 1976 January some 30 plates have been obtained, nearly half of which have been measured in each of two orientations on GALAXY; each measuring run takes about 16 hours.

A preliminary astrometric reduction of five plates extending from 1974.0 to 1975.5 indicates that the positional accuracy achieved using simple linear transformations, is about $\pm 0''.08$ per plate for 50 SAO stars uniformly distributed over the whole measured area. It is hoped, that, after a five year observing programme, parallaxes with standard errors of about $\pm ''.007$, at least for the brighter stars, will be obtained.

SEARCH FOR LARGE RADIAL VELOCITIES IN DIRECTION OF NGP USING THE FEHRENBACH TECHNIQUES

Alphonse FLORSCH
Observatoire Astronomique de Strasbourg, France

The Observatory of Strasbourg will participate in the future in the general radial velocity survey which is in hand at the Observatories of Haute-Provence and of Marseille with FEHRENBACH's Objective-prism astrograph. A programme of this type, providing a large number of data, is particularly suitable at the place where the "Data Center" is growing up.

A radial velocity survey is a vast programme which needs many years to be fulfilled. It is natural that one starts with the most interesting areas in the sky, one of them being the area near NGP. In 1976 twenty fields were covered, each by one plate, in that direction. The whole programme covers 70 overlapping fields in an area of 100 square degrees centered at NGP. The limiting magnitude shall be almost the 12th.

The meticulous measurement of the plates gives the radial velocities but a first superficial investigation is sufficient to find out the high velocity stars. That is what we shall do first. An investigation into ABT's catalogue of radial velocities in a wide area around the NGP shows that the rate of the number of high velocity stars (with $|v|>100$ km/s) to the number of normal stars increases rapidly with the magnitude, getting three times higher for stars fainter than the 10th magnitude than for brighter stars.

In the future, the survey will partially be done with the new devices which are already available at the Observatoire de Haute-Provence: the 60 cm Objective-prism mounted on a Schmidt telescope and the new automatic measuring device called "MESUCOR". The first results obtained by FEHRENBACH and his collaborators with this instrument are very promising, giving better than an accuracy of 7 km/s for the velocities.

THE GRAVITATIONAL FIELD OF THE GALAXY IN THE Z DIRECTION

R. F. Griffin
Cambridge Observatories

This is a progress report of a project designed to find the component, perpendicular to the Galactic plane, of the gravitational potential of the Galaxy. The principle is to measure the radial velocities and distances of a large number of K-giant stars near the North Galactic Pole. My student G. A. Radford is masterminding the project; collaborating with us are Drs. J. E. Gunn of the Hale Observatories and L. Hansen and K. Gyldenkerne of Copenhagen.

We have measured the radial velocities of all the HD stars of type K0 and later, and many of the G5 stars, within 15° of the Galactic Pole, using the Cambridge photoelectric spectrometer. In addition, we have observed all the stars classified as K giants by Upgren in his declination zones 25° to 31°, using the spectrometer on the Hale telescope. There are about 900 stars observed altogether, including about 200 Upgren stars, running down to twelfth magnitude or so, which are not in the Henry Draper Catalogue. To determine the distances of all these stars we are now trying to determine the absolute magnitudes by narrow-band photoelectric photometry in the Copenhagen system. Most of the observations have been made, thanks largely to the very generous grants of observing time given by the Hale Observatories earlier this year; but the reductions have only been completed for about 300 stars (including 244 K giants) which were observed last year at Kitt Peak, and the present, very preliminary, discussion is based on those stars alone.

The velocity dispersion shows an unexpected variation with distance from the Galactic plane. Within the uncertainties of its determination, it can be considered constant at 23 km/s up to 400 parsecs from the plane, and then to rise to 30 km/s at 600 parsecs. Using these velocity dispersions together with K-giant density distributions drawn from various sources, we infer a gravitational force towards the Galactic plane going from zero in the plane to a maximum at 300 parsecs, thereafter decreasing until it levels off at about 500 parsecs. The existence of this turning point, which has also been obtained by others, is in contravention of Poisson's Law, which states that dK_z/dz must be proportional to the local mass density and must therefore always be positive; so our preliminary result is certainly vitiated either by errors arising from inadequacy of the data — something that may be at least partly redeemed when the project is completed — or else by undetected falsehood in one or more of the premises of the discussion.

The presently-indicated value for the local mass density is 0.30 solar masses per cubic parsec — higher than the accepted value; if it is sustained by the complete investigation it will exacerbate the long-standing problem of the "missing mass".

BERKELEY STUDIES OF FAINT STARS AT HIGH LATITUDES

Richard G. Kron, Liang-Tai George Chiu, and Kate O. Brooks
Astronomy Department, University of California, Berkeley

Several Lick 3-m prime-focus plates of Selected Areas 57, 68, and 51 (taken by I. King) in B, V, and R have been measured for stellar magnitudes down to the plate limit by L. Hinrichs and King, and are currently being measured by Chiu for proper motions (several hundred stars per plate) with the Berkeley PDS microdensitometer. Prime-focus plates are also available from the Hale 5-m and Mayall 4-m telescopes, giving an overall baseline of 20 years. Work so far indicates that on the Lick plates stars brighter then V = 19 can be measured to within one micron standard error; the error becomes unacceptably large for stars fainter than V = 20. A large number of stars bluer than B-V = 0.4 show proper motion and are therefore excellent candidates for white dwarfs. For $0.4 \leq B-V \leq 0.8$, the proper motion stars are expected to be predominantly subdwarfs.

The frequency distributions of the stars in V and B-V for the three fields are being analysed by Brooks; the fields are advantageously placed for study of the density distributions in both the disk and the halo. These data should allow the halo stars to be studied out to a distance of 10 to 15 kpc, as well as a determination of the degree of flattening of the halo. Also, a study will be made of the z and ϖ density gradients in the disk, and the luminosity function of disk stars.

In a separate study by Kron and H. Spinrad, image tube scans have been obtained for ten of the faintest and reddest stars on the proper motion list of Murray and Sanduleak, stars which have been suggested to have low velocities. The strength of the MgH 5211 + Mg "b" blend is shown to be correlated with velocity dispersion for M dwarfs; according to this band strength, the Murray-Sanduleak stars are spectroscopically similar to old-disk M dwarfs. Only two of the ten stars have Hα emission. Generally, there is no evidence for a dominant, young population among these stars.

SWEDISH PROGRAMMES CONCERNING THE Z DISTRIBUTION OF STARS

Tarmo Oja
Astronomical Observatory, Uppsala, Sweden.

Swedish investigations on stellar space distribution generally are based on the Uppsala - Stockholm spectral classification system. In this system the properties of the stars are derived from objective prism spectra by means of narrow-band photometry. The main criteria are the hydrogen-line intensities, the intensity of the K line, the intensity of the G band, the break at the G band, and the intensity of the blue cyanogen band (for a description of the system see e.g. Ljunggren and Oja, 1961), when possible supplemented by a measure of the Balmer discontinuity (Rydström, 1976). The method yields absolute magnitudes with a dispersion of about $0\overset{m}{.}6$ and intrinsic colours $(B-V)_o$ with a mean error below $0\overset{m}{.}05$ for most kinds of stars.

Three investigations are relevant to the subject of this Joint Discussion.

Dr T. Elvius (at the Lund Observatory) and his collaborators are investigating the stellar distribution in 52 of Kapteyn's Selected Areas distributed over all latitudes. The aim is to establish the z distribution of stars not only at the position of the sun, but also at some distance from it. Up to now data have been published for 19 areas (Elvius and Lodén, 1960).

A region around the South Galactic Pole is investigated by Eriksson at Uppsala. The excess curve and the z distributions of several groups of stars have been established; the final result will be published very soon.

Häggkvist and Oja are carrying out an investigation towards the North Galactic Pole. Data will be determined for more than 12,000 stars; see e.g. Häggkvist and Oja, 1973.

References:
Elvius, T. and Lodén, K. 1960, Stockholm Obs. Ann. 21, N:o 2.
Häggkvist, L. and Oja, T. 1973, Astron. Astrophys. Suppl. 12, 381.
Ljunggren, B. and Oja, T. 1961, Uppsala Astr. Obs. Ann. 4, N:o 10.
Rydström, B. 1976, Uppsala Astr. Obs. Rep. N:o 8.

STUDIES OF A AND F STARS IN THE REGION OF THE NORTH GALACTIC POLE

R. W. HILDITCH   and   GRAHAM HILL
St Andrews, U.K.   DAO, Canada

A short account was given of results recently published (Hill et al. 1976; Hilditch et al. 1976a,b) of a spectroscopic and photometric study of 310 A0-F8 stars within 15° of the north galactic pole. Using these data, it has been found that δm_0 for the programme stars is constant out to 250 pc and corresponds, via Crawford's (1975) calibration, to the solar value of [Fe/H]. The distribution of radial velocities of A0-F8 stars is asymmetric and may be interpreted as showing a broad distribution of mean -7 kms^{-1} and dispersion $\sigma \sim 11$ kms^{-1} together with a superposed narrow distribution of mean 0 kms^{-1} and dispersion $\sigma \sim 3$ kms^{-1}. The narrow component originates from a group of 37 stars which appear to be kinematically and spatially associated with the Coma cluster. Only 14 of these 37 stars are recognised members of the Coma cluster. Removal of these 37 stars from the sample results in a distribution which matches closely the W component distribution of A0-F8 stars in Gliese's (1969) catalogue. Thus Oort's "well-mixed" hypothesis seems to be confirmed out to 250 pc. Much additional work on the space motions is required before the hypothesis of the 'Coma group' can be thoroughly tested. Results of this analysis and a preliminary value for the total mass density in the solar neighbourhood is expected to be completed during the next few months.

Extension of this programme to $12^m.5$ at the NGP has already been started at DAO (spectroscopy) and KPNO (photometry). In addition, a similar programme to study the SGP region was started in 1975 at CTIO and Las Campanas. To date, 60% of the Hβ data and 40% of the radial velocities have been obtained. The four-colour observations should start in 1976 September/October at SAAO.

References
Crawford,D.L., 1975. Dudley Obs. Report No.9, 17.
Gliese,W., 1969. 'Catalogue of Nearby Stars'. Veröff. Heidelberg Nr.22.
Hill,G., Allison,A., Fisher,W.A., Odgers,G.J., Pfannenschmidt,E.L., Younger,P.F. & Hilditch,R.W., 1976. Mem. R. astr. Soc., 82, 69.
Hilditch,R.W., Hill,G. & Barnes,J.V., 1976a. Mem. R. astr. Soc. 82, 94.
Hilditch,R.W., Hill,G. & Barnes,J.V., 1976b. Mon.Not.R.astr.Soc. 176,175.

THE DISTRIBUTION OF FIELD HORIZONTAL-BRANCH STARS IN THE GALACTIC HALO

A. G. Davis Philip[+]
Dudley Observatory

Since the time allowed per paper at this joint discussion is short only one aspect of the stellar distribution at high galactic latitudes will be presented here, namely the distribution of field horizontal-branch stars (FHB) in the galactic halo. First, the method by which FHB stars are found will be described. Second, the density distribution of FHB stars will be compared with that of the RR Lyrae stars.

As part of a general program to study the stellar density distribution perpendicular to the galactic plane a number of possible FHB stars has been discovered. In each survey area objective prism plates (at a dispersion of 280 Å/mm. to a limiting magnitude of V = 14) are taken with the Michigan Curtis Schmidt telescope at Cerro Tololo Inter-American Observatory or the Schmidt telescope at the Warner and Swasey Observatory. A set of direct plates are taken also to obtain photographic magnitudes for the stars with spectral classifications.

Four-color and H_β measures of the B and A stars allow the color excess in each region to be determined accurately. In the magnitude range 12 - 14 about half of the A-type stars measured had peculiar colors; c_1 indices that were 0.2 mag. or greater higher than the normal c_1 index for a star on the main sequence and m_1 indices that were 0.05 mag. or more lower than normal. These characteristics are identical to those measured for blue horizontal-branch (BHB) stars in globular clusters and to those of four well known FHB stars (-6° 86, HD 86986, 109995, and 161817, Oke, Greenstein, and Gunn 1966). Radial velocities have been obtained for 33 FHB stars (Philip 1973). The velocity dispersion for this group of stars is ± 113 km/sec which tends to confirm their classification as Population II stars. Recently, Danford (1976) has confirmed the classification of many of these FHB stars in a photometric and spectroscopic study of halo horizontal-

---

[+] Visiting astronomer, Cerro Tololo Inter-American Observatory, which is operated by the Association of Universities for Research in Astronomy, under contract with the National Science Foundation.

branch stars.

In two areas, the north and south galactic poles, complete surveys have been made of all the early-type stars in ~30 square degrees centered on each pole. There are not sufficient stars to do a conventional stellar density analysis but one can obtain an estimate of the density distribution of FHB stars by assuming a mean absolute magnitude $M_v = 0.5$, correct the apparent magnitude for the small amount of interstellar reddening and calculate the distance to each star. The stars can be divided into groups by distance from the galactic plane and the number of stars per $10^6 pc^3$ calculated. If the stellar densities are plotted versus the perpendicular distance from the galactic plane one finds that the main sequence A2 - A7 stars at the NGP have densities in the range of a few tenths and A0 stars in the range of a few hundredths stars/$10^6 pc^3$. (Upgren 1962, 1963). The RR Lyrae stars have densities of a few thousandths stars/$10^6 pc^3$. (Kinman, Wirtanen, and Janes 1966).

The stellar density distributions of FHB stars at the galactic poles are quite similar and fall in the range of a few tenths stars/$10^6 pc^3$. For distances closer than one or two kpc. the volume of the cone surveyed and the number of FHB stars found are too small for density estimates to be made. For distances greater than five kpc. the apparent magnitudes of the FHB stars are below the spectral plate limit. It is interesting to note that in the range of two to four kpc., where the density estimates are best, the number of FHB stars is approximately ten times that of the RR Lyrae stars.

With the new thin prism on the Michigan Curtis Schmidt telescope it is possible to identify early-type stars to V = 16 and thus FHB stars can be surveyed to distances of 10 kpc. If the ratio of FHB/RR Lyrae stars is ten to one out to these distances, and if the remainder of a globular cluster-like population (giants and main sequence Population II stars) then the density distribution of stars in the galactic halo will be much higher than previously estimated from the study of the distribution of Population I stars. Photometric data for over 75 FHB stars measured in the Stromgren four-color system will be made ready for publication in 1977. Additional measures are planned of stars in areas at $\ell = 0°$ and $180°$ so that complete surveys will be obtained.

REFERENCES

Danford, S.C. 1976, Thesis, Yale University.
Kinman, T.D., Wirtanen, C.A., and Janes, K.A. 1966, Ap. J. Supp. <u>13</u>, 379.
Oke, J.B., Greenstein, J.L., and Gunn, J. 1966, <u>Stellar Evolution</u>, R.F.
    Stein and A.G.W. Cameron eds., (Plenum Press: New York) p. 399.
Philip, A.G.D. 1973, <u>Spectral Classification and Multicolor Photometry</u>,
    Ch. Fehrenbach and B.E.Westerlund eds., (Reidel: Dordrecht) p. 230.
Upgren, A.R. 1962, A.J. <u>67</u>, 37.
Upgren, A.R. 1963, A.J. <u>68</u>, 475.

A NEW PROGRAM TO DETERMINE K(z) FROM MAIN SEQUENCE STARS

A. R. Upgren
Van Vleck Observatory, Wesleyan Univ., Middletown, Conn.

Recently Dessureau and Upgren (1975) redetermined the velocity distribution of giant stars in the north galactic pole direction using Upgren's (1962) catalogue and Oort's (1960) determination of K(z). The velocities were assumed to be represented by n Gaussian distributions with no further constraints imposed. The velocities are well represented by three such distributions whose properties disagree with those found by Oort. A larger number did not improve the stability of the solution. Without radial velocities, however, they could not redetermine the K(z) force itself.

Solutions for K(z) have usually had to be based on the giants, although the main-sequence A-stars have also been used (Perry 1969). Both groups are luminous, reasonably abundant at high latitudes and easily identified by their spectra, even in low-dispersion surveys. But the A-stars may not be numerous enough at large z-distances and the absolute magnitudes of giants appear to vary with z in a way which is difficult to measure with certainty. If the limiting magnitudes of recent objective-prism surveys could be extended by a few magnitudes, the main-sequence F-stars would be more suitable than either the A-stars or giants, since they then would also be identifiable at large distances from the plane. The new 1.0-1.5 meter Schmidt telescope of CIDA, the Venezuelan observatory, with its objective prisms reaches a limiting magnitude of about 15. As Stock, Osborn and Upgren (1976) show, spectral classes, radial velocities and proper motions could be determined for the F-stars to distances beyond one kpc (since their absolute magnitudes vary from +3 for F0V stars to +5 for G0V stars) with photographic plates taken with a conventional objective prism.

The abundance of these stars has been found by Bok and Basinski (1964); they obtained V magnitude of stars in a south polar region. Their group defined by $0.30 < B - V < 0.60$ closely corresponds to F0-G0 on the main sequence and they find about 19 stars within this interval per square degree. In a modest polar region of 100 square degrees, perhaps as many as 2000 such objects may be discovered.

The F-stars possess the further advantage over the giants of having absolute magnitudes whose variation with z is well known, providing only that the sequence for subluminous stars is well determined. For this purpose, the Van Vleck Observatory is measuring parallaxes for many dwarfs and subdwarfs in the F0-G0 range. The absolute magnitudes of these stars must be known with precision and at present, those of the high-ultraviolet excess stars (which become relatively abundant at large z) are not well known (Sandage 1970). Also, since the new parallaxes are of higher precision than existing ones, the volume correction (Lutz and Kelker 1973) can be successfully applied to provide more definitive main and subluminous sequences from which to calibrate absolute magnitudes.

REFERENCES

Bok, B.J. and Basinski, J. 1964, Memoirs of Mt. Stromlo Obs., No. 16.
Dessureau, R.L. and Upgren, A.R. 1975, Publ. Astr. Soc. Pacific 87, 737.
Lutz, T.E. and Kelker, D.H. 1973, Publ. Astr. Soc. Pacific 85, 573.
Oort, J.H. 1960, Bull. Ast. Inst. Neth. 15, 45.
Perry, C.L. 1969, Astron. J. 74, 139.
Sandage, A.R. 1970, Astrophys. J. 162, 841.
Stock, J., Osborn, W., and Upgren, A.R. 1976 this volume.
Upgren, A.R. 1962, Astron. J. 67, 37.

OBJECTIVE-PRISM RADIAL VELOCITIES AT HIGH LATITUDES

J. Stock, W. Osborn, and A. R. Upgren
CIDA, Merida, Venezuela, and Van Vleck Obs., Middletown, CT

Stock and Osborn (1972, 1973) have shown that objective-prism spectra taken with a conventional prism may be measured to produce radial velocities of sufficient accuracy for statistical purposes. They are determined by means of a third-order power series in both x and y coordinates. The same spectral-line measures can also yield positions such that proper motions can be determined if first-epoch positions are also available. For many stars, both tangential and radial velocities can be obtained with about the same error which is of the order of $\pm$ 20 km/sec. The field distortions caused by the prism are large but are constant and predictable to the degree that measured residuals are similar in size to those for direct images (Stock and Upgren 1968). A survey of a high-latitude zone between -30° and -35° in declination is underway and a catalogue of about 3000 stars has already been compiled by Stock. For each star, the catalogue lists an accurate 1950 position, spectral and luminosity type, apparent photographic magnitude, relative radial velocity and its weight, and the number of plates on which the star was measured.

The use of rapid scanning for objective-prism spectral types and radial velocities has been investigated using the PDS Microdensitometer of the Kitt Peak National Observatory for the measurement of the plates. If rapid scans can be made without loss of precision on the introduction of a systematic error, their great reduction of measuring time is of real advantage in the determination of these data for many thousands of stars.

REFERENCES

Stock, J. and Osborn, W. 1972, in "The Role of Schmidt Telescopes in Astronomy", Conference, Hamburg, Germany.
Stock, J. and Osborn, W. 1973, in "Spectral Classification and Multicolor Photometry", I. A. U. Symposium No. 50, page 290.
Stock, J. and Upgren, A. R. 1968, Publ. Obs. Ast. Nacional, Univ. de Chile, Vol. II, No. 1.

DISCUSSION OF THE CALCULATION OF THE DENSITY LAW (Dz)

C. TURON LACARRIEU
Observatoire de Genève, Switzerland,
On leave from Observatoire de Paris, France

Abstract : A matrix method involving eigenvector expansion is used to solve the "fundamental equation" of stellar statistics. This method is applied to M dwarfs and K giants.

I - INTRODUCTION

We investigate a new matrix method for solving the "fundamental equation" of stellar statistics. This equation is a linear integral equation of the first kind (Fredholm's type) and is difficult to solve because of the inherent ill-conditioning of the problem. Classical methods for the determination of D(r) have been reviewed for example by van Rhijn (1965) and Mihalas (1968). None of these "classical" methods allow a rigorous determination of the resultant stellar space density and of the uncertainties.

A new approach to this problem was made by Dolan (1974). He used a matrix method which allows a rapid and easy reduction but also requires an analytic form for the luminosity function. Its main advantage is that it provides an explicit measure of the uncertainty in D(r).

Solving this type of integral equation is a common problem in many fields of applied mathematics and physics and new methods using eigenvector expansions have been developped in the last few years : Varah (1973), Ekström and Rhoads (1974), Anderssen and Bloomfield (1974). The approximate and physically acceptable solution is obtained by a generalized Fourier expansion, the coefficients of which are weighted so as to minimize the effects of the higher order Fourier modes of the density law. This method is rigorous, allows a direct and rapid numerical computation of the solution and of its uncertainty and does not require any restriction regarding the shape of the luminosity function. It is even possible, and very easy, to introduce into the equation the variation of the luminosity function with the distance of the galactic plane.

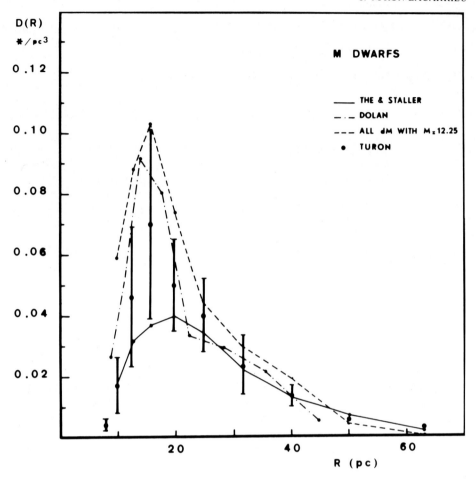

Fig. 1 - Density law for M dwarfs

II - APPLICATIONS

Until now, we have applied the method to two types of stars : the M dwarfs and the K giants.

1) The M dwarfs

The sample of Thé and Staller (1974) is used to derive a solution for the space density of M2 to M4 dwarfs near the South Galactic Pole. The same assumptions are made for the luminosity function (gaussian distribution with $\overline{M}_{pg}$ = 12.25 and σ = 0.5 magnitude) and the same magnitude interval (δm = 0.5) is used as in the paper of Thé and Staller in order to facilitate the comparison of the results. The results are shown in Fig. 1 with the solution given by Thé and Staller who used the Malmquist method and that of Dolan (1975). Dolan's results, values of D(r) and the uncertainties in them, are confirmed. It should

DISCUSSION OF THE CALCULATION OF THE DENSITY LAW (Dz)

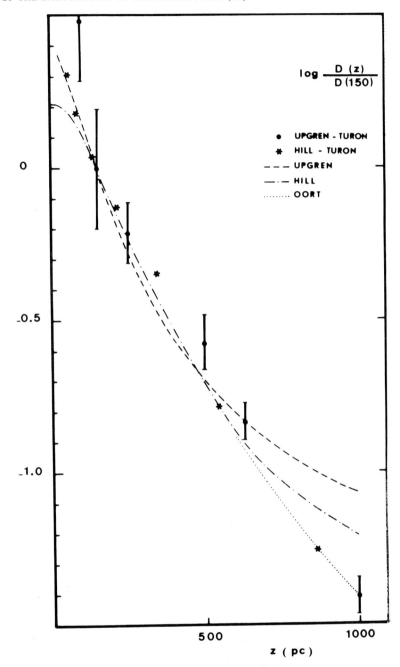

Fig. 2 - Density law for K giants

be noted that the error bars only represent the effect of the statistical uncertainties in the data and not the possible systematic effects such as incompleteness.

2) The K giants

The method has been applied to two samples of K giants : those of Hill (1960) and of Upgren (1962), both towards the North Galactic Pole. The results are shown in Fig. 2 with the solutions given by Hill and Upgren and with the dynamical solution given by Oort (1960). It is quite remarkable that the present solutions, for both samples, are in very nice agreement with Oort's solution, justifying his assumption that the z-distribution of the total stellar density is the same as that of the K giants. The slope of our solutions for z smaller than 500 pc is a little different from that of the other authors. This slope is confirmed by Grenon (1976) from a new sample of K giants which has been very carefully chosen and which is quite complete in the zone under consideration. These results will have some implications on the run of the force law Kz.

In this application, Oort's luminosity function (1960) has been used. An important advantage of this method is that it allows the use of a numerical form for the luminosity function.

Acknowledgments : I should like to thank Prof. L. Martinet for suggesting this work and for the hospitality of the Geneva Observatory.

REFERENCES

Anderssen, R.S., Bloomfield, P. 1974, Numerische Math. **22**, 157
Dolan, J.F.   1974, Astron. Astrophys. **35**, 105
Dolan, J.F.   1975, Astron. Astrophys. **39**, 463
Ekström, M.P., Rhoads, R.L. 1974, J. of Computational Physics, **14**, 319
Grenon, M.   1976, I.A.U. Present Joint Discussion
Hill, E.R.   1960, B.A.N. **15**, 1
Mihalas, D.   1968, Galactic Astronomy, W.F. Freeman and Company ed., San Francisco and London, ch. 4
Oort, J.H.   1960, B.A.N. **15**, 45
Thé, P.S., Staller, R.F.A. 1974, Astron. and Astrophys. **36**, 155
Upgren, A.R.   1962, Astron. J. **67**, 37
Van Rhijn, P.J.   1965, Galactic Structure, ch. 2, p. 27
Varah, J.M.   1973, SIAM J. Num. An. **10**, 257

To be published in Astronomy and Astrophysics

MOTIONS OF NEAR-POLAR K-GIANTS ALONG THE Z-COORDINATE

A.N. BALAKIREV

*Sternberg Astronomical Institute
Moscow University
Moscow, U.S.S.R.*

The motion of G8-K5 III,IV stars near galactic poles ($|b|>60°$) was investigated. The main attention was paid to motion in the z-direction. The spectral types, visual magnitudes, colors and radial velocities (v_r) were taken mainly from Schild (1973), Heard (1956), Blanco et al (1968) and Wilson (1953). To reduce the values v_r by Heard to those by Wilson, the correction of -1 km/sec was made. The stars with variable v_r and spectroscopic binaries were excluded from the investigation.

The study of motion was made by least-squares solution of the system of conditional equations taken at first (calculation 1) in the form

$$-v_r = U_\odot \cos l \cos b + V_\odot \sin l \cos b + W_\odot \sin b \qquad (1)$$

where $U_\odot$, $V_\odot$, $W_\odot$ are the solar motion components relative to the investigated stars. In order to ascertain whether the value $W_\odot$ changes with z-coordinate, we made calculation 2, having replaced in (1) the value $W_\odot$ by value $(W_\odot + (dW_\odot/dz) z)$. The z-coordinates were defined as the heights above the galactic plane, assuming the Sun's height $z_\odot = 10$ pc. The distances r were found mainly as luminosity-distances. The intrinsic colors and absolute magnitudes were taken from Bulon (1963) and the ratio $R = A_V / E_{B-V} = 3.0$ was adopted. The results of calculations for different groups of stars are presented in the Table, where n is the number of stars. The components of the solar motion found by Vyssotsky and Janssen (1951) (group F) and those relative to G8-K5 III, IV stars from Woolley et al (1970) (group D) are also listed. The stars of these groups belong to the disk population. Although the values $U_\odot$ and $V_\odot$ are determined with uncertainty for the investigated stars, still the difference between those values for the groups A, B and C on the one hand and the groups D and F on the other hand appears to be real. This distinction may be accounted for by the presence of the old disk and halo objects among the investigated stars.

Group of stars	Calculation 1				Calculation 2	
	$U_\odot$ km/sec	$V_\odot$ km/sec	$W_\odot$ km/sec	n	$dW_\odot/dz$ km/sec/kpc	n
A ($b > 60°$)	7.8 ± 7.6	41.4 ± 10.3	7.0 ± 1.8	157	15 ± 32	89
B ($b < -60°$)	2.6 ± 8.4	13.1 ± 7.4	4.8 ± 2.3	87	30 ± 50	79
C ($|b| > 60°$)	8.6 ± 5.7	26.4 ± 6.3	6.5 ± 1.5	245	7 ± 15	169
D	8.8 ± 3.6	13.7 ± 2.7	6.3 ± 2.2	49		
F	10	10	6.2			

The quantity $dW_\odot/dz$ is determined with great uncertainty, although it is quite possible that the small positive value of it is real. This should mean that the subsystem of investigated stars is contracting towards the galactic plane, the farther the stars outward from the plane the greater the speed of contraction. But further we shall assume that $dW_\odot / dz = 0$.

The residual radial velocities (Δv_r) contain information on peculiar W-velocities of the investigated stars (w). The knowledge of w makes it possible to determine the dynamic parameter C in the form (see Kuzmin, 1952):

$$c^2 = (\sigma_w^2 / \sigma_z^2) = - (d^2\Phi / d^2z)_o \qquad (2)$$

where Φ is the galactic potential for the unit mass, σ_z^2 is the dispersion of z and σ_w^2 is the dispersion of w of investigated stars. The parameter C is related to the constants of differential galactic rotation A and B by Poisson's law

$$4\pi G\rho = c^2 - 2(A^2 - B^2) \qquad (3)$$

where G is the gravitational constant and ρ is the total mass density near the Sun.

To find C we shall use the values Δv_r instead of true values w. Such substitution results in the increase of the value σ_w^2 relative to reality and therefore, according to (2), in the increase of the value C. But the nearer the star is to the pole, the more accurate is the assumption that Δv_r is equal to w. Using the stars with $|b| > 75°$ (group 1, 60 stars) and with $|b| > 84°$ (group 2, 18 stars) we found C equal to 157 ± 14 km/sec/kpc and to 103 ± 17 km/sec/kpc, respectively. Having substituted A = 15 km/sec/kpc, B = -10 km/sec/kpc and C = 103 ± 17 km/sec/kpc in (3) we found $\rho = 0.15$ $M_\odot/pc^3$. Such total mass density does not correspond to the observed mass density near the Sun, $\rho_{obs} \approx 0.09 M_\odot/pc^3$, universally recognized at present. The investigated stars appear to be a mixture of disk population, old disk population and halo objects. But the formula (2) is strictly correct only near the galactic plane, where the condition

$$d^2\Phi / d^2z = \text{const} = -C^2 \qquad (4)$$

is satisfied. Therefore only the disk population stars may be used for such study. Besides, such mixture results in the increase of velocity dispersion and in the increase of value C. But there is no method for classification of each star under investigation according to population type.

We try to do this with the help of modulus w - modulus z ($|w|-|z|$) graph (Fig.1). In the region where the condition (4) holds, the stars move along the z-coordinate according to the law (see Parenago,(1954)

$$\ddot{z} = (d^2\Phi / d^2z)\, z = -C^2 z \qquad (5)$$

where $\ddot{z}$ is the acceleration. Then the trajectory of the star's motion in the coordinate plane w - z is an ellipse with semiaxes Z^0 and $W^0 = C\, Z^0$, where Z^0 is the amplitude in the z-direction and W^0 is the amplitude of the w-velocity. Suppose that the quantity C is close to that found by Oort (1932) (C^0 = 73 km/sec/kpc), which corresponds to the total mass density $\rho = \rho_{obs.}$ = 0.09 M☉/pc^3 exactly. This condition leads to some trajectories in the plane w - z (see Fig.1). The curve 1 is such a trajectory with amplitude Z^0 = 500pc and the curve 2 is that with amplitude Z^0 = 360pc. If the investigated stars move according to the law (5) (i.e. are the disk population stars) the stars below the curves may not have z-coordinates higher than the corresponding Z^0.

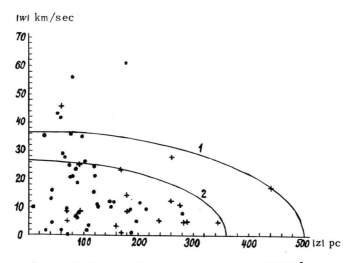

Figure 1. The $|w|-|z|$ graph for stars with $|b|>75°$; crosses are the stars with $|b|>84°$, points are all the other stars.

Using the stars of group 2 below the curve 1 we obtain $C = 79\pm13$ km/sec/kpc and using those stars below the curve 2 we obtain $C = 76 \pm 14$ km/sec/kpc. Hence the flatter the stars we study, the nearer to C^0 is the quantity of the dynamic parameter. The value $C = 76 \pm 14$ km/sec/kpc corresponds to $\rho = 0.11 \pm 0.02$ $M_\odot / pc^3$. That is in good agreement with $\rho_{obs.}$, the observed mass density near the Sun.

References

BLANCO, V.M., DEMERS, S., DOUGLASS, G.G., FITZGERALD, M.P.:
    1968, Publ. U.S. Naval Obs., second ser., XX, Washington.
BULON, J.:
    1963, Publ. Obs. Haute-Provence, 6, 187.
HEARD, J.F.:
    1956, Publ. David Dunlap Obs., 2,107.
KUZMIN, G.G.:
    1952, Publ. Tartu Astron. Obs. 32,5.
OORT, J.H.:
    1932, Bull. Astron.Inst.Neth. 6,249.
PARENAGO, P.P.:
    1954, Kurs zvezdnoj astronomii, Gostechizdat, Moscow, p.402.
SCHILD, R.E.:
    1973, Astron.J., 78,37.
VYSSOTSKY, A.N., JANSSEN, E.A.:
    1951, Astron.J., 56.58.
WILSON, R.E.:
    1953, General Catalogue of Stellar Radial Velocities, Carnegie Inst. of Washington, Washington, D.C.
WOOLLEY, R., EPPS, E.A., PENSTON, M.J., POCOCK, S.B.:
    1970, Roy. Obs. Ann., No.5.

The full text of this paper is published in Astron.Zhur. SSSR 1976, 53, 1 (see Soviet Astronomy 1976, 20, 64).

DISCUSSION

Bok: Will King tell us what he considers the composition of the massive halo.

King: I simply do not know. What I think I do know is that the massive halos that have been suggested for galaxies have a different structure from the galactic halo of which we normally speak. According to the best tracers, RR Lyrae stars and globular clusters, the density gradient in the halo is inverse-cube or even steeper. The so-called "massive" halos, on the other hand, seem to require an inverse-square law. This is why I feel that any "massive" halo is a different population from the conventional halo we are discussing.

Biermann: A question for Kron: could you elaborate on subdwarfs in Brook's work at Berkeley?

Kron: Among the faint stars in the survey, the stars with $0.4<(B-V)<0.8$ are expected to be mainly subdwarfs.

Buscome: It is reassuring that Kharadze and Bartaya find many supergiants F, G and K stars at high galactic latitudes as also Stock had done.

Bartaya: I wish to comment on the peculiar distribution of the Am stars which was mentioned today. I would like to recollect now that in Orion Ic stellar association whose age is estimated as being only a couple of million years, Smith has discovered Am stars among A-type stars on the evolutionary track in front of the main sequence. If we consider this and if we remember the presence of Am stars in older stellar clusters, it seems quite probable that among Am stars we have representatives of different subsystems.

Kharadze: The papers we have listened to during this day show that the problem of the Galactic Polar Structure seems to have become quite ready for a complex attack by widely coordinated facilities of many observatories and individual scientists. This should be kept in mind by the members of all three commissions participating in this Joint Discussion.

THE ROLE OF PROPER MOTIONS IN DETERMINING THE LUMINOSITY AND DENSITY FUNCTIONS

Willem J. Luyten
University of Minnesota, Minneapolis, Minnesota USA

The papers presented this morning form a good illustration of the old adage that you should never publish a paper that is completely correct, for then you can write only one paper. But if you publish a paper with a lot of errors, you are sure to be criticized, and you can write a second paper to answer your critics, and if the heat proves to be too much you can write a third paper withdrawing everything.

First a bit of history. It is difficult to pin-point when exactly this question of the star density in high latitudes first came up, but I feel sure that it is implied, to say the least, in the early work on faint blue stars by Malmquist (1927, 1938) and by Humason and Zwicky (1947). This morning Ivan King showed us a table of the changes in the frequency of occurrence among the stars in high galactic latitude as one goes to fainter and fainter magnitudes. This is exactly what I did in 1960 when I published a color-magnitude diagram for 4000 stars down to 19th photographic magnitude near the South Galactic Pole (Luyten 1960). These data had been obtained with the Palomar 48-inch telescope using Haro's three-image method. I also made up a similar diagram calculated from what I thought were then the best available data on the luminosity and density functions. The conclusion of my analysis was that there seemed to be rather fewer M stars than had been expected, and a great many more stars with the color of F and G stars than expected. This idea was not popular at the time; hence this paper has been carefully ignored until these ideas were used by others twelve years later without reference to the 1960 paper.

Later, Klare and Schaifers (1966) published results of an objective prism survey giving data for 1571 stars at high galactic latitudes. They made two fatal errors, however; one, they not only mentioned but actually used proper motions in order to separate giants and dwarfs, and two, they concluded that their results agreed with my luminosity function. Hence this paper too has been carefully ignored. It is surprising that in 1968 and again in 1976 other similar surveys described themselves as a first unbiased survey of faint red stars.

That word "unbiased" appears to be the key word in many of the recent papers. Spectroscopists and photometrists appear to be firmly agreed that anything done with or derived from proper motions is extremely biased. Now, what are the facts? Photoelectric observers are apt to claim that their observations of magnitudes and colors are subject to errors of $\pm 0\overset{m}{.}005$ (or even less). Hence if systematic errors of $0\overset{m}{.}2$ turn up, these amount to 40 times the mean error. When I did the Bruce Proper Motion Survey and the Palomar Proper Motion Survey (by hand), my motions had errors of $0\overset{''}{.}025$ annually. Forty times that would be a proper motion of one arc second per year, and I believe that anyone who has even heard of proper motions would know that such systematic errors are out of the question. Another salient comparison may be made as follows. If we should make a systematic error of, say, ten per cent, in the size of a proper motion, this eventually comes through as a thirty-three per cent error in the star density derived. But if, as in the present topic of discussion, a systematic error of ten per cent, i.e. $0\overset{m}{.}1$, is made in the (B-V) of an M dwarf, the estimate of the absolute magnitude will be off by nearly $1\overset{m}{.}5$, the distance by a factor of nearly 2 and the resulting star density by a factor of 8.

Spectroscopists also have their troubles. You all remember the case of T202, which was first classified as a white dwarf and later, by the same person, as a quasar, which means that by his own figures he first underestimated the luminosity by a factor of 10^{14}. A second case is AT Cancri, first announced as a cepheid because it is variable in light, then classified spectroscopically as a white dwarf. I measured the proper motion - but it doesn't have any, and a 12th-magnitude white-dwarf variable would be very interesting; so I strongly suspect this too to be a quasar. The third one is CD $-42°$ 14462, announced as the brightest single white dwarf at the St. Andrews Conference. A telephone call to Cape Town brought the detailed photometry, and hence at the Brighton IAU meeting it was announced again as a definite 9th-magnitude white dwarf. More recently a parallax of $0\overset{''}{.}001$ was determined for it; so again it is definitely not a white dwarf.

The proper way to conduct these investigations, apparently, is to take two or three papers with miniscule samples of stars, and one paper with a good solid systematic error in the colors, while ignoring all the basic data, especially those on proper motions, and then call this the total evidence. Next, as one of our columnists put it, "with the unerring academic eye for the false solution," you derive a new luminosity function. In this case it demands 29 stars nearer than two parsecs - we know four: the three components of Alpha Centauri and Barnard's star. Hence there must be twenty-five more, all brighter than $m = 12$! Does anyone believe such nonsense?

Recently Graham Hill processed a pair of Palomar plates on the North Galactic Pole with a 26-year interval and found 3000 motions of stars down to $m_{pg} = 21$ with $\mu \geq 0\overset{''}{.}035/\text{yr}$. Previously I had done a plate at R. A. $13^h 36^m$ and dec. $+6°$ with an interval of 23 years, on which some 2000 similar stars were found. Data for more than 5000 stars are shown in the table, where the abscissae indicate my very rough colors and where the ordinates express the quantity $\underline{H} = \underline{m} + 5 + 5 \log \mu$, which can also be written $\underline{H} = \underline{M} + 5 \log \underline{T}$, where $\underline{T}$ is expressed in a.u./yr. Since we have no parallaxes for these stars, $\underline{H}$ is the best statistical approach to absolute magnitudes; and for these proper-motion stars we have, roughly, $\underline{H} = \underline{M} + 5 \pm 2$. If the people who see the sky filled with nearby

Color index	0.0	+0.5	+1.0	+1.5	
$\underline{H}_{pg}$ = 7.0		2	2	2	1
8.5		6	10	3	
10.0	3	66	87	16	
11.5	19	121	246	77	5
13.0	36	113	263	155	38
14.5	43	99	215	274	118
16.0	42	86	216	467	320
17.5	14	28	88	308	613
19.0	4	5	8	50	599
20.5	1		1	5	220
22.0				1	46
23.5			1	1	4
25.0					1

	P321	$12^h 34^m$ + 30°		5149 stars
	P558	$13^h 36^m$ + 6°		

M dwarfs with very small proper motions were correct, then the three circled squares should contain by far the largest numbers. There is no evidence for this, and the two most heavily populated squares lie about 3 magnitudes further down; i.e., they represent M dwarfs with reasonable motions.

I have now analysed proper motions in more than 800 Palomar Survey fields; one thing which emerges from this is that the number of stars per unit area with proper motions larger than a given amount <u>definitely</u> decreases with galactic latitude. The simplest explanation of this, I believe, is that stars in high galactic latitude have larger tangential velocities because the main motion is parallel to the galactic plane. If now one wants to add large numbers of stars in high galactic latitude with small tangential velocities, then the end result will be equidensity surfaces in the shape of prolate ellipsoids with the long axis perpendicular to the galactic plane.

Recently I completed and published a new catalogue of stars with proper motions larger than 0".5 annually. It contains 3600 entries, 2000 of which have come from the Bruce and Palomar Surveys. I do not now intend to derive a new luminosity function, because as of now the number of new accurate parallaxes being determined for very faint stars, especially at the U.S. Naval Observatory in Flagstaff, Arizona, is so large that in five or ten years we shall be in a much better position to make such a solution. In fact, in about ten years we should be able to get an almost definitive determination for the luminosity function in the solar neighborhood.

By now I have published about 30,000 proper motions for stars near the North and South Galactic Poles, with estimated colors for 27,000. In addition I have data on magnetic tape and computer printouts for another 60,000 stars, which I shall be glad to make available.

Recently an article appeared which gave a rehash of the colors of all of nine stars. It took up 4 pages and had 28 references. On that basis I should be entitled to ask the same journal to publish 3000 4-page papers on my proper motions and another 3000 4-page papers on my colors, for a total of 24,000 pages, including 168,000 references, i.e., if I followed the mutual-admiration society of these Messiahs of the Missing Mass in copiously referring to each other but never to the basic data. Is this what our science has come to? This is not astronomy or astrophysics, it is astrofantasy.

Ours is the age of automation and the computer. Since I had nothing to do either with the design of or the fabrication of my machine, I am not boasting when I say that the engineers at Control Data Corporation have achieved success brilliantly and superlatively in automating the proper-motion survey. The best we ever did was to process a pair of plates on each of which the computer counted 585,000 star images. The scanning took three hours, the computerizing took $2^h 23^m$ of central-processor time and 15 minutes for the peripherals. Thus in less than 6 hours we determined $\underline{x}$, $\underline{y}$, and diameter to one micron for 1,170,000 stars, which works out as better than 50 stars per second; this is 200 times faster than GALAXY. In addition, we made a least-squares solution and the printout showed right ascension to $0^s.1$, declination to $1"$, red magnitude to $0^m.1$, size of proper motion to $0".001/yr$ and direction to $1°$ for 400 stars.

What we need is automation of the photometry for these 30,000 stars near the galactic poles, and this should not be too difficult. I know whereof I speak, for we have several times processed a pair of blue and red Palomar plates and determined colors. And this goes so fast that we could easily repeat it five or six times and thus virtually eliminate any machine-introduced errors and end up with only the plate error. So I suggest that the photometrists get busy - but no systematic errors, please.

Astronomy, like all other fields of human endeavour, has had its funny theories and observations. My generation especially will remember the rotation of what were then known as "spiral nebulae". But how many remember the new theory of gravitation proposed to explain these motions? I think it was due to J. H. Jeans and E. W. Brown, and their force of gravitation was dependent not only upon the distance but also on the angle of direction. Then there was the theory for the origin of the Solar System by Lyttleton and Hoyle, which ended up in a game of cosmic billiards. Here the Sun was supposed to form a binary; and a third star came in, first collided with the other component, glanced off, then hit the Sun, which at that precise moment exploded as a nova. I took a dim view of this.

Some of you may remember, some ten or fifteen years ago, an advertisement in The New Yorker for the Taittinger Champagne Co. It was a very striking one and showed only the black silhouette of the rather distinctively shaped bottle. Then, at the top, it said: "This is the finest champagne in the world.*" The asterisk refered the reader to a footnote which said: "This is probably an understatement."

In my fifty-five years of being in astronomical research this enterprise of the plethora of nearby M dwarfs with no proper motions is the most absurd I have ever experienced. And this is certainly an understatement.

REFERENCES

Humason, M. L., and Zwicky, F. 1947, Astrophys. J., 105, 85.
Klare, G., and Schaifers, K. 1966, Astron. Nachr. 289, 81.
Luyten, W. J. 1960,'A Search for Faint Blue Stars. XXII. The Star Density in High Galactic Latitude,' University of Minnesota.
Malmquist, K. M. 1927, Lund. Medd., Ser. II, No. 37.
Malmquist, K. M. 1938, Stockholm Obs. Ann. 12, No. 7.

Concluding Remarks Part II

A. Blaauw
Leiden Observatory

In research in the galactic polar caps we may roughly distinguish two categories:

I. Research on problems concerning the galactic population at large distances from the plane, including the density and velocity distribution and the field of force at large z.

II. Research on problems concerning the population in the immediate neighbourhood of the sun; at high galactic latitudes it is easier to discriminate the nearest stars than at low latitudes - see, for instance the investigations of the M dwarf population.

Most of the papers presented in part II of this Joint Discussion deal with the first category, and of these, the distances reached are mostly within a few Kpc. Classical investigations, some of them dating from more than thirty years ago, have provided our current knowledge of the force $K(z)$ perpendicular to the galactic plane up to about 1500 pc and the large scale features of the density distribution. What emerges from recent and current work, is attempts to improve the knowledge of $K(z)$ and to extend it to greater distances; and refinement of the density and velocity investigations by more and more detailed discrimination of population types. The dominating problem is that of the separation of the stars according to two basic parameters: age and chemical composition. Kinematic and space distributional properties for subgroups according to these two parameters will be the necessary elements on which a theory of the (local) evolution of the Galaxy is to be built. The vague notion, that metal abundance may be taken to be an equivalent of age is abandoned and replaced by the recognition that at any epoch, star formation may have occurred with a considerable spread in the resulting stellar metal abundances.

The investigations reported show that insight in these aspects is now building up, but it is in most cases limited to the layers within a few Kpc. These investigations therefore may hopefully lead to understanding of the later stages of the local galactic evolution. For the earlier stages, comprising most of the lifetime of the galaxy, we need much more information about the stars beyond a few Kpc. Here, for instance, the exploratory work of W. Becker and his associates for stars down to 19th photographic magnitudes would require precise photoelectric follow-up, as a first step toward the finer segregation of age and abundance groups, pending the measurement of kinematic properties which for the moment still seems hard to attain. But we have already heard about interesting probes to very large distances like those reported by Kron et al on the basis of B, V, R photometry down to V = 19 m and by Davis Philip on very distant faint Horizontal Branch stars. With the coming into operation of the new generation of large telescopes (Kitt Peak, CT10, ESO, AAT, FCH) a breakthrough to these distant domains in space and time may well be within reach.

CURRENT RESULTS AND SUGGESTIONS FOR FUTURE WORK

M. McCarthy, A.G.D. Philip, I. King, and U. Steinlin

In the days following our Joint Discussion the Co-editors were joined by two of the participants, Dr. Ivan King and Dr. Uli Steinlin, in an effort to evaluate results reported as well as the problems posed for future work. As a result of these discussions at Grenoble the present listing was made. This list was presented to members of the Organizing Committee but since it was impossible in the last days at Grenoble to convene this group or to speak collegially with all participants it must remain the responsibility of the above-named authors. It cannot be a complete resumé nor can it presume to represent adequately the varied opinions expressed at Grenoble on 25 August. We hope here only to attempt a synthesis of certain evident results and to outline certain prospects for exploring further the exciting problems of structure and evolution in the galactic polar caps.

1. Current results

There is widespread convergence of results from studies of faint M stars which tends to confirm the conventional luminosity function for the solar neighborhood.

The diversity of luminosity functions (besides the purely local one) as outlined earlier in classical papers, is now clearly confirmed. The luminosity function must be described in terms of a mixture of stellar populations with different concentrations towards the galactic plane.

Recent advances and developments in instrumentation, automation and data reduction techniques have yielded substantial improvements in the measurement of stellar radiation and an increased possibility for the systematic control of errors. Through observations from satellites important regions in the far ultraviolet have become available for photometric and spectroscopic studies. Meanwhile the new infrared techniques yield much needed information especially for the study of late-type stars.

2. Problems to be solved

The luminosity function of each component of the galactic population at higher z distances must be determined separately. To determine the spatial, kinematic and chemical characteristics of these components will be a major task. We must consider how this can be verified by spectroscopic, photometric and kinematic measurements.

The individual M dwarfs found in the direction of the polar caps are undoubtedly authentic. What is needed is a higher accuracy concerning their absolute magnitudes. How can this be best obtained?

Many important problems in radial velocity and parallax remain to be solved, but present techniques and approaches are promising. A moral of our experience is that future progress will depend very much on the accuracy and reliability of our fundamental data.

It is not clear whether studies of the local halo are related to the larger problem of the possibility of a massive halo of the galaxy. Does it exist and how can it be verified?

Photometric methods for a reliable multi-dimensional classification of faint late-type stars which go beyond the limited possibilities of standard three- or four-color photometry must be sought. How can these be best developed?

JOINT DISCUSSION NO. 5

STELLAR ATMOSPHERES AS INDICATOR AND FACTOR OF STELLAR EVOLUTION

(Edited by R. Cayrel)

Organizing Committee

R. Cayrel (Chairman), C. Jaschek, L. Mestel, and B.E.J. Pagel.

CONTENTS

5. STELLAR ATMOSPHERES AS INDICATOR AND FACTOR OF STELLAR EVOLUTION

Introduction-Observational Evidence

B.E.J. PAGEL / Introductory Talk   103

R. CAYREL and G. CAYREL / Observational Evidence for Atmospheric Physical Characteristics Relevant to Stellar Evolution   105

R. FOY / Behaviour of Microturbulence with Evolution   115

B.E.J. PAGEL / Observational Evidence for Atmospheric Chemical Composition Peculiarities Relevant to Stellar Evolution   119

Stellar Atmospheres as Boundary Condition of Internal Structure

P. DEMARQUE / Sensitivity of Internal Structure to the Surface Boundary Condition   137

R.N. THOMAS / Boundary Conditions with Mass Loss: General Considerations   143

C. DE LOORE / Mass Loss in Stars of Moderate Mass by Stellar Winds and Effects on the Evolution   155

D. MIHALAS / Boundary Condition with Mass Loss: The Radiatively-Driven Wind Model   175

Alteration of Surface Chemical Composition of Stars by Transport Phenomena

G. MICHAUD / Stratification of Elements in a Quiet Atmosphere: Diffusion Processes   177

G. VAUCLAIR and S. VAUCLAIR / Competition Between Diffusion Processes and Hydrodynamical Instabilities in Stellar Envelopes   193

Alteration of Surface Chemical Composition of Stars by Nuclear Reaction

M. SCHWARZSCHILD / Mixing Between Burned Core Material and Surface Layers 205

A.M. BOESGAARD / Decay of Light Elements in Stellar Envelopes 209

R. CAYREL / Conclusion 217

INTRODUCTORY TALK

B.E.J. Pagel
Royal Greenwich Observatory

Welcome to this Joint Discussion on stellar atmospheres and interiors. For stellar structure, the atmospheres are of importance for two reasons, first as a source of information on effective temperature, gravity (or mass: luminosity ratio), chemical composition and dynamical effects, and secondly as an outer boundary condition for stellar models.

In the last few years there have been substantial advances in stellar atmospheres as a result of both improved theories and new measurements. In the particular field of abundance determination, non-LTE theory has at last come of age and provided a certain number of corrections that are both realistic and significant, for example in reconciling the neon abundance in B stars with that in nebulae and solar flare particles. Solar abundances have been improved by the provision of better oscillator strengths and better treatment of line broadening, generally with the effect of leading to still closer agreement with carbonaceous meteorites. Nevertheless, he would be a bold man who claimed to know the initial solar abundance parameters Y and Z to better than, say 25 per cent, even if we grant that the photosphere is a true sample of the initial interior abundances - an assumption that is now being questioned (again) in view of the solar neutrino problem.

Another phenomenon revealed by quantitative stellar spectroscopy is the presence of velocity fields including the notorious "microturbulence". The existence of such small-scale velocity fields has been questioned on grounds (among others) of the effect of a non-LTE source function in pushing up the flat part of the curve of growth. In G and K giants and supergiants, however, one needs microturbulent parameters approaching 2 km s^{-1} and 5 km s^{-1} respectively even if one uses a curve of growth in which all re-emission in lines is neglected; so the velocity fields have to be real and they can be credibly interpreted as a field of acoustic waves coming up from the convective zone.

The most striking evidence for the existence of these velocity fields comes from stellar chromospheres in the form of the Wilson-Bappu effect in Ca^+ and its analogues in $H\alpha$ and Mg^+. The slow but steady increase in width with luminosity, and the decline in K-line emission with age among main-sequence stars, are powerful signs of an intimate relationship between stellar atmospheres and basic features of internal structure and evolution. A fully coherent theory of the Wilson-Bappu effect still does not exist, but a reasonable scaling law can be derived on the assumption that the FWHM is proportional to the Doppler width with something like 10^{-5} of the radiative flux being propagated upwards as mechanical energy with increasing velocity amplitude through the chromosphere. This fraction of 10^{-5} is comparable to the power needed both to heat the solar corona and to drive mass loss according to the formula suggested by Reimers, although this mass-loss rate is still very uncertain, of course.

These problems are also related to the effect of atmospheric phenomena on the outer boundary condition for stellar models, going back to the old arguments as to whether the dog wags its tail or the tail wags the dog. From textbooks one has the impression that for hot stars there is really no problem, but in cool stars with convective envelopes the adiabatic constant, and therefore the whole structure, depends on a proper treatment of the atmospheric opacity and other details. I look forward to hearing today about the current state of this problem with special reference to the effects of mass loss. So let battle commence and I trust that we shall leave at the end of the day with a better understanding of both stellar atmospheres and interiors.

OBSERVATIONAL EVIDENCE FOR ATMOSPHERIC PHYSICAL CHARACTERISTICS RELEVANT TO STELLAR EVOLUTION

R. and G. CAYREL
Observatoire de Paris

1. Introduction

As a star burns its nuclear fuel, its radius R and its luminosity L are modified. Its mass may as well be affected if the mass loss rate has a time scale comparable to the nuclear time scale ; this is likely to occur for stars of very high luminosity. Currently, the change in radius R and luminosity L of an evolving star is described in the so-called theoretical Herzsprung-Russel diagramme with in abscissa the logarithm of the effective temperature defined by :

$$\sigma T^4_{eff} = \frac{L}{4\pi R^2} \quad (1)$$

(σ Stephan's constant, π usual meaning) and in ordinate Log L.

The path described by the star in this reference frame is known as the "evolutionary track" of the star. If we want to restrict our attention to the stellar atmosphere itself, it must be noted that a stellar atmosphere of given chemical composition is determined not by effective temperature and luminosity, but effective temperature and <u>gravity</u>. Unfortunately, both in the language and in practice, the concepts of luminosity and gravity are often mixed up. For example, the "luminosity classes" in MK classification are in fact gravity classes, and gravity criteria in many classification systems are most of the time calibrated in terms of absolute magnitudes. If mass loss does not occur during stellar evolution this confusion between luminosity and gravity is not too great of a problem, because there is a one-to-one relationship between luminosity and gravity along each evolutionary track :

$$g = \frac{G M}{R^2} = \frac{G M}{L} \quad 4\pi\sigma T^4_{eff} \quad (2)$$

But, if mass loss does occur at the same stage of evolution, a low gravity (for example) may either be the result of a large radius or of a low mass, and the confusion does matter. This is not all an academic question, as we shall see later in the case of our neighbour Arcturus.

2. Evolution of the main sequence from photospheric data
2.1 Theory

By far the most significant evidence of stellar evolution displayed by the atmosphere of a star is the decrease in gravity caused by the increase of radius undergone as the star moves off the main sequence towards the giant branch. This effect is shown in Figs. 1 and 2.

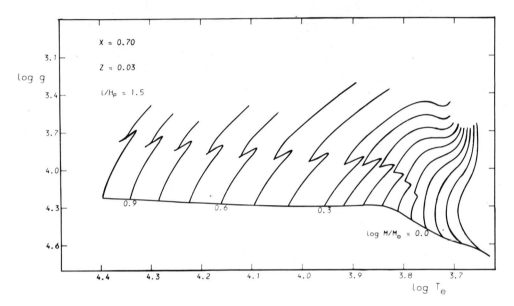

Fig.1 : Evolutionary tracks in the (log T_{eff}, log g) plane according to Hejlesen (courtesy I.A.U. Colloquium no.17).

Figure 1 shows the evolution of stars following the fate of individual objects of various masses (evolutionary tracks) ; whereas Fig.2 shows where stars of all masses fall in the diagram (log T_{eff}, log g) after prescribed times (Isochrones). Figure 3 shows the same type of diagramme computed taking into account overshooting from the convective core (Maeder, 1976). The locus of stars at zero age, referred to as the zero age main sequence (ZAMS), is of particular interest, as it defines the starting point of the evolution. The location of the ZAMS on the (log T_{eff}, log g) phase depends upon the initial chemical composition of the star, namely upon the fraction by mass Y of helium, and the fraction by mass Z of elements heavier than hydrogen and helium, lumped together. As usual we designate by X the fractional image of hydrogen. One has obviously :

$$X + Y + Z = 1 \quad (3).$$

Before anything can be derived from the knowledge of (log T_{eff}, log g) for a given object of its evolutionary status, Y and Z must be known.

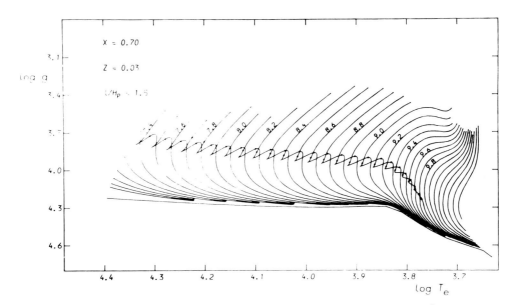

Fig. 2 : Isochrones in the (log T_{eff}, log g) plane according to Hejlesen (courtesy I.A.U. colloquium no.17).

Figure 4 shows the sensitivity of the ZAMS to the value of Z. It must be noted that the effect of a change in Z is not only an upwards or downwards translation of the ZAMS but also that a star of given mass moves in effective temperature when Z changes. It is interesting to note that a star of one solar mass (big dots on Fig. 4) has its ZAMS effective temperature moved from 5600° to 6900°K when Z varies from 0.03 to 0.04 respectively, its gravity remaining practically constant at log g = 4.5. If the star is such that the diffusion processes discussed in Michaud's and Vauclair's papers do not occur thanks to efficient stirring mechanisms, it is possible to infer from a detailed spectroscopic study of the atmosphere of the star and therefore to account properly for this Z effect from purely atmospheric observations. More troublesome is the effect of a change in Y as shown on Fig.5. The ZAMS seem little affected by a change in Y (or X = 1 - Y - Z, fraction of hydrogen by mass) as long as the mass of the star is ignored. However, a star of one solar mass, with Z = 0.02, has its ZAMS effective temperature moved from 5200°K to 6750°K when Y varies from 0.18 to 0.38, respectively. The difficulty here is that Y is generally unknown by lack of helium lines in the photospheric spectrum of stars later than the spectral type B0. If there is still the theoretical possibility of deriving a helium abundance from chromospheric lines one must admit that this is not yet a very practical nor reliable procedure. Therefore we stress the point that before any " spectroscopic" mass can be claimed (even on the ZAMS), a reliable determination of Z and Y must be available.

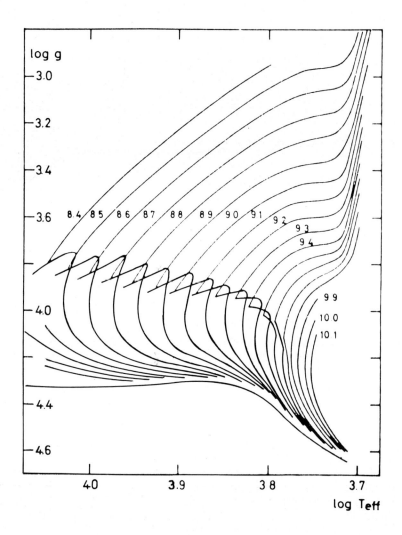

Fig. 3 : Isochrones in the (log T_{eff}, log g) plane according to Maeder, with convective overshooting in the core (courtesy Astron. & Astrophys.)

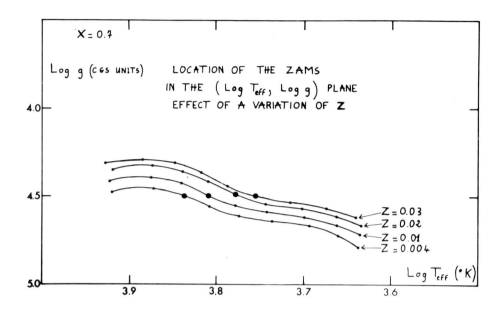

Fig. 4 : Variation of the ZAMS in the $(\log T_{eff}, \log g)$ plane in function of the heavy elements content.

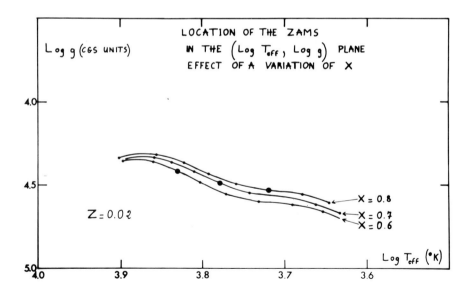

Fig. 5 : Variation of the ZAMS in the $(\log T_{eff}, \log g)$ plane in function of the helium content.

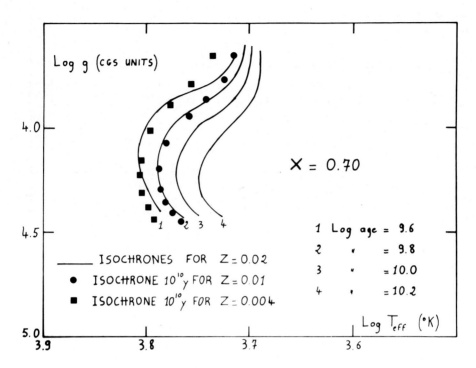

Fig. 6 : Dependence of isochrones upon the content Z in heavy elements. (Evolutionary tracks according to Hejlesen).

Equally interesting is the effect of Y and Z on the age of an evolved star. Figure 6 shows the isochrones for various ages corresponding to $Z = 0.02$. The open dots represent the isochrone 10^{10} years for $Z = 0.01$. One notes that this isochrone practically coincides with the isochrone ($Z = 0.02$, 6.3×10^9 years at log age = 9.8). This demonstrates the sensitivity of age determination upon actual value of Z. An error by a factor of 2 on Z produces for a sub-giant an error of 0.2 dex on its age. The more metal deficient is the object, the older it is for a given position in the ($\log T_{eff}$, $\log g$) plane.

A similar effect exists for the content in helium. Figure 7 illustrates this effect.

One notes that the more helium poor is the object, the younger it is for a given location in the ($\log T_{eff}$, $\log g$) plane. Figures 4 to 7 are all based on Hejlesen evolutionary tracks (Perrin et al.1977).

2.2 Observations

It is possible to locate a given star in the ($\log g$, T_{eff}, $\log g$) by using detailed analyses of stellar atmospheres. This has been done recently by M.N. Perrin (Thesis, Dec. 1975, Univ. Paris VII, and M.N. Perrin et al., 1977). Much more frequently the observations are done

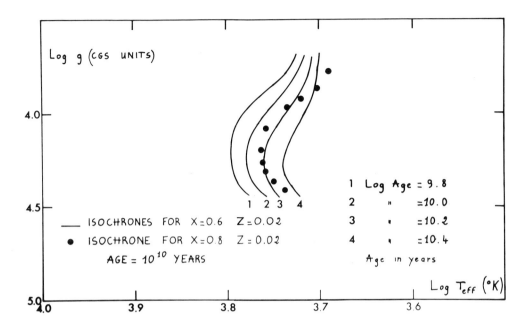

Fig. 7 : Dependence of isochrones upon the content Y in helium. (Evolutionary tracks according to Hejlesen).

on a classification system, (for example the Stromgren Crawford uvby β system, the Barbier-Chalonge-Divan (φ_b, D, λ_1) system, the DDO photometric system, the Cambridge system, the Geneva 7-color photometric system, the Danish system ,etc...).

The internal accuracy of such systems may be very high but the main problem is their calibration. A recent and very good review of this question may be found in the proceedings of the conference "Multicolor photometry and the theoretical H R diagram " cited in the reference list.

Of particular interest in the cited reference are those papers on the calibrations of the uvby β system respectively by Crawford, by Breger and by Philip and Matlock ; the calibration of the DDO system by Janes and McLure, and a paper on empirical effective temperatures, bolometric corrections and fundamental stellar properties by Code. The calibration of the Cambridge system can be found on a series of papers by Williams (1975) and in a paper by Bell (1970). The danish system has been calibrated by Hansen and Kjaergaard (1971) and the consistency of their calibration has been investigated by Olsen (1971). On the average it is difficult to obtain log/T_{eff} with an accuracy greater than 0.01 to 0.02 dex and log g better determined than 0.1 dex and often worse. The internal accuracy of some photometric systems is of course much greater. For example the uvby β system has a discrimination power almost one order of a magnitude larger for spectral types near F0 IV-V.

It is nevertheless discouraging to see that in the case of Arcturus, for which top quality spectrophotometric data are available, a controversy on the actual log g value is still going on, with extreme values of 0.9 ± 0.35 (Mäckle et al., 1975) to 1.7 ± 0.2 (Ayres and Johnson (1976) and log g = 1.8 (Martin Mullor, 1977) letting an uncertainty of over an order of magnitude on the mass of Arcturus.

The pattern of isochrones in the ($\log T_{eff}$, log g) plane makes that the errors on age and mass determination corresponding to a given error on $\log T_{eff}$ and log g vary very much according to the location in the plane. The funelling effect of evolutionary tracks along the Hayashi limit makes age and mass determinations extremely unaccurate for K giants. The most favourable region is the one of stars evolved by one or two magnitude above the ZAMS, i.e. stars with log g between 4.5 and 3.6 and an effective temperature not less than 5500°C.

2.3 Interplay between internal structure, stellar atmospheres and basic stellar data.

So far we have seen that taking for granted the validity of internal structure computation it is possible to determine from stellar atmosphere characteristics *alone* , the evolutionary stage (or age) and mass of a star, provided :

i) some assumption is made on the helium content of the object ,

ii) the Z value found in the atmosphere is representative of the whole star ,

iii) mass loss has been insignificant during evolution.

If the star is member of a multiple system the mass of the star may be available from astrometric observations and assumption (i) can be relaxed.

But it must be stressed that having M, g and T_{eff} does not allow us yet to check observationally the validity of internal structure computations.

If the distance of the star is known then the luminosity of the star is known. Unfortunately that does not give any further way of checking the internal structure as the radius or the luminosity can be derived from g , m, and T_{eff} already. If an apparent diameter of the star is known R can be derived directly without using the effective temperature determined via the stellar atmosphere. If one is tempted to completely by-pass the data derived from the stellar atmosphere one needs to determine L from the absolute magnitude, R from apparent diameter plus distance, and   from the orbital motion (companion star or planet is needed). There are only 4 stars with such data left : the sun, Procyon, Sirius and Spica (cf Code in Philip and Matlock ref.)

In conclusion, it is not possible to check the validity of internal structure computations by observing unrelated individual stars, as the number of parameters which can be determined is always smaller than (or equal to) the total numbers of parameters from which the star depends on (M, Y, Z, age).

It is possible to check internal structure computation only when one has a cluster of stars, because then the assumption that all stars in the cluster have the same age and the same chemical composition decreases the number of free parameters for all stars, except one of them.

3. Evolution from chromospheric lines

3.1 Evolution from chromospheric width.

In 1957, Wilson and Bappu made the amazing discovery that the width of the Ca II K_2 chromospheric emission lines was merely in a one-to-one relationship with the absolute visual magnitude of the star. A discussion of this question is given in another paper of this joint discussion (by B.E.J. Pagel). I shall then just mention here the fact, as a way of telling from a chromospheric line if the star is near zero age or how much evolved it is, from its departure in luminosity from the ZAMS. A recent paper by O.C. Wilson (1976) updates this remarkable discovery. Preston has shown that a similar effect exists with the shape of the core of H .

3.2 Evolution from chromospheric intensity of Ca II K_2 line

On the main sequence and at a given effective temperature the intensity of Ca II K_2 emission line has a wide spread. Wilson (1963) has shown that this spread might be due to a time variation of the emission, the emission being stronger in young stars then in older stars. Wilson and Wooley (1970) have been able to confirm this hypothesis using kinematical data and Wilson and Skumanich (1966) using a correlation with the c index of Strömgren classification. Skumanich (1972) has also claimed a correlation with the Li resonance line strength and rotation and has suggested an inverse square root time variation of the emission.

4. Evolution from turbulence in the atmosphere

The subject is dealt with in the next paper by R. Foy and will not be approached here.

5. Conclusion

Atmospheric physical parameters supply ample evidence for stellar evolution, mainly through the decrease of the surface gravity as the star evolves off the main sequence.
When one attempts to use this criterion quantitatively one is faced with several problems. The main one is the fact that the helium content is not known by direct spectroscopic determination for most star locations on the HR diagram. A second one is the difficulty on obtaining a proper calibration of spectroscopic gravity criteria in terms of actual values of effective temperature under gravity. Finally there are very few stars for which all fundamental parameters mass, radius, luminosity and chemical composition (including helium) are known with satisfactory accuracy.

We shall conclude by saying that atmospheric physical parameters are an essential tool in connecting observations with internal structure computations as they supply three of the fundamental parameters, (Z, T_{eff}, g). The ultimate goal is of course to combine these three parameters with other independent data (mass or absolute magnitude) in order to interpret or to check the theory of stellar evolution.

REFERENCES
AYRES, T.R., JOHNSON, H.R. 1976 Bull. Am. Astron. Soc. $\underline{8}$,303.
BELL, R.A. 1970, Monthly Not. Roy. Astron. Soc. $\underline{148}$, 25.
HEJLESEN,P.M. I.A.U. Colloquium No.17. L'âge des étoiles (Ed. G. Cayrel et A.M. Delplace, Meudon,1972).
MAEDER,A. 1976, Astron. Astrophys. $\underline{47}$,389.
MACKLE, R., GRIFFIN, R., GRIFFIN, R. and HOLWEGER, H. 1975, Astron. Astrophys. $\underline{38}$,239.
OLSEN,E.M. 1971, Astron. Astrophys. $\underline{15}$,161.
PERRIN,M.N., HEJLESEN,P.M. ,CAYREL DE STROBEL,G., CAYREL,R. 1977, Astron. Astrophys. (in press).
PHILIP,A.G.D. and HAYES,D.S. Multicolor photometry and the theoretical HR diagram, Dudley Observatory Reports, No.9 (1975) Albany, New York.
PRESTON,G.W., KRAFT,R.P. and WOLFF, S.C. 1964, Astrophys.J. $\underline{140}$,235.
WILLIAMS, P.M. 1975, Monthly Not. Roy. Astron. Soc. $\underline{170}$,343.
WILSON,O.C. 1963, Astrophys. J. $\underline{138}$,832.
WILSON,O.C. and BAPPU,M.K.V. 1957, Astrophys. J. $\underline{125}$,661.
WILSON,O.C., SKUMANICH,A. 1964, Astrophys.J. $\underline{140}$, 1401.
WILSON,O.C., WOOLEY,R. 1970, Astrophys. J., $\underline{171}$,565.

BEHAVIOUR OF MICROTURBULENCE WITH EVOLUTION

Renaud FOY
Observatoire de Paris, France

It is reasonable to suppose that the state of the convective zone underlying the atmosphere of cool stars correlates with the atmospheric kinematics. Following Iben (1967), as a star evolves from the red giant tip to the helium burning core phase, the convection rapidly decreases. Then the microturbulence can be expected to decrease too.

To observationaly check this, I have analysed in detail a sample of 21 G and K giants. I have found that the microturbulent velocities ξ lie

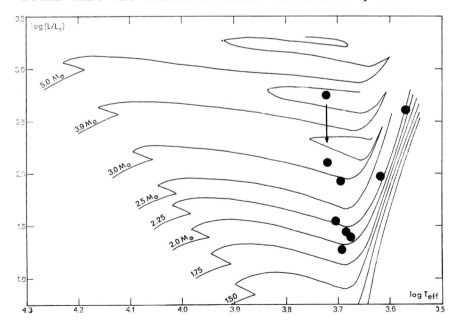

fig 1. Theoretical HR diagram for eight Hyades giants.

around ξ = 1.5 km/s. But few giants have a markedly lower value of ξ, that is to say $\xi \lesssim$ 1.0 km/s. Careful discussions of available data allow one to conclude that these low microturbulence stars are likely in the helium burning core phase.

For example, on the grid of evolutionary tracks shown in figure 1, I have plotted the eight Hyades giants of the sample. A difficulty occurs only for HD 71369. Assuming it is evolving, as the other ones on the red giant branch would imply that its mass is larger than 4 $\mathcal{M}_\odot$, which is very much too large for a Hyades star; therefore HD 71369 is not evolving on the red giant branch, and so has to be in the helium burning core phase. Then its mass is near 2 $\mathcal{M}_\odot$, which quite agrees with that deduced from the dynamical parallaxe, namely 1.5 $\mathcal{M}_\odot$. Its location then moves as shown by the arrow of figure 1. This result is supported by the one magnitude brighter luminosity of HD 71369 with respect to the other Hyades giants with similar R-I coulour indices, as shown by Eggen (1972).

But HD 71369 has another peculiarity: it has a markedly lower microturbulence ξ = 0.8 km/s than the seven other stars, for which a mean value is $\langle\xi\rangle$ = 1.55 ± 0.14: this difference is highly meaningful.

Therefore, in the case of HD 71369, we have equivalence between giant with an helium burning core and low microturbulence giant. Five other stars lead to the same conclusion.

The identification of the helium burning core phase will ever be open to criticism in the case of field stars. Therefore, I have also studied how the microturbulence varies along the red giant branch where stars have a very much greater probability to be observed. This work has been done with my student Luc Vigneron.

We have redetermined physical atmospheric parameters of 74 stars for which line equivalent width data are available in literature. From this work, the effective temperature appears to be not correlated with ξ. But the gravity is strongly correlated with ξ : ξ increases as g decreases(fig2).

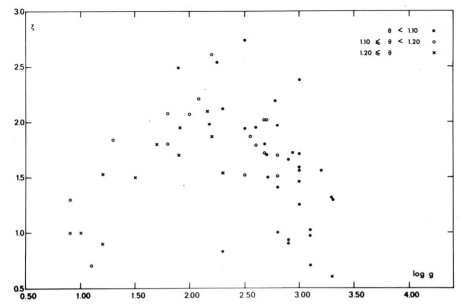

fig 2. Microturbulent velocity ξ versus logarithm of the gravity for cool giants.

As a star evolves along the red giant branch, both its gravity and its effective temperature decrease: note that the temperature also decreases along this sequence. I think that if we project evolutionary tracks on the ξ/log g plane, the red giant branch would be on the right decreasing part and the helium burning core branch would be on the left increasing part, since, with the exception of HD 71369 (log g = 2.2; ξ = 0.8), the giants which are very likely in the helium burning core phase lie on this second branch.

This correlation between microturbulence and degree of evolution is also shown in the theoretical HR diagram of figure 3. This diagram shows a segregation in the sense that the most evolved stars have a larger microturbulence than the less evolved ones.

I wish to emphasize the three main conclusions from the present study.

Firstly, the microturbulence in cool stars is a real phenomenon and not a spurious effect, as claimed, for instance by Worrall and Wilson (1972).

Secondly, the microturbulence is a good indication of the degree of evolution of giants. It allows us to distinguish between field stars on the

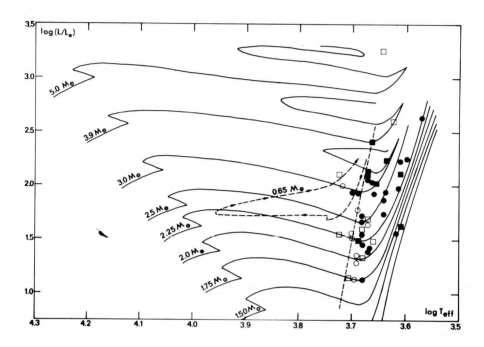

fig 3.Theoretical HR diagram for reanalysed cool giants. Open and filled symbols respectively refer to low and high values of the microturbulent velocity.

horizontal branch and the red giant branch.

Thirdly, the microturbulence in stellar photosphere of cool stars varies as the strength of the underlying convective zone: this supports the idea that microturbulence is induced from convection, presumably by the mechanism of overshooting.

References

Eggen,O.J., 1972, Publ. Astron. Soc. Pacific 84,406
Iben,I.Jr., 1967, Ann. Rev. Astron. Astrophys. 5,571
Worrall,G., Wilson, A.N., 1972, Nature 236,15

OBSERVATIONAL EVIDENCE FOR ATMOSPHERIC CHEMICAL COMPOSITION
PECULIARITIES RELEVANT TO STELLAR EVOLUTION

B.E.J. Pagel
Royal Greenwich Observatory, Herstmonceux, Sussex.

ABSTRACT

Abundance peculiarities in successive stages of stellar evolution are reviewed. Main-sequence stars show anomalies in lithium and, on the upper main sequence, the Am, Ap and Bp effects, which may be largely due to separation processes, and helium and CNO anomalies to which nuclear evolution and mixing could have contributed. Red giants of both stellar Populations commonly show more or less extreme variations among the C, N, O isotopes, sometimes accompanied by s-process enhancement, due to mixing out in various evolutionary stages. Detailed anomalies expected from galactic evolution are also briefly considered. Novae show strong effects in C, N, O and synthesis of heavier elements is displayed by the supernova remnant Cassiopeia A.

1. INTRODUCTION

Forty years ago, known abundance peculiarities were confined to the cool carbon and S stars, and it is still quite a good approximation to regard the vast majority of stars observed as having the same composition apart from minor variations in the metal: hydrogen ratio. When examined in detail, however, hardly any two stars are identical, particularly with regard to delicate features like Li and $^{12}C/^{13}C$. So it is convenient to regard the Sun (for which we have the most data) as a standard, describing departures from it as "abnormalities". These are generally interpreted as a consequence of four types of process:

A. Nuclear reactions and mixing in the star itself, having a direct bearing on its evolution.

B. Variations with time and position in the composition of the interstellar medium (ISM) from which the star was formed, resulting from galactic evolution which has been influenced in turn by the evolution of supernovae and other stars, now dead, which have

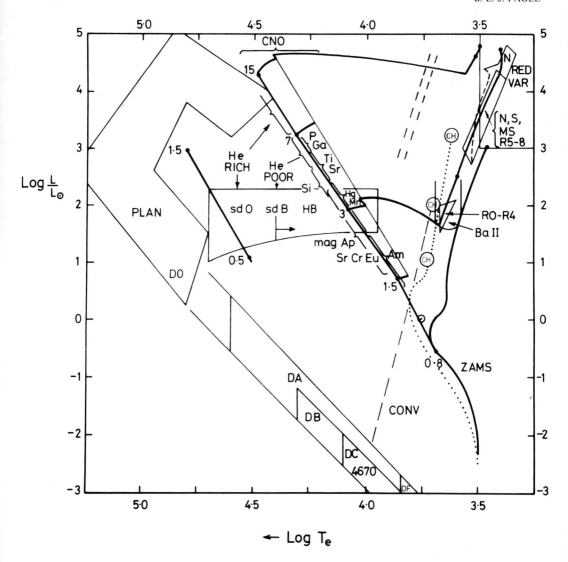

Figure 1. Approximate position of stars showing various abundance anomalies in the HR diagram. Full lines, ZAMS and partial evolutionary tracks adapted from Paczynski (1970) with masses marked along the ZAMS. The helium main sequence, planetary nebula, horizontal-branch and white dwarf regions are also shown, following Greenstein and Sargent (1974), and white dwarf classes after Weidemann (1975). The dotted line shows a schematic main sequence and evolutionary track for Population II, while various broken lines show roughly the cepheid instability strip, the transition to surface convective zones, and the He shell-flashing locus for Population I after Scalo (1976).

enriched the ISM.

C. Diffusive separation processes.

D. Selective or unselective accretion from the ISM.

In what follows I shall describe some abundance peculiarities observed in successive stages of stellar evolution from the main sequence onwards with special reference to category A above, with a brief aside on category B. Category C will be discussed by G. Michaud in these Proceedings.

2. STARS NEAR THE MAIN SEQUENCE

Apart from the light elements, especially Li, few anomalies might have been expected to occur near the main sequence, but in fact there is quite a number. (The light elements are discussed by Mrs A.M. Boesgaard in these Proceedings.) Most of these anomalies occur on the upper part of the main sequence (Figure 1) where convective envelopes are absent and the majority of stars are rapid rotators. Among the sharp-lined stars, one tends to find either small-amplitude variability or the various abundance anomalies associated with Bp, Ap and Am stars, although there are a few well-known stars with low v sin i and apparently normal abundances like τ Sco, γ Peg (a variable), ι Her and α Lyr. Conversely, some stars manage to spin quite fast and yet have anomalous abundances. I shall not discuss the Am and Ap stars because there will be a contribution on the subject by Michaud; see also recent reviews (Preston 1974; Baschek 1975).

Among early B stars, however, there are also He and C,N,O abundance anomalies whose status is not at all clear (see Figure 1). Helium-poor stars, cooler than about 20,000 K, can probably be explained by diffusion, but what of the intermediate helium-rich stars with He/H $\simeq$ 1 that are generally found somewhere near the main sequence at higher temperatures? The interpretation of these stars has been discussed by Hunger (1975). One well-known example is σ Ori E, a spectrum variable with variable Hα emission, v sin i = 150 km s^{-1} and a period of 1.19 days. Osmer and Peterson (1974) analysed σ Ori E and a group of more slowly rotating He-rich stars from the Michigan spectral survey and pointed out that between the He-rich stars and the Bp (He-poor) stars there seems to be a well-defined boundary in the T_{eff} -log g plane that is populated by the three interesting transitional cases 3 Cen A (with ^{3}He), ι Ori B (with ^{3}He) and a Cen (He variable). They concluded that all these peculiarities are due to diffusion with high temperatures and radiation-pressure driven mass loss favouring levitation of He to the surface, and suggested that a similar effect may operate in sd O stars. Hunger and his collaborators, on the other hand, while accepting that this may be the right explanation for σ Ori E and a few other stars, find lower values for the gravities and masses of most of the sharp-lined stars and suggest that these are well-mixed objects evolving to or from the helium main sequence. To complicate the issue further, Hesser et al. (1976) have studied the light curve of variable emission from σ Ori E and find a double eclipse pattern reminiscent of the U Gem stars. They therefore suggest that there is a low-mass collapsed companion with an accretion disk, that originally supplied the excess helium to the primary by mass transfer.

Another somewhat mysterious set of abundance anomalies are displayed by C,N,O in slowly rotating stars of spectral types between O9 and B6 and all luminosity classes that also tend to be radial-velocity and spectrum variables. These were discovered in classification work by the Jascheks, Walborn and others and the few abundance analyses that have been carried out reveal striking variations in N abundance with no very consistent pattern in other elements (see review by Baschek 1975). The existence of these anomalies can cause some difficulty in using galactic stars as standards to find the underabundance of nitrogen in stars of the Small Magellanic Cloud (Osmer 1973), which is believed to be primarily due to differences in the enrichment history of the ISM. The interpretation of the C, N, O anomalies is still unclear and more quantitative data are needed, but they could be associated with pulsationally driven mass loss (Baschek and Scholz 1974).

Another possible cause of CNO abundance anomalies, suggested by Paczyński (1973), is mixing by meridional circulation in rapidly rotating stars between about 3 and 10 solar masses. Preston and Paczyński (1974) accordingly looked for evidence of depletion of carbon in rapidly rotating stars of types B3-B5, but failed to find it. Alternatively, some authors have suggested that after-effects of this process might be present in red giants, where it would be distinguished from the effects of convective mixing-out of CNO products predicted by Iben (1965) by a steep increase in N/C ratio with mass. Statistical investigations of CN strength in red giants (Harmer and Pagel 1973; Demers 1975) do not generally show such an effect, as most of the variation in CN strength can be accounted for by the atmospheric parameters (effective temperature, gravity and metal abundance), but there are one or two cases where it could have operated, in particular the K0 II-III giant 37 Com which has weak CH, $^{12}C/^{13}C$ = 3.4 and relatively broad lines (Yamashita 1967; Tomkin et al. 1976). If so, however, it must be either a fairly unusual effect or one operating mainly in supergiants.

3. RED GIANTS OF POPULATION I

The possibility of surface abundance changes in stars evolving up the red giant branch was first pointed out by Iben (1965) as a consequence of "mixing-out" by penetration of the outer convective envelope to regions previously affected by the CNO cycle. If it penetrates deeply enough, it reaches the zone where the CNO cycle has reached equilibrium and almost all carbon has been changed into nitrogen, which would lead to an enhancement of the N/C abundance ratio at the surface. Except in a few cases there is not too much evidence that this happens; among a few bright stars analysed by Greene (1969) only the supergiant ε Peg showed N/C appreciably enhanced (by a factor of 4). Statistical studies of Population I red giants likewise show few CN anomalies, as has already been mentioned, the main exception being some of the so-called "super metal-rich" stars like μ Leo which are not markedly rich in iron (Blanc-Vaziaga et al. 1973; Peterson 1976a), but do have exceptionally strong CN.

If the outer convective zone penetrates somewhat less deeply, it mixes out a region where the C,N,O cycle has not reached equilibrium, but there is a peak in ^{13}C, due to the first reaction of the cycle, at the outer edge of the H-burning core, which permits enrichment of the atmosphere from $^{13}C/^{12}C$ = 1/90 to about 1/30, or from 1/40 to about 1/20 (Dearborn et al. 1976); larger initial values than 1/40 are unlikely in view of interstellar observations. In the last three years or so a great deal of new and exciting work on this topic has been carried out on $^{12}C/^{13}C$ ratios deduced from red CN bands in yellow and red giants supplemented by the 2μ CO band in the reddest stars, mainly by Lambert and his group at Texas (Dearborn et al. 1975; Hinkle et al 1976; Tomkin et al 1976). The results are summarized in an HR diagram by Tomkin et al. (1976). The majority of stars have $^{12}C/^{13}C \geq 18$, explicable by mixing, but considerably more ^{13}C is found in some of the more luminous stars, with more than about twice the mass of the Sun, and also in some metal-deficient stars like Arcturus and γ Leo. These pose an interesting problem for stellar evolution, since most of them are not bright enough to be expected to have undergone helium shell flashes (log $L/L_\odot \geq 3$) and various exotic mechanisms have been proposed to account for these results: meridional mixing (already mentioned); severe mass loss in the main sequence or early subgiant stage; mixing at the helium flash; thermal instability in the H-burning shell; and the engulfing of low-mass companions. None of these mechanisms seems quite sufficient by itself and the further exploration of this problem, preferably with better data on elemental abundances of C, N and O, will be a topic of "burning interest" as Tomkin et al. have put it.

Somewhat related to these effects is the oxygen isotope ratio $^{17}O/^{18}O$. Since the CNO cycle is a tri-cycle owing to the long life of the ^{17}O (p,α) ^{14}N reaction, the simple mixing mechanism can cause traces of ^{17}O to be mixed out and significantly affect $^{17}O/^{16}O$ in the atmospheres of the more massive stars (Dearborn and Schramm 1974). In agreement with this, features of $C^{17}O$ have been observed in a few red supergiants, IRC +10216 (for which $^{16}O/^{17}O \simeq 400$ compared to a terrestrial ratio of 3000), α Sco and α^1 Her (Rank et al. 1974; Maillard 1974).

The other abundance anomalies affecting Population I red giants are more conspicuous, so that they have been known for a long time, but affect only a minority of the stars. These fall into four main categories (Figure 1):

Table 1
Some Abundances in Cool Giants

	M,MS	S	SC,CS	C
[Fe/H]	$\sim 0^1$			
C/O	$\leq 0.95^2$	0.95 to 0.99	0.99 to 1.01	≥ 1
$^{12}C/^{13}C$	7 to 20^3	25^3 (χ Cyg)		3.5 to $40^{4,5}$
[N/Fe]				$\sim 0^5$
[O/Fe]				$0:^{5,6,7}$
[Zr/Ti]	-0.3 to $+0.4^8$	$+0.4$ to $+1.2^{8,9}$		$+0.9$ to $+1.9^9$ (WZ Cas $\sim +0.5^{9,10}$)
Tc?	some^{11} (Miras)	some^{11}	RZ Peg^{12}	some^{11}
[Li/Ca]	-2.1 to $+0.4^{13,14}$	$\leq 3.3^{9,14}$	$\sim 1.0^{16}$ (SC)	-1 to $+2^9$ WZ Cas $\sim +5^{17}$

References to Table 1

1. Huggins 1973.
2. Scalo and Ross 1976.
3. Hinkle et al. 1976; Tomkin et al. 1976.
4. Fujita 1970.
5. Kilston 1975.
6. Thompson 1974.
7. Querci and Querci 1976.
8. Boesgaard 1970.
9. Catchpole and Feast 1976.
10. Utsumi 1970.
11. Peery 1971; Cohen 1973.
12. Peery et al. 1971.
13. Merchant 1967.
14. Boesgaard 1970b.
15. Bretz 1966.
16. Catchpole and Feast 1971.
17. Hirai 1969.

(a) and (b) The BaII stars and early R stars which coincide with K giants in the HR diagram.

(c) Cool giants, mostly variables, of the sequence M, MS, S, SC, CS, C (or N) which is primarily a sequence of increasing carbon abundance relative to oxygen (Scalo and Ross 1976).

(d) Hydrogen-deficient carbon stars like R CrB and the closely related extreme helium stars, appearing with high luminosities over a wide range of temperatures.

The first problem in trying to understand these stars is to place them in the HR diagram, and the positions shown are largely inspired by the recent discussion by Scalo (1976). The majority of all these types seem to be old disk stars (Eggen 1972a,b) with masses slightly above that of the Sun, although a few are younger and presumably much more massive (e.g. the Ba star ζ Cap and the Large Magellanic Cloud Carbon stars). Most S and N stars lie to the right of the broken line in Figure 1 which is the predicted locus of helium shell flashes, whereas the BaII and early R stars are well below and to the left of this line, but they are suitably placed to have undergone helium core flashes. According to Williams (1975), a few per cent of all G and K giants have a noticeable enhancement of barium.

As regards composition, the Ba stars are comparatively easy to analyse and quite a lot is known; one recalls the important work of Burbidge and Burbidge (1957) which helped to establish the s-process. The most extensive data available now are for ζ Cap (Tech 1971) for which, within the errors, all s-process elements from Sr up to Pb are enhanced by about a factor of 10, whereas Sc, Cu, Zn and Nb, as well as Ge, Eu and Gd, are not noticeably enhanced, so that these elements presumably have no significant contribution from the s-process. Most known BaII stars have enhanced carbon features, but this does not seem to be a necessary condition, as is shown by o Vir (Williams 1972, 1975) and HD 101013 (Branch and Bell 1975). $^{12}C/^{13}C$ ratios seem to be quite large.

The properties of carbon stars have been reviewed in a very good article by Wallerstein (1973). Among early R stars most is known about HD 156074 (Greene et al. 1973) which shows some symptoms of the CNO cycle: $^{12}C/^{13}C \simeq 4$, N is enhanced, O somewhat depleted and other elements fairly normal as though nitrogen were acting as a neutron poison; RU Cam seems to be quite similar.

Abundances in the cool M,S and N stars are much more difficult to find because the abundance of molecules in their spectra prevents decent equivalent-width measurements and makes the structures of the atmospheres both hard to calculate and excessively dependent on the input assumptions. Nevertheless there seems to have been some progress in recent years; in particular, the presence of C_2 implies that carbon stars typically do not show mainly equilibrium products of the C, N, O cycle. The $^{12}C/^{13}C$ ratios no longer look so remarkable in view of the results for more ordinary giants, while at the same time it has turned out that N stars have s-process elements enhanced just about as much as S stars if not more. Table I contains some recent results of abundance determinations in cool giants, based on work by Scalo and Ross, Kilston, Boesgaard, Catchpole and Feast and others. (Square brackets denote change in logarithm of abundance ratio relative to the Sun.) The basic pattern suggests that the atmospheres show varying degrees of enhancement of carbon, s-process elements and lithium, presumably due to repeated incursions of the outer convective envelope into the inter-shell region, but it is difficult to be much more specific.

Important clues to the nature of BaII and carbon stars, as well as of planetary nebulae, are likely to come from the so-called "Rosetta Stone" FG Sge (Langer et al. 1974), which is rapidly evolving to the right of the HR diagram at a luminosity of about $10^4 L_\odot$ and suddenly became a barium star between 1965 and 1969 when its spectral type was near F0. Here apparently we can see the effects of helium shell flashes, mixing and mass loss taking place partly in real time, so to speak, and the future evolution of this star will obviously be followed with great interest.

The last type of nuclear abundance anomaly shown by Population I giants is that of the hydrogen-deficient carbon stars (Warner 1967), probably closely related to the hotter extreme helium stars (Hunger 1975). The composition and kinematics of helium stars suggests that one had a normal star of slightly above a solar mass in which the hydrogen and carbon originally present were changed into helium and nitrogen through hydrogen burning and then further carbon was added to what are now the surface layers by helium burning, while in the cooler but otherwise similar R CrB-like stars one finds Li but no excess of s-process elements. The impression that one has from these stars, which must be evolving quite rapidly, is that one may be looking directly into an inter-shell region after removal of the hydrogen-rich outer layers (Schönberner 1975).

4. STARS OF POPULATION II

The most marked abundance anomaly shown by stars of extreme Population II, that is to say those in globular clusters and nearby field stars of extremely high velocity, is of course the overall deficiency of heavy elements from carbon upwards by factors between 10 and 1000 and the corresponding weak line blanketing revealed by photometric indices like ultra-violet excess. Much effort has been devoted to the search for abundance anomalies in individual elements that could be attributed to a different history of galactic enrichment affecting the material from which these stars were born and the existence of such effects is still a matter of doubt and controversy which has been extensively discussed in the literature (Pagel 1973; Peimbert 1973a; Unsöld 1974; Peterson 1976b). To summarize the situation briefly, extragalactic HII regions show interesting abundance anomalies in nitrogen which may be a secondary nucleosynthesis product (Talbot and Arnett 1974); similar effects may occur in metal-weak stars, but

this is still not certain and it certainly does not occur in all cases
(Harmer and Pagel 1970; Sneden 1974). Odd-numbered elements like V
and Mn seem to be slightly overdeficient, but the effect is marginal;
among s-process elements, Ba seems definitely overdeficient when
[Fe/H] < -2, but by factors of 20 or less, and Sr, Y and Zr by
much smaller factors, which implies that no star that can be seen
now was formed until after the iron-group elements had been
subjected to a substantial amount of s-processing. The ratio of
carbon to iron is fairly constant, but that of oxygen does seem to
vary, in the sense that metal-deficient objects are significantly less
deficient in oxygen (Peimbert 1973b; Lambert et al. 1974), which could
be a sign that the progenitors of oxygen evolved more rapidly than those
of iron (cf. Chevalier 1976). However, most of these remarks apply to
nearby field stars, and the situation in globular clusters and/or near
the galactic centre could be different (Hesser et al. 1976).

Apart from these difficult problems, the evolved stars of
Population II show a similar gamut of abundance peculiarities as do
those of Population I, including He-poor stars sometimes showing other
peculiarities (the B subdwarfs, Searle and Sargent 1972), He-rich stars
(the O subdwarfs) and a variety of C, N, O anomalies among the red
giants, both above and below the expected locus for helium shell
flashing and both with and without enhancement of s-process elements.

An attempt to classify the abundance peculiarities is given in
Table II. Among field stars one sees both enhanced carbon features
(CH stars) and a tendency among the more luminous extremely deficient
stars like, for instance, HD 122563 to have nitrogen enhanced at the
expense of carbon (Sneden 1973, 1974). This could be related to a similar effect seen in asymptotic giant branch stars of the extremely metal-weak globular clusters M 92 and NGC 6397, but it would be odd if all the
field stars were actually AGB stars. One then thinks of the
possibility of mixing out along the first ascent of the red giant branch,
but in that case what are we to make of the planetary nebula in M 15
where N is extremely deficient (Peimbert 1973)? Detailed quantitative
analyses of stars on both giant branches of, say, 6397 would be helpful
in settling this question.

The CH stars appear at various luminosities (see Figure 1) and
possibly cover a continuous range from $M_v \simeq -2$ to +2. The brightest ones
(which are the three CH stars in ω Cen) are presumably products of
helium shell flashing followed by mixing out, but the remarkable
discovery of subgiant CH stars by Bond (1974) raises a serious problem
in connection with this, and Bond himself suggested that some stars
undergo large-scale mixing after either a shell or a core flash which
brings them back down to the neighbourhood of the main sequence.

Apart from CH stars, a great variety of CN and s-process anomalies
have been found in globular clusters. An accurate discription is
difficult owing to the observational limitations; for instance in a
large number of cases we only know that CN is strong from DDO photo-

Table II

Abundance anomalies among Population II Red Giants and Subgiants

Brief Description	Where Observed
A: mainly strong carbon features:	
CH stars with ^{13}C,N markedly enhanced	Field[1]; ω Cen (within 1^m of RG tip)[2,3]
CH stars without ^{13}C,N markedly enhanced	Field[4], $M_v \simeq 0$
Subgiant CH stars	Field[5], $M_v \simeq +2$
CN strong	ω Cen (mainly red edge)[6,7]; M22, M71, 47 Tuc, 6352[7]
CN strong, no C_2	ω Cen (tip and red edge)[3,8]; M5, M10[9]; 47 Tuc[10]
s-process, no C_2; various CN,CH	ω Cen (tip and red edge)[3,8]
B: weak carbon features:	
Weak CH, strong NH	M 92 (AGB)[11]; 6397 (AGB)[12]
C deficient, N enhanced	Field, $M_v \leqslant 0$, [Fe/H] $\leqslant -1.6$ [13]

References to Table II

1. Climenhaga 1960; Wallerstein 1969.
2. Harding 1962; Bell and Dickens 1974; Bond 1975.
3. Dickens and Bell 1976.
4. Wallerstein and Greenstein 1964; Harmer and Pagel 1973.
5. Bond 1974.
6. Norris and Bessell 1975.
7. Hesser, Hartwick and McClure 1976
8. Mallia 1976.
9. Zinn 1973b.
10. Bell et al. 1975.
11. Zinn 1973a; Butler et al. 1975.
12. Mallia 1975.
13. Sneden 1973, 1974.

metry. Where spectra exist, observers are not always unanimous as to whether CH and s-process elements are enhanced, although they agree on CN and C_2. And finally, the apparent confinement of strong CN stars to the less metal-weak globular clusters can itself be a sort of selection effect because in extreme cases like M 92 CN would still be invisible even if it were enhanced by a very large factor. Apart from these difficulties, an intricate variety of effects seems to occur, particularly in ω Cen which is noted for the great width of its red giant branch in (B-V). Freeman and Rodgers (1975) give evidence that some or all of this effect arises from an inhomogeneity in initial metal abundance analogous to that in elliptical galaxies, while Norris and Bessell (1975) and Mallia (1976) suggest that it is due to different kinds of mixing process exemplified by the anomalies listed in the Table which, according to spectrum synthesis studies by Dickens and Bell, can be largely attributed to an enhancement of nitrogen. Quite possibly both effects occur, the incidence of anomalies being itself a function of primeval composition, but it must also be noted that the behaviour found in 47 Tuc is quite different from that of field stars of similar metallicity (Hesser, Hartwick and McClure 1976) so that more than one composition parameter would have to be involved.

5. FINAL STAGES OF EVOLUTION

The later stages of stellar evolution comprise planetary nebulae, white dwarfs and supernovae. A related question is that of the abundances in common novae, which may undergo a special kind of C,N,O nucleosynthesis resulting from mass transfer on to a white dwarf (Starrfield et al. 1974).

The composition of planetary nebulae is of special interest as an indication of the nature of the outer layers (presumably above the hydrogen shell source) ejected by red giants when they reach the appropriate stage. In general the abundances of He, O and Ne are found to be more or less as expected from stars of the usual Population I range of compositions (Kaler 1970; Osterbrock 1974), but there are signs of enhancement of N (Peimbert and Torres-Peimbert 1971; Boeshaar 1975) which suggest that at some stage the mixing-out process may become more effective than has so far been suggested by observations of most of the red giants themselves, although in M giants and carbon stars an enhancement by a factor of 2 or 3 could easily pass unnoticed. In agreement with theory, the nitrogen enhancement in planetary nebulae seems to diminish or disappear at low metal abundances; in the extreme case of K 648 in the globular cluster M 15, N is deficient by at least a factor of 40, possibly matching the metal-deficiency of the cluster as a whole, while O and Ne are deficient by only a factor of 10 (Peimbert 1973). There are some difficulties in the determination of abundances from emission lines, since nitrogen is only detected from [NII] which is very sensitive to the presence of condensations (Kirkpatrick 1972; Webster 1976); but this effect can be allowed for in a semi-empirical way. Unfortunately the elements observable in planetary nebulae and in red giant atmospheres are still sufficiently orthogonal that no particular class of red giant can be excluded as a source of planetaries on grounds of composition.

White dwarfs have extremely strange compositions at their surfaces which have been very well reviewed by Weidemann (1975). These are suggestive of diffusion and perhaps other processes and will not be discussed further here. Novae, on the other hand, are extremely interesting in showing marked overabundances of C,N and O (Antipova 1974); a recent example is Nova Cygni 1975 (Figure 2), which had an unusually high temperature at maximum and rather weak absorption lines among which, however, NII was by far the dominant species after hydrogen (Andrews 1975; see Figure 2). From the model of Starrfield et al. (1974), involving energy release from the fast CNO cycle operating on C and O in the white dwarf component when it accretes mass from its companion, it is possible (though not certain) that interesting isotopic effects should occur. These were looked for in the 4215 band of N Her 1934 by Lambert and Sneden (1975), who found some signs of ^{13}C and possibly ^{15}N, although these odd isotopes are certainly not dominant. The observations provide some constraints on the models, though not very strong ones: a model with pure ^{13}C is not allowed, for example, but pure ^{12}C could be allowed if the ^{13}C is made by some other means.

"Last scene of all, in this strange and eventful history," is the supernova outburst wherein one can expect to see effects of nucleosynthesis beyond the C, N,O group and up to the iron group and r-process elements. The only evidence I am aware of for abundance anomalies in supernovae themselves (as opposed to their remnants) comes from the study of the Type I supernova 1972e in NGC 5253 by Kirshner and Oke (1975) who identified several prominent emission bumps on the spectrum two years after maximum with [FeII]. Assuming an excitation temperature of 5000 K, the total luminosity in [FeII] implied a mass of about 10^{-2} $M_\odot$ in the form of Fe^+, which, with an estimated mass for the whole envelope of ~ 1 $M_\odot$, gives [Fe/H] $\simeq 1$, a marginally significant enrichment which has inspired the suggestion that iron is supplied to the ISM by Type I supernovae rather than Type II (Chevalier 1976).

Among supernova remnants, only the young ones can be expected to show abundance anomalies because after a few thousand years or so they sweep up more than their own mass of interstellar material. This leaves us with just two objects: the Crab Nebula, in the filaments of which only He (and perhaps marginally N) are noticeably overabundant (Davidson 1973), and the 300-year old supernova remnant Cas A, which is the most interesting case. The nebulosity is a highly reddened incomplete shell broken up into two components: fast-moving condensations or knots with emission lines of [OI], [OII], [OIII], [SII], [SIII] and [ArIII], but nothing else detected, and a smaller number of slowly moving condensations or flocculi having Hα, strong [NII] and weak [OII] (Minkowski 1968; Peimbert and van den Bergh 1971; Searle 1971).

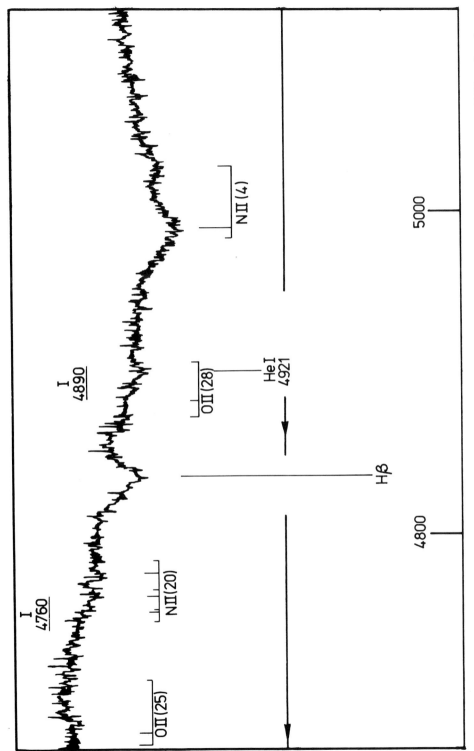

Figure 2. Spectrum of Nova Cygni 1975 taken on August 29.9 with the 0.75 m telescope at RGO, showing relative strength of NII lines.

The abundance analysis involves considerable uncertainties, but it seems clear that O, S and Ar are strongly enhanced in the fast-moving and rather dense knots (Peimbert 1971), presumably as a consequence of nucleosynthesis, while the almost stationary flocculi are most readily interpreted as swept-up remnants of a fossil HII region with enhanced nitrogen abundance that was formed from the hydrogen envelope of the supernova and ionised during the outburst, as there are no ionizing stars in the vicinity. Arnett (1975) has compared the inferred abundances in the Crab and Cas A with evolutionary models of massive supernovae and finds reasonable qualitative agreement, particularly with the inference from Type II light curves that the outburst is preceded by the ejection of a hydrogen-rich circumstellar shell, modified by CNO processing, that subsequently gives rise to the knots and filaments through Rayleigh-Taylor instability (Falk and Arnett 1973). Thus, in at least one case, the observer's supernova and the theoretician's supernova have at last come close enough together for fairly detailed comparisons to be attempted and there is direct evidence that supernovae can indeed supply heavy elements to the ISM.

6. CONCLUSION

To summarise the upshot of this talk briefly, both the observational determination and the theoretical prediction of abundance peculiarities have come sufficiently far that quite detailed things can be inferred about the effects of stellar evolution in certain cases, particularly the existence and extent of surface mixing not only after helium shell flashes but all the way up the giant branch. On the other hand we do not know where all the details fit in, e.g. what is the precise significance of seeing s-process elements and Li in certain cases while we do not see them in others, why do we see both peculiar and normal (or at least less peculiar) stars in the same part of the HR diagram and why do some stars show abundance peculiarities at much lower luminosities than one might have expected? Perhaps by the next I.A.U. Assembly we shall have a more systematic collection of information which will help to throw light on these questions.

REFERENCES

Andrews, P.J. 1975, Private communication.
Antipova, L.I. 1974, Highlights of Astronomy, 3, ed. G. Contopoulos, p.501.
Arnett, W.D. 1975, Astrophys. J., 195, 727.
Baschek, B. 1975, Problems of Stellar Atmospheres and Envelopes, ed. B. Baschek, W.H. Kegel and G. Traving, Springer-Verlag, p. 101.
Baschek, B., and Scholz, M. 1974, Astron. and Astrophys., 30, 395.
Bell, R.A. and Dickens, R.J. 1974, Mon. Not. R. Astron. Soc., 166, 89.
Bell, R.A., Dickens, R.J., and Gustafsson, B. 1975, Bull. Amer. Astron. Soc., 7, 535.

Blanc-Vaziaga, M.J., Cayrel, G., and Cayrel, R. 1973, Astrophy J., 180, 871.
Boesgaard, A.M. 1970a, Astrophys. J., 161, 163.
　　　　　　　1970b, Astrophys. J., 161, 1003.
Boeshaar, G.O. 1975, Astrophys. J., 195, 695.
Bond, H.E. 1974, Astrophys. J., 194, 95.
　　　　　1975, Astrophys. J. Let., 202, L47.
Branch, D.R., and Bell, R.A. 1975, Mon. Not. R. Astron. Soc., 173, 299.
Bretz, M.C. 1966, I.A.U. Symposium No. 26, p.304.
Burbidge, E.M., and Burbidge, G.R. 1957, Astrophys. J., 126, 357.
Butler, D., Carbon, D., and Kraft, R.P. 1975, Bull. Amer. Astron. Soc., 7, 239.
Catchpole, R.M., and Feast, M.W. 1971, Mon. Not. R. Astron. Soc., 154, 197.
　　　　　　　　　　　　　　　 1976, Mon. Not. R. Astron. Soc., 175, 501.
Chevalier, R. 1976, Nature, 260, 689.
Climenhaga, J.L. 1960, Pub. Dom. Astrophys. Obs., 11, 307.
Cohen, J.G. 1973. Publ. Astron. Soc. Pacific, 85, 187.
Davidson, K 1973, Astrophys. J., 186, 223.
Dearborn, D.F., Eggleton, P.P., and Schramm, D.N. 1976, Astrophys. J., 203, 455.
Dearborn, D.F., Lambert, D.L., and Tomkin, J. 1975, Astrophys. J., 200, 675.
Dearborn, D.F., and Schramm, D.N. 1974, Astrophys. J., 194, 67.
Demers, D., 1975, Bull. Amer. Astron. Soc., 7, 518.
Dickens, R.J. 1972, Mon. Not. R. Astron. Soc., 159, 7P.
Dickens, R.J., and Bell, R.A. 1976, Astrophys. J., 207, 506.
Eggen, O.J. 1972a, Astrophys. J., 174, 45.
　　　　　1972b, Astrophys. J., 177, 489.
Falk, S.W., and Arnett, W.D. 1973, Astrophys. J. Let., 180, L65.
Freeman, K.C., and Rodgers, A.W. 1975, Astrophys. J. Let., 201, L71.
Fujita, Y., 1970, Interpretation of the Spectra of Cool Carbon Stars, Tokyo University Press.
Greene, T.F. 1969, Astrophys. J., 157, 737.
Greene, T.F., Perry, J., Snow, T.P., and Wallerstein, G. 1973, Astron. and Astrophys., 22, 293.
Greenstein, J.L., and Sargent, A.I. 1975, Astrophys. J. Sup., 28, 157.
Harding, G. 1962, Observatory, 82, 205
Harmer, D.L., and Pagel, B.E.J. 1970, Nature, 225, 349.
　　　　　　　　　　　　　　1973, Mon. Not. R. Astron. Soc., 165, 91.
Hesser, J.E., Hartwick, F.D.A., and McClure, R.D. 1976, Astrophys. J. Let., 207, L113.
Hesser, J.E., Walborn, N.R. and Ugarte, P. 1976, Nature, 262, 116.
Hinkle, K.H., Lambert, D.L., and Snell, R.L. 1976, Preprint.
Hirai, M. 1969, Pub. Astron. Soc., Japan, 21, 91.
Huggins, P.J. 1973, Astron. and Astrophys., 28, 217.
Hunger, K. 1975, Problems of Stellar Atmospheres and Envelopes, p.57.
Iben, I. 1965, Astrophys. J., 142, 1447.
Kaler, J.B. 1970, Astrophys. J., 160, 887.
Kilston, S. 1975, Pub. Astron. Soc. Pacific, 87, 189.

Kirkpatrick, R.C. 1972. Astrophys. J., 176, 381.
Kirshner, R.P., and Oke, J.B. 1975, Astrophys. J., 200, 574.
Lambert, D.L., and Sneden, C. 1975, Mon. Not. R. Astron. Soc., 170, 533.
Lambert, D.L., Sneden, C., and Ries, L.M. 1974, Astrophys., 188, 97.
Langer, G.E., Kraft, R.P., and Anderson, K.S. 1974, Astrophys. J., 189, 509.
Maillard, J.M. 1974, Highlights of Astronomy, 3, 269.
Mallia, E.A. 1975, Mon. Not. R. Astron. Soc., 170, 57P
        1976, Astron. and Astrophys., 48, 129.
Merchant, A.E. 1967, Astrophys. J., 147, 587.
Minkowski, R. 1968, Nebulae and Interstellar Matter, ed. B.M. Middlehurst and L.H. Aller, University of Chicago Press, p. 652.
Norris, J., and Bessell, M.S. 1975, Astrophys. J. Let., 201, L75.
Osmer, P.S. 1973, Astrophys. J. Let., 184, L27.
Osmer, P.S., and Peterson, D.M. 1974, Astrophys. J., 187, 117.
Osterbrock, D.E. 1974, Astrophysics of Gaseous Nebulae, San Francisco: W.H. Freeman, p. 132.
Paczynski, B. 1970, Acta Astron. 20, 47.
        1973, Acta Astron., 23, 191.
Pagel, B.E.J. 1973, Space Sci Rev., 15, 1.
Peery, B.F. 1971, Astrophys. J. Let., 163, L1.
Peery, B.F., Keenan, P.C., and Marenin, I.R. 1971, Pub. Astron. Soc. Pacific, 83, 496.
Peimbert, M. 1971, Astrophys. J., 170, 261.
        1973a, I.A.U. Symposium No. 58, p. 141.
        1973b, Les Nébuleuses Planétaires, Mem. Soc. R. Sci. Liege, 5, 307.
Peimbert, M., and Torres-Peimbert, S. 1971, Astrophys. J., 168, 413.
Peimbert, M., and van den Bergh, S. 1971, Astrophys. J., 167, 223.
Peterson, R. 1976a, Astrophys. J. Sup., 30, 61.
        1976b, Astrophys. J., 206, 800.
Preston, G.W. 1974, Ann. Rev. Astron. Astrophys., 12, 257.
Preston, G.W., and Paczynski, B. 1974, Ann. Report Dir. Hale Obs. 1973-74, p.133.
Querci, M., and Querci, F. 1976, Astron. and Astrophys., 49, 443.
Rank, D.M., Geballe, T.R., and Wollman, E.R. 1974, Astrophys. J. Let., 187, L111.
Scalo, J.M. 1976, Astrophys. J., 206, 474.
Scalo, J.M., and Ross, J.E. 1976, Astron. and Astrophys., 48, 219.
Schönberner, D. 1975, Astron. and Astrophys., 44, 383.
Searle, L. 1971, Astrophys. J., 168, 41.
Searle, L., and Sargent, W.L.W. 1972, Comments Astrophys. and Space Phys., 4, 59.
Sneden, C. 1973, Astrophys. J., 184, 839.
        1974, Astrophys. J., 189, 493.
Starrfield, S., Sparks, W.M., and Truran, J.W., 1974, Astrophys. J. Sup., 28, 247.
Talbot, R.J., and Arnett, W.D. 1974, Astrophys. J., 190, 605.
Tech, J.L. 1971, Nat. Bur. Stand. Monograph, No. 119.
Thompson, R.I. 1974, Highlights of Astronomy, 3, 255.

Tomkin, J., Luck, R.E., and Lambert, D.L. 1976, Preprint.
Unsöld, A. 1974, Proc. First Europ. Astron. Meeting, Athens, 3, 84 Springer-Verlag.
Utsumi, K. 1970, Pub. Astron. Soc. Japan, 22, 93.
Wallerstein, G. 1969, Astrophys. J., 158, 607.
            1973, Ann. Rev. Astron. Astrophys., 11, 115.
Wallerstein, G., and Greenstein, J.L. 1964, Astrophys. J., 139, 1163.
Warner, B. 1967, Mon. Not. R. Astron. Soc., 137, 119.
Webster, B.L. 1976, Mon. Not. R. Astron. Soc., 174, 157.
Weidemann, V. 1975, Problems of Stellar Atmospheres and Envelopes, p. 173.
Williams, P.M. 1972, Mon. Not. R. Astron. Soc., 155, 17P.
            1975, Mon. Not. R. Astron. Soc., 170, 343.
Yamashita, Y. 1967, Pub. Dom. Astrophys. Obs., 12, 455.
Zinn, R. 1973a, Astrophys. J., 182, 183.
            1973b, Astron. and Astrophys., 25, 409.

Postscript — On the question of C,N anomalies in metal-weak globular clusters, R P Kraft (private communication) has recently found some anomalous stars on the giant branches of M 92 and M 15, as well as on the AGB.

SENSITIVITY OF INTERNAL STRUCTURE TO THE SURFACE BOUNDARY CONDITION

Pierre Demarque
Yale University Observatory

It is now nearly fifty years since Eddington and Milne had a lively controversy on the importance of the surface boundary condition on the internal structure of stars [see Eddington (1930) and Milne (1930)]. We remember that Eddington believed that the internal structure of stars is basically determined by the physical processes occurring in the deep interior and that what happens in the surface layers has little effect on the total stellar luminosity. On the other hand, Milne emphasized the importance of the properties of the outer layers and the effect these could have on the run of pressure and temperature in the deep interior of the stars. We know now that both Eddington and Milne were correct. Eddington's considerations apply to the hot stars, the early-type stars which have surface layers in radiative equilibrium. Milne's arguments are relevant to the cool stars, the late-type stars which have deep convective envelopes. In the former case, one can safely assume in calculations of stellar structure that the density and the temperature both approach zero simultaneously at the surface (the so-called "zero" surface boundary conditions). For late-type stars, most of the convective envelope is adiabatic and its structure is determined by the adiabatic equation:

$$P = KT^{\gamma/\gamma-1} \tag{1}$$

which requires the parameter K to be determined by the run of P and T near the stellar surface (Schwarzschild 1958).

Now let us consider the properties of these convective zones. One can show that for a star with a deep convective envelope in a diabatic equilibrium, the radius is chiefly determined by the specific entropy in the adiabatic region [see Larson (1973)]. For a perfect gas, the specific entropy s is given by:

$$s \propto c_v \ln(P/\rho^\gamma) \propto c_v \ln(T/\rho^{\gamma-1}) \tag{2}$$

where $\gamma = c_p/c_v$, the ratio of specific heats.

For a star in hydrostatic equilibrium, of given mass M and radius R, one finds that dimensionally:

$$\rho \propto \frac{M}{R^3} \quad \text{and} \quad P \propto \frac{GM^2}{R^4}$$

so that:

$$R \propto \left[\frac{\exp(s/c_v)}{GM^{2-\gamma}} \right]^{\frac{1}{(3\gamma-4)}} \quad (3)$$

For a polytrope of index n = 1.5 (i.e. γ = 5/3), using the tables of Chandrasekhar (1939), one can then write:

$$R = 2.36 \frac{\exp(s/c_v)}{GM^{1/3}} \quad (4)$$

This expression gives the radius of configurations with a deep adiabatic envelope with accuracy of 10 to 15 percent.

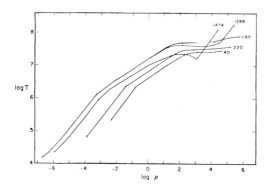

Figure 1. Interior characteristics of models for red giant stars with 0.85 $M_\odot$. The labels refer to the model number on the evolutionary sequence. The onset of helium burning occurs between models 1130 and 1288. The surface convection zone is for each model bounded by the first two tick-marks (starting at the low-temperature end). Note that in model 1474, which has a smaller radius than 1288, the specific entropy in the convection zone has also begun to decrease [from P. Demarque and J.G. Mengel (1971). Courtesy of the University of Chicago Press].

Figure 1 illustrates this result in terms of models for red giants with increasing luminosities along an evolutionary sequence which reaches the onset of helium burning between models 1130 and 1288. At

model 1474, the star begins to move down the giant branch toward the
horizontal branch. The radii of each of the configurations of Figure 1,
which have deep convective envelopes, is determined by the specific
entropy in the convective region. This is illustrated by Figure 1
which shows that as the luminosity and therefore the radius increase,
the specific entropy increases also, since one moves toward higher
temperatures and lower densities in the (log T-log ρ)-plane. And
once the core flash has begun and the radiative layers below the convection zone are cooling down (as in model 1474), the radius decreases
in turn.

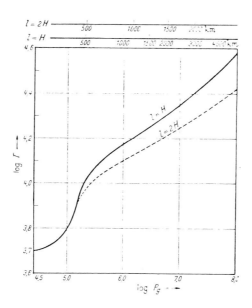

Figure 2. The run of gas pressure P_g and temperature T in
the solar convection zone for two values of the
mixing length ℓ in terms of the local pressure
scale height H [from E. Vitense (1953). Courtesy
of Springer-Verlag].

The calculation of the radius of the stellar model would thus seem
to reduce to the problem of determining the specific entropy in the
adiabatic envelope. One might think that a radiative model for the
atmospheric layers which gives the quantities P and T at the point of
the onset of convection, could by substituting into equation (1),
provide all the information that one requires to fix the stellar
radius. The problem is, however, complicated by the presence of a
superadiabatic layer at the top of the convection zone and is best
illustrated in the case of the sun. Although the sun is believed to
have a relatively shallow superadiabatic zone compared to red giants,
it has the advantage of readily observable surface features and a
measurable limb darkening. Karl Schwarzchild (1906) was aware of this
problem a long time ago, and it has not yet been fully resolved: al-

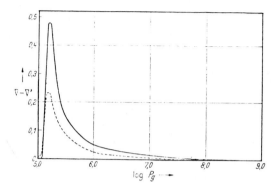

Figure 3. The "degree of superadiabaticity", plotted in ordinate, in the solar convection zone for two values of ℓ. Note how thin this superadiabatic layer is. [from E. Vitense (1953). Courtesy of Springer-Verlag].

though there is evidence for convection in the solar photosphere, the law of limb darkening observed on the solar disk is that which is characteristic of radiative equilibrium. In other words, we are seeing a layer which is unstable with respect to convection, but which undergoes a very inefficient kind of convection because of the low densities and the large radiative losses, and in which the temperature gradient is as a result intermediate between the local radiative and adiabatic gradients. For lack of a better theory, it is customary in studies of stellar interiors to use the mixing length formalism to describe this layer, after the work of Vitense (1953). Figure 2 and 3 show the models of Vitense for the solar convective zone. Note how sensitive the run of temperature and pressure are on the choice of the mixing length in Figure 2. Note also that it is in the thin superadiabatic region shown in Figure 3 on the outer part of the convective zone that the specific entropy of the whole adiabatic envelope is determined, i.e. that the boundary condition which fixes the radius is set.

Figure 4 illustrates the effect of the choice of the mixing length on main sequence position in the H-R diagram, Figure 5 shows the well known result that the structure of red giant envelopes are even more sensitive to the choice of the mixing length.

The problem that we face in the determination of what one might call an effective mixing length for late-type stars is compounded by other major gaps in our understanding of the relevant physical processes. The structure of the superadiabatic region is sensitive not only to the choice of the mixing length, but also to the opacity. This situation can be particularly complicated since various molecular species can be found in the atmospheres of cool stars which can affect the opacities in an important way.

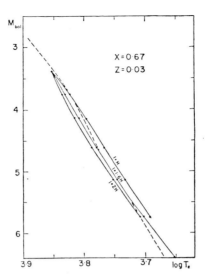

Figure 4. Effect of the choice of ℓ on the position of the main-sequence in the theoretical H-R diagram for stars in the mass range 1.3-0.8 $M_\odot$. [from P. Demarque and R.B. Larson (1964). Courtesy to the University of Chicago Press].

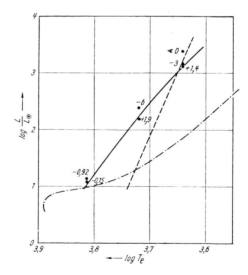

Figure 5. The large effect of the choice of the mixing length ℓ on the position of the giant branch is illustrated in this figure for a star of 1.3 $M_\odot$. The continuous line was obtained with $\ell=H$, the dot-dashed line with $\ell=2H$. [from R. Kippenhahn, St. Temesvary and L. Biermann. Courtesy of Springer-Verlag].

The details of the outer radiative layers can also affect the structure of the surface convection zone and in turn modify the radius. For example, a recent experiment by Prather (1975) in which he varied the function $q(\tau)$ of the grey solution of the equation of transfer from Milne's $q=2/3$ to the Krishna Swamy empirical fit to the sun meant a shift of 0.01 in log T_{eff} on main sequence interior models. This shift corresponds to a change in metallicity from $Z=0.01$ to 0.02. For red giants the sensitivity is greater still. And it is quite possible that non-LTE effects are important in this context.

In summary, one can say that proper surface boundary conditions for interior models of late-type stars require a detailed understanding of the structure of the stellar atmosphere. Much progress still remains to be made on several problems which are separated here for convenience, but which are obviously closely interrelated: 1) the problems of the treatment of convection and of the uncertainties in convective efficiency, and the related problem of convective overshoot; 2) the problem of the opacities and of the molecular equilibrium in late-type stellar atmospheres; 3) the problem of the radiative transfer itself and its implications for the construction of atmospheric models of great spectral complexity.

REFERENCES

Chandrasekhar, S. 1939. <u>Introduction to the Study of Stellar Structure</u>, University of Chicago Press.
Demarque, P. and Larson,R.B. 1964. Astrophys.J., <u>140</u>, 544.
Demarque,P. and Mengel,J.G. 1971. Astrophys.J., <u>164</u>, 317.
Eddington,A.S. 1930. Monthly Notices Roy.Astron.Soc. <u>90</u>,284,808. also Observatory, <u>53</u>, 208,342.
Kippenhahn,R., Temesvary, St. and Biermann,L. 1958. Z.Astrophys.,<u>46</u>,257.
Larson,R.B. 1973, Fundamentals of Cosmic Phys., <u>1</u>, 1.
Milne,E.A. 1930, Monthly Notices Roy. Astron. Soc., <u>90</u>,17,678. also Observatory, <u>53</u>,119, 238.
Prather, M.J. 1975 quoted by L.H.Auer and E.B. Newell, Dudley Obs. Report No. 9, p.14.
Schwarzschild,K. 1906, Nachr. Kon. Ges. d. Wiss. Gottingen, <u>195</u>,41.
Schwarzschild,M. 1958, <u>Structure and Evolution of the Stars</u>, Princeton University Press, p.89.
Vitense, E. 1953. Z. Astrophys.,<u>32</u>, 135.

BOUNDARY CONDITIONS WITH MASS-LOSS : GENERAL CONSIDERATIONS

Richard N. Thomas
Institute d'Astrophysique, Paris

I. INTRODUCTION

   I begin by stating very explicitly and unambiguously that I completely disagree with any idea that a stellar atmosphere can be in any sense a boundary of the star, or that the atmosphere sets boundary conditions on stellar structure. Such an idea is what underlies all the single-layer, LTE atmospheric models, which misdirected atmospheric studies for so many years by introducing a false separation between "normal" and "extended" atmospheric phenomena. On the contrary, I assert that a more correct physical picture for the atmosphere than as a boundary comes from the characterization of star and atmosphere as :

   (1) The STAR is : a concentration of matter and energy, C(M,E), in its parental environment, the interstellar medium (ISM) ; with boundary conditions on stellar structure being set by the way one models (a) storage modes for matter and energy in the star and ISM, and (b) energy generation in the star.

   (2) The ATMOSPHERE is : (a) <u>functionally</u>, a transition-zone between star and ISM (Pecker, Praderie, Thomas, 1973) ; (b) <u>structurally</u>, a set of regions, each of which has distinctly different characteristics and reflects some aspect of the transition ; (c) <u>diagnostically</u>, the immediate place of origin of the fluxes, the analysis of which gives direct information on these regions and indirect information on thermodynamic structure of the star. In consequence, boundary conditions are imposed <u>on</u> the atmosphere by star and ISM, <u>not</u> imposed <u>by</u> the atmosphere on anything.

   Equally explicitly, I emphasize that we do not know, a priori, either : (i) which of the several possible alternatives for thermodynamic structure of the star actually exists at each phase of the star's evolution ; (ii) or, in consequence, which variety of atmospheric regions is most important observationally for any given class of star. Thus our decisions on stellar structure and atmospheric regions must be empirical/observational, not a priori theoretical. We <u>observe</u> atmospheric regions, and we <u>infer</u> sub-atmospheric storage modes.

In this spirit, an alternative title to my remark, to parallel that of the preceding paper, might be : "Sensitivity of Atmospheric Structure to Lower (or Sub-atmospheric) Boundary Conditions". I focus on 4 points, and a summarizing slide prepared jointly by NASA, Pecker, and myself : (A) Non-Equilibrium thermodynamic alternatives on structure of, and fluxes from, the star as a C(M,E) in the ISM ; (B) Survey of atmospheric phenomena which guide an empirical choice among alternatives (A), hence boundary conditions on models ; (C) Pictorial caricature of the representation of the star as a system with matter-fluxes (OPEN system) and nonthermal kinetic energy storage modes ; (D) Mathematical representation of thermodynamic model and boundary conditions ; Summarizing-Slide, The Sun as a proto-example of the relation between sub-atmospheric storage modes and sequence of atmospheric regions.

II. POINTS OF FOCUS

A. NonEquilibrium Thermodynamic Alternatives on Structure of, and Fluxes from, the Star $\equiv$ C(M,E) :

Consider the models defined by various combinations of the (KIND, TYPE, DEGREE) alternatives of Table 1. We exclude those containing ISOLATED and EQUILIBRIUM because we observe the star. But any other choice must be empirical/observational, following the sequence of successive approximation : (1) choice of a combination and the implied boundary conditions ; (2) construction of a model, and of a diagnostic methodology consistent with the model to interpret observations, based on (1) ; (3) comparison of model to observations, seeking consistency or anomaly. To the degree of accuracy needed for the present discussion, and using our limited knowledge of general nonEquilibrium thermodynamic modeling, such successive approximation converges rapidly to show us those combinations of Table 1, hence which boundary conditions, we must study.

The great majority of stellar atmospheric models in the literature are based on the (closed, thermal storage, both linear and nonlinear nonEquilibrium) alternatives. In addition to chemical composition, these models depend upon just <u>two</u> parameters : F_R --- radiative flux, or T_{eff} ; gravity, g. The <u>lower</u> boundary condition in the atmosphere is a prescribed value for each of the two parameters. The <u>outer</u> boundary condition is no incident radiation on the star. The condition of radiative equilibrium ensures only thermal storage modes, because the computed T_e-gradients are too small to make thermal conduction be of importance. IF the models are <u>non</u>linear nonEquilibrium, they can provide a smooth transition from a static star to a static ISM. Any matter-flux comes wholly by thermal escape ; values computed from the outer-atmospheric properties of such models show these matter-fluxes to be negligible, in any observational way. Thus, these models are self-consistent : physically, and mathematically. But, those models for which density decreases monotonically outward (essentially all models) are unstable against arbitrarily-small outward radial velocities. The introduction of such small rad-

TABLE 1

NonEquilibrium Thermodynamic Alternatives For $C(M,E)$

Kind of C		TYPE storage modes	DEGREE of nonEquilibrium
ISOLATED	: no fluxes	THERMAL only	EQUILIBRIUM
CLOSED	: only energy fluxes-- F_{rad}, F_{mech}	THERMAL + nonTHERMAL	LINEAR nonEquilibrium
OPEN	: energy and matter fluxes-- F_{rad}, F_{mech}, F_{mat}	nonthermal kinetic energy : convection, pulsation, rotation, magnetic, hydromagnetic	nonLINEAR nonEquilibrium

TABLE 2

Atmospheric Phenomena Guiding Choice Among Table 1 Alternatives

PHENOMENA	OBSERVED RANGE
$F_M(obs) \sim (10^4 - 10^7) \, F_M(thermal)$	WR-OB-Θ-M
$F_M(obs) \to U \sim q$ at $R/R_o < 1.04$	WR-OB-Θ-M
C-C($T_e > T_{eff}$) ioniz. level	WR-OB-Θ-M
BaC excess	O-B$_e$-T Taur-Me
Balmer Line Emission	B$_e$-T Taur-Me
Nonthermal line widths	WR-OB-Θ-M

ial velocities is the kind of thing which would result from admitting nonthermal sub-atmospheric storage modes.

Those presently-existing atmospheric models which are superposed upon sub-atmospheric models having nonthermal storage modes in the form of convection are not physically-consistent because of the imposed boundary conditions. As upper boundary conditions on the convection zone --- lower boundary conditions on the atmosphere --- they impose the condition F_M(matter-flux) = 0. That is, they impose the condition that the star be a CLOSED rather than an OPEN system, without asking the stability of such a condition against $F_M \neq 0$ and thus against its consistency.

Indeed, those CLOSED system models that have been enlarged to include kinetic-energy, as well as radiative-energy, fluxes provide counter-examples to demonstrate this inconsistency. The usual solar models having a convectively-induced acoustic mechanical energy flux produce chromospheres-coronas as a region of the atmosphere. Parker demonstrated that such models cannot match, statically, the ISM : but must produce a nonthermal matter-flux = "wind". Thus the star/Sun becomes an OPEN system ; and one must return to the convective nonthermal kinetic energy storage modes to include both matter- and nonthermal kinetic energy-fluxes. These must be part of the UPPER convective-zone, LOWER atmospheric boundary conditions. It is not so important whether such changes would introduce significant effect on the properties of the convective zone. It is important that without such fluxes, the stars would have no chromospheres-coronas ; with them, they have chromospheres-coronas, and all the attendant atmospheric features.

In summary : CLOSED stellar models can be constructed which are physically and mathematically self-consistent IF they have only thermal storage modes. But they are unstable against sub-atmospheric radial matter-, or nonthermal kinetic energy-, fluxes.

SOME nonthermal sub-atmospheric kinetic energy storage modes require, in parallel, chromospheres-coronas and matter-fluxes. In such situations, the proper lower atmospheric boundary conditions on models must be the non-restricted (by $F_M = 0$) upper boundary conditions on the sub-atmospheric nonthermal kinetic energy storage modes.

It is therefore essential to survey atmospheric phenomena, especially correlations between chromospheric-coronal and matter-flux phenomena, to ask whether all nonthermal kinetic energy and matter-fluxes produce such associations, so that the lower boundary conditions on them should be the upper boundary conditions on sub-atmospheric nonthermal kinetic energy storage modes ? Or, can we have nonthermal matter-fluxes without having chromospheres-coronas ? Current models of so-called radiation-presure origin of matter-fluxes indeed adopt this latter possibility. Unfortunately, just those hot (OB)stars, for whose description these radiation-pressure models were contrived, have been observed to have chromospheres-coronas. It is not generally agreed whether this conflict between observations and model-prediction comes from incomplete equations or inadequate boundary conditions for these models. I comment later on this point.

B. Survey of Atmospheric Phenomena guiding an Empirical Choice among Table 1 Alternatives :

The goal of such a survey is to decide whether observed atmospheric phenomena, across a sufficiently-broad range of spectral classes, suffice for us to conclude that nonthermal kinetic energy fluxes, and nonthermal matter-fluxes, are associated and characterize stars generally. If so, they imply equally-generally the existence of nonthermal storage modes in the sub-atmosphere, which couple directly to the atmosphere in the sense of providing both storage and fluxes there. If so, such coupling must be represented by the proper boundary conditions at the sub-atmosphere, atmosphere interface or transition. If so, we have then identified the "Universal" nonEquilibrium thermodynamic character of stars : they are OPEN, NonTHERMAL storage mode, NonLINEAR NonEquilibrium Concentrations. (The latter property follows immediately from recognizing the nonlinear character of distribution functions for photons, internal energy states, and particle velocities.)

It is necessary to define the meaning of chromosphere-corona characteristics associated with matter-fluxes. Nonthermal matter-flux characteristics are clear : macroscopic velocity fields comparable with, or exceeding, chromospheric-coronal, as well as photospheric, thermal velocities. If, in a macroscopic flow, having velocity gradients, velocities comparable with the local thermal velocity appear, mechanical energy dissipation begins --- and the local kinetic temperature of the gas rises. Such a rise in T_e above T_{eff} is the essential character of a chromosphere : because a rise to a value slightly smaller than T_{eff} can be induced wholly by nonlinear photon + atomic phenomena in a static atmosphere.

So, I present the following list, making no pretence that it is not highly selective, in no way complete. I could equally-well have used the similar list presented by Lamers in the Commission 44 session on stellar mass-loss. Refer to Table 2.

Table 2, by itself, is a bit cryptic ; hence, a brief explanation. The first line is clear ; observed matter-fluxes exceed by a very large factor those which would be produced by only thermal escape from the atmosphere. The second entry corresponds to a computation (Thomas, 1973), which considers the observed data on $F_M = 4\pi R^2 \rho U$ for a wide variety of stellar types. Because the photospheric thermal velocity, q, is much less than the escape velocity ; because the thermal velocity does not change much through the atmosphere unless/until the systematic velocity $U \sim q$; and because until this point $U \sim q$ the photospheric, radiative equilibrium density distribution does not change much ; we can use essentially isothermal density distributions to ask at what R/R(photosphere) $U \sim q$ from the observed F_M. Wholly empirically, for all stars for which F_M has been inferred, observationally, in some way, we obtain the result of line 2. This suggests that all these stars have chromospheres, which begin very near the photosphere. The third entry asks in what stellar classes we observe ionization levels significantly above those which would result for $T_e > T_{eff}$ (LTE or nonLTE calculation. The long-known observations of OVI in WR stars in the visual spectrum couple with recent satellite observations of OVI in OB stars (Lamers and Rogerson, 1975) in the far UV to demonstrate that T_e is at least some 2.10^5 K in the atmospheres of these hot stars, thus that they have chromospheres-coronas, for which the symbol C-C stands in Table 2. The fourth entry states a well known fact ; of which one possible explanation is a deep-lying chromosphere, such as suggested by Herbig (1962) long ago, for T Taur. The fifth entry is again a well-known observational fact which, we have suggested (Dumont et al., 1973), follows directly from a chromospheric interpretation of the BaC emission in these same stars. The fifth item is closely linked, suggesting that the deeper-lying the chromosphere, the greater the line-width, in either inhomogeneous (Thomas, 1973) or homogeneous (Dumont et al., 1977) atmospheres. Computations exist thus far only for the cool stars ; they are projected, for the hotter stars.

In short, a variety of atmospheric phenomena suggest that chromospheres-coronas, and matter-fluxes, arise from coupling between subatmospheric nonthermal storage modes and the atmosphere ; thus, that the star is an OPEN, NonTHERMAL storage mode, system. Let me return to Mrs. Gaposchkin's unpublished ejaculation, long ago, on seeing the first solar rocket spectrum : "the UV Sun is a WC6 star" (i.e. the Sun is also a symbiotic star) ; and re-phrase it, to be "the atmospheres of WC6 stars are visually deep-lying solar chromosphere-coronas, along with all their cousins". Boundary conditions, and equations to which the boundary conditions are applied, must reflect these empirical non-Equilibrium thermodynamic conclusions.

The boundary conditions for this (Open, nonthermal) model are, in parallel to those of the (Closed, thermal) model : prescribed values of F_M, F_{mech}, F_R, g at the sub-atmosphere, atmosphere interface for the lower boundary conditions ; no incident F_M (non-accreting) on the star, F_R (incident) equal to the statistical stellar contribution to the ISM for the upper boundary conditions. The condition of radiative equilibrium is replaced by including terms in both systematic and "turbulent" (viscous, or velocity gradient, terms ; acoustic terms ; "true" turbulence) velocities in the storage/transport equations describing the spatial/temporal evolution of the structure of star and atmosphere. One cannot, as is customary in the radiation pressure models, simply ignore the "turbulent" velocity terms, deleting them from the equations ; for this a priori exludes the production of chromospheres-coronas. Moreover, one cannot introduce the entropy as a state-parameter, as did Demarque in the preceding summary of boundary conditions on the interior, because in the nonlinear nonEquilibrium region, the entropy is undefined. In either of the (Closed, thermal) or (Open, nonthermal) models, one must admit the possibility of non-spherical-symmetry, for which F_M, F_{rad}, etc. depend on angular as well as radial coordinates. Whether one chooses to simply take these parameters as given, or as coming from a solution of the sub-atmospheric nonthermal mode problem, is a choice which must reflect our degree of progress in modeling the stellar sub-atmosphere + interior. In the following illustrations, I take these F_M etc. as given.

C. Pictorial Caricature of a Representation of the Star as an (Open, nonthermal) System by an Imperfect Wind-Tunnel :

Consider the wind-tunnel analogy to the solar wind (Clauser, Germain, 1961), for which the equations, without "turbulent" terms, are essentially : ($q = \gamma p/\rho$ in 1a ; kT/M in 1b)

$$\left\{\frac{q^2}{U^2} - 1\right\} U \frac{dU}{dr} = \begin{cases} -\frac{q^2}{A} \frac{d A(r)}{dr} & \text{(1a) WIND-TUNNEL} \\ \frac{1}{r}\left[w_o^2 \frac{r_o}{r}(1-\beta) - 2q^2\left(1 - \frac{d \ln T_e}{d \ln r^2}\right)\right] & \text{STAR (1b)} \end{cases}$$

and a schematic diagram is :

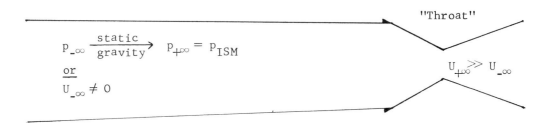

The caricature is both physical and mathematical. It is physical because it replaces the star-as-a-storage-pot by the storage section of the wind-tunnel (far left of the diagram), and replaces the atmospheric acceleration-to-escape-velocity by the converging-diverging nozzle (right part of diagram). The caricature is mathematical, because it ignores the 2-dimensional character of the diagram in writing the wind-tunnel equation.

The slope of the storage-converging section is very gradual, to represent the slow decrease of $(r\, w_o^2/r)/q^2$ in a "normal" = thermal-storage atmosphere. The smallest value of this ratio is some 400, in real stars ; for most hot stars, it exceeds 2000. In consequence, for such "normal" models, where $U_{-\infty} = 0$, and we can ensure that the "gravitational-decrease" of $p_{-\infty}$ brings it smoothly to $p_{+\infty}$, the wind-tunnel does not "flow" ; the atmosphere is static.

In Parker's solar chromosphere-coronal model, $p_{-\infty}$ is too large to decrease, gravitationally, to $p_{+\infty}$; so there is a flow ; because the "storage-section" is taken as the "hot" chromosphere-corona. The "perfect" wind-tunnel corresponding to Parker's model produces a continuously accelerating flow to supersonic velocity without shock-waves ; this requires both $U = q$ and $\sqrt{2}\, q = w$ (the escape velocity) to occur at the "throat". (For the moment, we ignore radiation-pressure, whose ratio to gravity is β).

But such a "perfect" wind-tunnel requires that $p_{-\infty}$ and $p_{+\infty}$ be specified, and $A(r)$ designed to produce such shock-free, continuously-accelerating flow. Generally, if $A(r)$ is also given a priori --- as in the stellar case, because gravity is fixed independently of the details of atmospheric structure, and q is constant in this outer atmospheric region --- such a "perfect" flow does not occur. If $w^2 (= w_o^2 r_o/r)$ decreases to $2q^2$ before U increases to q, the flow is not continuously accelerated but decelerates after $w^2 = 2q^2$. Such alternative does not produce the large observed wind-velocities in the hot stars, which much exceed the escape velocity ; so, observationally, this alternative is not interesting. On the other hand, if U reaches q before w^2 reaches $2q^2$, shocks occur, and the resulting mechanical dissipation of energy heats the gas and raises q. The resulting behavior of U and q must be determined from a set of equations that are expanded over those above to include all forms of these energy dissipation terms --- which are essentially those included in "turbulence", plus heat-conduction, etc. Now, this alternative corresponds to the observed presence of chromospheres, even in these hot stars. Note that we can have chromosphères-coronas without a heating produced by the velocity in the matter-flux reaching the thermal velocity --- in this case the region of subsonic acceleration is extended, over the case of constant q. But we cannot have chromospheres-coronas without mechanical heating of some kind. Neither the "perfect" wind-tunnel, nor the accelerated-decelerated alternative, produce such heating : particularly, they do not produce such heating at such low atmospheric levels as correspond to Table 2 ; and especially they cannot produce such heating if there are no terms in the equations to describe such production.

So, we criticize the "perfect" wind-tunnel models, with or without radiation pressure, on three points. Physically, they do not produce chromospheric heating below the "critical point $w^2 = 2q^2$. Mathematically, they apply the wrong boundary conditions ($U = q$ where $w^2 = 2q^2$) to an incomplete set of equations.

Adding radiation pressure, $\beta \neq 0$, in these equations and under these boundary conditions does not change these essential criticisms, when they are taken in the context of observations. To move the "critical-point" deep-enough into the atmosphere to match the observed mass-fluxes requires values of some 0.95 for β. To say that there are some, very hot, stars for which such large β can be reached in a static atmosphere is not useful in a general sense, because Bo and O5 stars, which are not this hot, show chromospheres ; and these chromospheres must be explained. So, I prefer the above explanation : "imperfect wind-tunnels"; and the associated boundary-conditions applied to the complete equations.

D. Mathematical Representation of Thermodynamic Model and Boundary-Conditions in the One-Dimensional, Plane-Parallel Approximation :

The one-dimensional, plane-parallel approximation suffices for the present summary ; because our only objective is to show that an association, like the observed one, between matter-flux, nonthermal kinetic energy flux, and chromosphere-corona existence would be predicted by the model of a star as an open system with nonthermal kinetic energy storage modes and the associated boundary conditions. We have already shown that the observed matter-fluxes imply the beginning of chromosphere-corona well within the plane-parallel approximation limit. The one-dimensional approximation is highly-restrictive, especially under the variety of nonthermal storage modes possible, but only in the details, not the existence, of the stated association ; so it suffices for simplicity of illustration.

Then we can write the three equations which describe the storage/evolution of matter, thermal/microscopic energy, and nonthermal/macroscopic energy --- and their coupling --- as (Cannon and Thomas, 1977) :

Matter : $\dfrac{d(U\rho)}{dx} = 0 \quad ; \quad (U\rho) \neq 0$ (2)

Microscopic energy :

$$\left\{ \dfrac{d[U(\epsilon + p + p')]}{dx} - \dfrac{Ud(p + p')}{dx} \right\} \left\{ \dfrac{d}{dx}\left(\dfrac{\lambda dT_e}{dx}\right) + 4\pi \int (J_\nu - S_\nu)\dfrac{d\tau_\nu}{ds}\,d\nu \right\}$$ (3)

Macroscopic energy + Matter :

$$\rho(q^2 - U^2)\dfrac{dU}{dx} U^{-1} = \rho g + \dfrac{d}{dx}(p_r + p')$$ (4)

ϵ is the internal energy of the particles ; p' is the "generalized turbulent" pressure/energy (eg. Moyal, 1952), as is p the thermal pressure/energy, and p_r the radiation pressure/energy. In the static, thermal-storage-mode situation, the left-hand-sides of each of the above vanish, except for the q term in (4), which reduces to dp/dx. Thus, the left-hand-side gives the nonthermal-storage-mode presence and effects. Equation (4) represents the correction to equation (1), coming from the inclusion of these "generalized-turbulent" terms.

In the absence of the nonthermal modes, the terms in U and p' disappear ; and the boundary conditions are the values of g and of F_R, noting that

$$dF_R/dx = -4\pi \int (J_\nu - S_\nu)\dfrac{d\tau_\nu}{ds}\,d\nu$$ (5)

and that the thermal conductivity term (first term on right) in (3) is negligible. So one solves (3) with this given F_R (in the simplest opacity cases, like the grey case, T and ρ de-couple in this solution) to obtain $T_e(\tau)$; then uses this $T_e(\tau)$ in (4) to obtain $\rho(\tau)$. (2) is irrelevant, because U = 0 by assumption. One can, however, infer the outward variation of any small value of U <u>imposed from below</u>, by setting U = the imposed value of U, coupled with the static-atmosphere-inferred value of ρ at the given atmospheric level. Then, the variation of U follows that of this static atmosphere solution for ρ, until U becomes large enough to cause the left-hand sides of (3) - (4) to change the static-atmosphere solutions for T_e and ρ.

In essence, this last procedure corresponds to the situation with nonthermal storage modes. Hence, the boundary conditions for this case are simply values of F_M (or, alternatively, U) and F_{mech} added to the values of F_R and g. If <u>only</u> F_M is imposed, then the value of U increases to near q ; heating begins ; and T_e rises. But also, the solution of equation (4) begins to "oscillate". That is, because the right-handside is positive throughout this plane-parallel region, dU/dx > 0 so long as q > U. But when U > q, dU/dx < 0 and the flow decelerates, until q > U, when the process repeats. This is the well-known trans-sonic instability ; and the oscillation is accompanied by the production of many small shock-waves (eg. cf. the photographs in Charters and Thomas, 1945). So, to treat this problem completely, one must restore the complete set of time-dependent, 2- or 3-dimensional equations. But, in these regions, we have already produced the chromosphere, and the matter-flux has a significant (U ~ q) value. Hence, the stated association : matter-flux and chromospheres. If, in addition a non-zero F_{mech} produces an acoustic-wave component of p', chromospheric heating begins before U ~ q, as earlier mentioned in section C ; so that we have the further association with nonthermal mechanical energy fluxes. So, we see, pictorially as well as arithmetically, how this (open, nonthermal storage mode) model, <u>plus</u> an adequate set of boundary conditions can produce the kind of association observed.

Summarizing-Slide : The Sun as a Proto-Example of the Relation Between Sub-Atmospheric Nonthermal Storage Modes and Sequence of Atmospheric Regions. (See Figure 1.)

The slide follows a figure from work by Pecker and myself, and is reproduced courtesy of NASA (Pecker and Thomas, 1977). Going clock-wise from the right hand side of the figure, the Sun is divided into sectors. Each sector shows the kind of atmospheric region which would result if the sub-atmospheric nonthermal storage were as indicated in the sector. Thermal storage gives only photosphere ; nonthermal storage with one type of mode (convection, pulsation, or rotation) gives chromosphere-corona, matter-fluxes etc. but no magnetic "activity" ; two types of storage modes produce the "active" Sun etc. The implication of the increasingly-detailed structure coming from increasingly elaborate storage modes is clear. It is equally clear that such a sequence of phenomena and atmospheric regions hardly follows from considerations based on radiation pressure alone. It is equally clear that these "pictorial" considerations must be developed in detail ; and the relation between

BOUNDARY CONDITIONS WITH MASS-LOSS: GENERAL CONSIDERATIONS 153

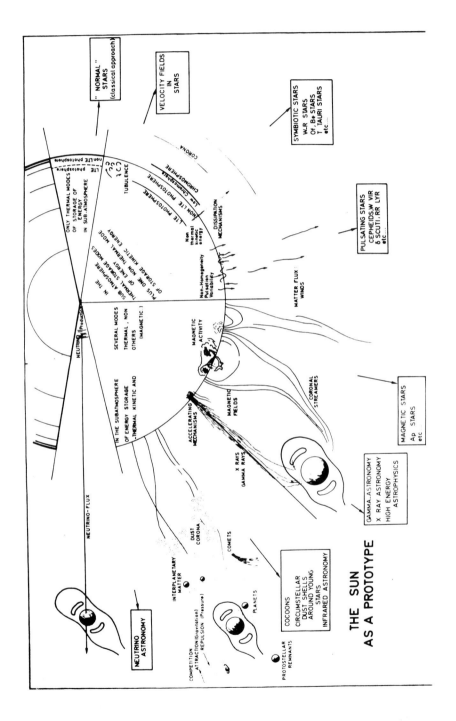

Figure 1. The Sun as a proto-example of the relation between sub-atmospheric non-thermal storage modes and sequence of atmospheric regions.

atmosphere, sub-atmosphere, and interior examined in detail. Until the boundary conditions have been actually applied in detail, it is hardly worthwhile to debate whether the interior --- and its evolution --- is or is not altered as significantly as is the structure and evolution of the atmosphere.

REFERENCES

Cannon, C.-J. and Thomas, R. N. 1977, Ap. J. Feb. 1977
Charters, A. C. and Thomas, R. N. 1945, J. Aero. Sci. 12, 468
Dumont, S., Heidmann, N., Kuhi, L. V., Thomas, R. N. 1973, Astron. Astrophys. 29, 199
Dumont, S., Heidmann, N., Thomas, R. N. 1977, in preparation
Herbig, G. H. 1962, Adv. Astron. Astrophys. 1, 47
Lamers, H. J. and Rogerson, J. B. 1975, The Liege Annual Symposium
Moyal, J. E. 1952, Proc. Camb. Phil. Soc. (Pt 2) 48,•329
Pecker, J.-C., Praderie, F., Thomas, R. N. 1973, Astron. Astrophys. 29, 289
Pecker, J.-C. and Thomas, R. N. 1977, Space Science Rev. in press.

MASS LOSS IN STARS OF MODERATE MASS BY STELLAR WINDS
AND EFFECTS ON THE EVOLUTION

C. de Loore

Astrophysical Institute

Free University of Brussels
Belgium

1. INTRODUCTION

Some 30 years ago it became clear that the solar corona is a plasma with a temperature of the order of 10^6 K. As the underlying layers have only temperatures of 5000 K a mechanism had to be discovered, capable to explain this high temperature. A solution to the problem was found when it was realized that mechanical energy losses, by shock dissipation of wave energy can heat up a plasma to such high temperatures. This mechanical energy is formed in the deeper layers of the atmosphere and transported outwards. Dissipation becomes significant in regions where the density is sufficiently low.
Wave propagation in a compressible medium in the presence of gravity and magnetic fields has been treated as a general problem, among others by Ferraro and Plumpton (1958). Three basic parameters are present : compressibility of the medium, gravity and magnetic field.
Here solutions will be discussed for cases where magnetic fields are not included.
Biermann (1946) and Schwarzschild (1948) suggested that gravity modified sound waves are, by shock dissipation responsible for the heating of the outer atmosphere. A few years later Lighthill (1952) and Proudman (1952) derived quantitative expressions for quadrupole sound generation. Starting from these data, de Jager and Kuperus (1961) and Kuperus (1965) were the first to construct models of the corona and the transition region to the chromosphere which at least qualitatively explain the high temperature of the corona and the sudden temperature rise in the transition region; later considerable improvements were made of the physical theory by Ulmschneider (1967, 1971a,b) and Stein (1968), and improved models were calculated by de Loore (1971). The computation of stellar coronae is based on these ideas developed for the calculation of the solar corona

and transition region.

2. MECHANICAL FLUXES

In convection regions turbulence is caused by large scale convective motions. Starting in these regions waves will penetrate the radiative layers. Generally gravitational waves, or sound waves are considered (see de Jager, 1975). Gravitation waves propagate through changes in the local gas pressure as a consequence of the change of the air mass above a certain horizontal plane in a hydrostatic equilibrium medium. They can propagate only for frequencies below the Brunt-Väisälä frequency

$$\omega_{bv} = (\gamma-1)^{1/2} g c^{-1}$$

Sound waves (compression waves) propagate as a consequence of increase and decrease of the gas pressure by compression or dilatation of the gas. The phase velocity λ/P depends on the period, so the waves will disperse. Vertical propagation is inhibited for frequencies below the limit frequency

$$\omega_s = \gamma g/(2c)$$

The diagnostic diagram (Figure 1), i.e. the dispersion relation $F(\omega,k)$, shows the possible waves for an isothermal atmosphere. The solutions in the upper left corner represent the gravity modified sound waves; the gravity wave solutions occupy the lower right part.
Whitaker (1963) proposed gravity waves for the coronal heating because sound waves with wavelengths comparable to the dimensions of the granules (Bahng and Schwarzschild, 1961) would not be able to propagate through the region of temperature minimum. However, it was pointed out by Souffrin (1966) that the short radiative relaxation time in the low photosphere (order of 1^s) would rapidly eliminate these oscillations. In the region of the temperature minimum the cut-off frequency for sound waves is 0.0233 s^{-1}, and the most representative frequency of sound waves generated by the Lighthill mechanism is $\omega \sim 0.2$ s^{-1}. Hence the mechanical flux generated by sound waves can to a large extent penetrate into the chromospheric regions. During the sixties (Osterbrock, 1961; Kopp, 1968) the heating of the chromosphere was almost generally attributed to acoustic waves with typical periods of ~ 5 min (however see Kuperus (1965, 1969)). Now it is known that the 5 minute oscillations (Leighton, 1960; Leighton et al. 1962) are produced by nonradial pulsations. More than five dis-

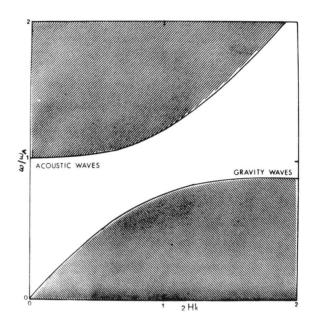

Figure 1. Diagnostic diagram showing the dispersion relation $F(\omega,k)$. The coordinates are normalised. The upper left part represents the acoustic wave solutions, the lower right part the gravity wave solutions (after de Jager, 1975).

crete stable modes of the 5 min oscillations could be resolved by Deubner (1976) in the average k,ω diagram. They agree with the predicted modes of trapped acoustic waves in the subphotospheric regions and especially with the solutions of linear nonradial oscillations of the solar envelope, obtained by Ando and Osaki (1975). Ulmschneider (1970,1974) argued in favour of the short-period nature of the waves, by comparing the theoretical dissipation rate of acoustic shock waves for different mechanical fluxes and periods with the computed chromospheric radiation losses. A study of the radiatively damped acoustic waves in the solar atmosphere is being performed by Ulmschneider et al. (1976a), Kalkofen and Ulmschneider (1976) and Ulmschneider and Kalkofen (1976). Theoretical computations of the spectrum of acoustic flux produced in the solar convection zone were performed by Stein (1968). He made computations for mechanical fluxes for several different turbulence spectra and found a flux maximum at periods of ~ 30 s.
Computations of mechanical fluxes for stars of various types were made by : Kuperus (1965), for stars of solar composition, with T_{eff} ranging from 4400 K to 7000 K; de Loore (1967,1970) for stars of solar composition, with T_{eff} ranging

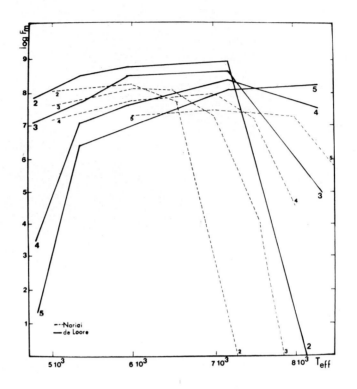

Figure 2. Comparison between the mechanical fluxes computed by de Loore and Nariai.

from 2500 K to 16630 K; Nariai (1969) for helium rich stars; Böhm and Cassinelli (1971) for helium rich stars, with T_{eff} from 5790 K to 30000 K. The results of the computations of de Loore (1970) and Nariai (1969) are shown in figure 2. Curves of constant mechanical fluxes are shown in figure 3. All this is rather uncertain and subject to some criticism. First of all, as was demonstrated by Stein (1968) by computing the mechanical energy for different turbulence spectra, the acoustic power output is highly sensitive to the high frequency tails. His computations yield an uncertainty factor of the order of one magnitude. In the second place the result is highly dependent on the turbulent velocity amplitudes (the acoustic emission is a function of the fifth power of the turbulent Mach number). However, actual refined computations of stellar mechanical fluxes by Ulmschneider et al. (1976b) are in rather good agreement with our previous calculations, with observations and with other refined calculations (Stein 1968).

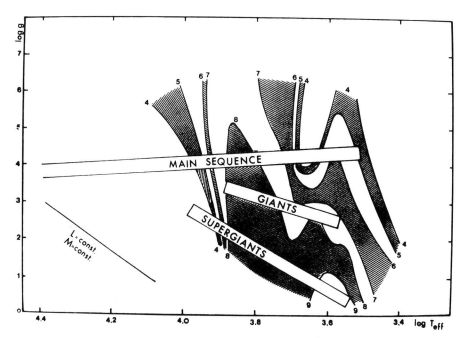

Figure 3. Curves of constant mechanical flux as computed by de Loore (according to Kippenhahn, 1972).

3. HEATING OF THE CORONA

It is assumed that the mechanical flux F_m generated in the convection zone is known (e.g. from computed models of this zone). Spatial dissipation of this flux becomes only then important when the shock front is attained; this dissipation can be described by a local absorption coefficient.
The balance between the different energy terms, dissipation, radiative energy losses and stellar wind energy losses is closed by conductive energy losses from corona to chromosphere. The heat conductivity of a fully ionized gas is a function of the temperature; hence the temperature can be calculated. The hydrostatic equilibrium equation furnishes the density. De Jager and Kuperus (1961), and Kuperus (1965) did not include the effects of stellar winds and started the integration at the photosphere. Inclusion of a velocity kinetic energy term complicates the problem extremely. For this reason one assumes the initial value of the flow speed to be a supplementary parameter. Iteration of this value leads to a given outer boundary condition. The behaviour of the various fluxes for the solar transition region is given in figure 4. Results of de Loore (temperature and density) for some stellar coronas are shown in figure 5. This leads

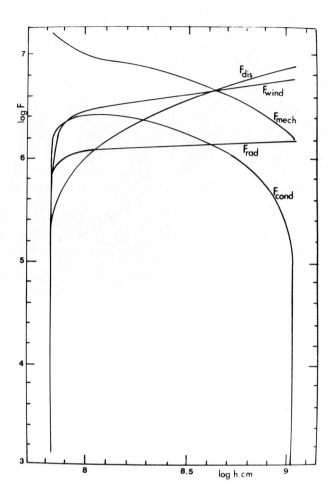

Figure 4. Behaviour of the various fluxes for the solar transition region.

essentially to the conclusion that coronal temperature and density in the range covered by the calculations seem to be monotonic functions of the mechanical energy flux.
A new discussion of the mechanical energy fluxes for main-sequence stars and supergiants was given by Lamers and de Loore (1974,1976). In order to get a better insight in the mechanism that drives microturbulence in supergiants the energy flux associated with the observed microturbulence was calculated for different spectral types. It was assumed that the microturbulent motions are outward propagating sound waves.

$$F = \frac{1}{2} \rho a^2 c$$

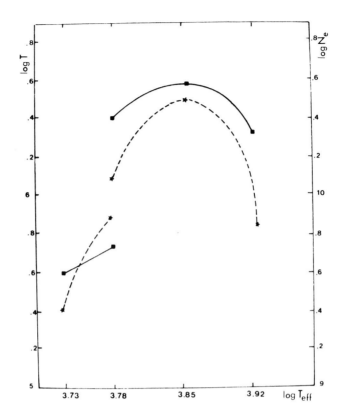

Figure 5. Temperature (full curves) and density (dashed curves) for some stellar coronas as a function of the effective temperature. For the lower curves log g = 5, for the upper curves log g = 4 (after de Loore, 1970).

with ρ density

c the local sound speed

a is assumed equal to the "observed" turbulent velocity, as adopted in the model atmospheres from Kurucz, Peytremann and Avrett (1974) and in those for the F and G supergiants from Parsons (1967). These fluxes can then be compared with computed mechanical fluxes; for stars with convection zones values of de Loore (1970) were used, for hot stars the mechanical fluxes produced by radiation-driven sound waves of Hearn (1973) were used. It turns out that the overall agreement between "observed" and predicted fluxes is reasonable, but should not be exaggerated. In any case, it turns out that the flux associated with microturbulence shows a minimum around type A; this could mean that microturbulence in stars earlier than A is created by another mechanism than

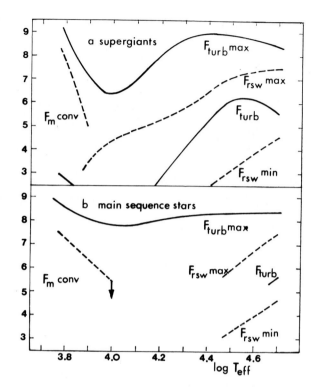

Figure 6. Predicted and observed mechanical fluxes versus T_{eff} for early-type supergiants and main-sequence stars. For cool stars the mechanical fluxes generated by convective turbulence are used (F_m(conv)). For hot stars Hearn's mechanism of radiation driven sound waves is considered (dashed lines). The solid lines are derived from observed values of the microturbulent velocity component, and with the assumption that the turbulence propagates outward with sound speed.

in late types. Figure 6 shows the results of the computations.

4. X-RAY FLUXES EXPECTED FROM STELLAR CORONAS

On the basis of earlier work of de Loore (1970), de Jager and de Loore (1971) derived values for the expected X-ray fluxes of some nearby stars, the dominant emission mechanism being free-free emission by electrons. The ratio between the expected X-ray flux of a star F_{x*} and that of the sun $F_{x\odot}$ is given by

$$\frac{F_{x*}}{F_{xo}} = \left(\frac{R_*}{R_\odot}\right)^2 \left(\frac{Ne_*}{Ne_\odot}\right)^2 \left(\frac{H_*}{H_\odot}\right)\left(\frac{d_\odot}{d_*}\right)^2$$

with R and d radius and distance respectively, Ne the electron density at the basis of the corona, and H the coronal scale height.

For the sun the observations of Friedman (1959) were used, i.e. 0.13 erg cm^{-2} s^{-1} at sunspot minimum and 1 erg cm^{-2} s^{-1} at sunspot maximum, corresponding with 0.2 and 1.5 photons cm^{-2} s^{-1} at earth distance.

Some good candidates for X-ray research are given in table 1 (de Jager and de Loore, 1971).

	spectral type	expected photon flux	
Procyon	F5	0.016	0.12
α Cen	G2	0.008	0.06
β Cas	F2	0.0007	0.005

Table 1. The expected minimum and maximum soft X-ray photon fluxes for some F and G stars.

Minimum flux coronae in dwarfs and giants were calculated by Mullan (1976), using a method of Hearn (1975) : it is assumed that the corona has a strictly radial magnetic field and adjusts itself such that the sum of radiative, conductive and stellar wind fluxes, for a given pressure at the base of the corona is minimal. It is assumed that the flux loss is compensated by an input of mechanical energy from the star. By assuming that the fraction of the total stellar luminosity which is used for the coronal heating is the same as for the sun (certainly a very strong assumption!) stellar coronal temperatures and densities can be predicted. Coronal temperatures ranging from 280 000 K (for M6) to 4 980 000 K (for O5) MS stars are found. Red dwarfs turn out to have coronas 3-10 times cooler.

5. OBSERVATIONS OF STELLAR CORONAS

The Astronomical Netherlands Satellite (ANS), in a systematic search for stellar coronas examined some 30 stars, in order to find possible soft X-ray fluxes. Most of these stars showed no detectable flux, and only upper limits could be given. (Mewe et al. 1975, 1976). Figure 7 shows these upper limits for main-sequence stars, compared with observations. Two stars showed a detectable soft X-ray flux, α CMa (Sirius), and α Aur (Capella).

Observations made with OAO-3 (Dupree, 1975) also indicate

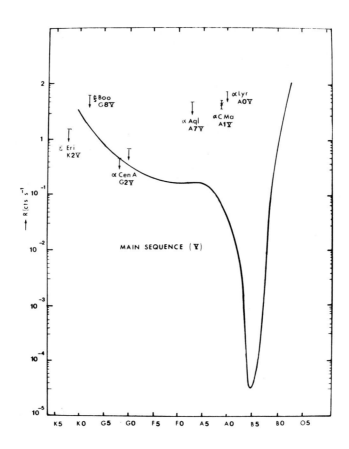

Figure 7. Predicted count rate (full line) for the soft X-ray channel of ANS for main-sequence stars, compared with recent observations (mostly upper limits) after Mewe et al. (1976).

the presence of hot outer layers for α Aur and for α Aqr. The soft X-ray flux from Sirius is probably generated in the white dwarf companion, Sirius B. Indeed Sirius A has spectral type A1, T_{eff} ~ 10.000 K, hence too hot for allowing a convection zone. As pointed out by Böhm (1972) the helium convection zone persists to very high temperatures, and in dense stars, a large part of the energy must be transported by convection. So for Sirius B the conditions dor the production of a large mechanical flux and possibly for a chromosphere and corona are probably present.

ANS observations gave also indications for the presence of a corona in α Pup (Mewe et al., 1975a). Indications for a corona in β Gem (Ko III) were presented by Gerola et al. (1974), with a temperature of 140 000 K - 260 000 K, fairly

Star	Spectral type	Coronal temperature-indication	Reference
α Aur	G5 III + G0 III	600000	ANS - Mewe et al., 1975 OAO-3 Dupree, 1976
α CMa	A1 WD		ANS - Mewe et al., 1975
ζ Pup	O4f	200000	Lamers and Morton, 1976
β Gem	K0 III	140000-260000	Gerola et al., 1974
α CMi	F5 IV V	$3 \cdot 10^5 - 6 \cdot 10^6$ $10^5 - 1.3 \; 10^6$	Evans et al., 1975
α Cyg	A2 Ia	320000	Hearn, 1975
α Car	F0 Ib	Mg II emission	Evans et al., 1975
α Boo	K1 III	asymmetries $T > 10^6$	Dupree, 1976 Gerola et al., 1974
α Tau	K5 III	Ly α - Mg II	Dupree, 1976
α Aqr	G2 Ib	asymmetries Mg II	Evans et al., 1975
ε Peg	K2 I	asymmetries Mg II	McClintock et al., 1976
τ Sco	B0 V	$1 - 5 \cdot 10^5$	Rogerson and Lamers, 1975

Table 2. Observational evidence on stellar coronas.

well in agreement with the calculations of de Loore (1970). Furthermore the discovery of O VI and Si III emission lines in the uv spectrum of Procyon (Evans et al.) indicates a transition region to a corona with a temperature between 3.10^5 and 6.10^6 K, confirming a suggestion of de Jager and de Loore (1971) that Procyon may have a corona.

6. MASS LOSS BY STELLAR WINDS

For a wide variety of stars mass loss has been observed, as summarized in table 3.

Type	$\dot{M}$ in $M_\odot$ yr^{-1}	Reference
WR	$10^{-6} - 10^{-5}$	Underhill, 1969
OB SG	10^{-6}	Morton, 1967 ; Hutchings, 1976
Bo V	10^{-8}	Lamers & Rogerson, 1975
B,A SG	$10^{-9} - 10^{-8}$	Hutchings, 1968
A2 Ia	3.10^{-10}	Kondo et al., 1975 ; Lamers, 1975
P Cyg	$10^{-5} - 10^{-4}$	Hutchings, 1968 ; De Groot, 1971
	3.10^{-6}	Kuan & Kuhi, 1975 ; Wright & Barlow, 1975
F2 Ia	10^{-8}	Sargent, 1961
F8 SG	10^{-5}	Sargent, 1961
Sun	2.10^{-14}	Hundhausen, 1972
Mo III	$2.5\ 10^{-9}$	Reimers, 1975
M1 III	4.10^{-9}	Reimers, 1975
M3 III	10^{-8}	Reimers, 1975
M5 III	3.10^{-8}	Reimers, 1975

Table 3. Observed mass loss rates for different types of stars.

An upper limit for the mass loss rate $\dot{M}$ was derived by Thomas (1973) following an approach of Williams (1970). These authors considered that the upper limit is determined by the ratio of the nuclear energy output in the sphere where core burning occurs, to the escape energy for that sphere. The values of Thomas (1973) are shown in table 4.

Spectrum	WR	B0	A0	F0	G0	K0	M0
dwarf		-3.8			-7.5		-8.4
giant					-5.6	-5.0	-4.2
supergiant	-1.5	-2.3	-2.7	-2.9	-2.7	-2.3	-1.5

Table 4. Upper limit for the mass-loss rates $\dot{M}$ for different types of stars according to Thomas (1973) (the table shows log $\dot{M}$).

Mass loss calculations for F and G stars were performed by de Loore (1968). The results are given in table 5. The values in the table were found by reducing the straightforwardly computed values (obtained from the computed density and flow velocity) by an efficiency factor, determined as the ratio between the observed solar mass loss rate of $2 \cdot 10^{-14}$ and the straightforwardly computed values of $1.1 \cdot 10^{-11}$ M_o yr^{-1}.

Te	log g	v	$\dot{M}$ (in M_o yr^{-1})
G5	5350	5	19 $4.4 \cdot 10^{-15}$
G0	5940	5	17 $8.4 \cdot 10^{-15}$
G2	5800	4.45	12 $2.2 \cdot 10^{-14}$
F0	7130	4	9 $4.8 \cdot 10^{-13}$
A5	8320	4	7 $9.6 \cdot 10^{-14}$

Table 5. Calculated mass losses for F and G stars. In the table are given the spectral types, effective temperatures, gravity acceleration, outflow velocity in km s^{-1} at temperature maximum and the mass flow in M_o yr^{-1}.

7. MASS LOSS IN RED GIANTS AND SUPERGIANTS

A semi-empirical estimate of the mass loss by stellar wind in red giants and supergiants has been made by Fusi-Pecci and Renzini (1975). They computed for stellar envelope models the quantity $\dot{\mu} = R \cdot L_{ac}/(GM)$, with R and M stellar radius and mass respectively and L_{ac} the acoustic energy output in the convection zone. The real mass loss rate $\dot{M}$ is then related to $\dot{\mu}$ as

$$\dot{M} = \eta \cdot \dot{\mu}$$

with η an efficiency factor, in the case of the sun $\eta_0 = 8 \times 10^{-4}$.

For red horizontal branch stars and $\eta \sim \eta_0$ the mass loss rate is $\sim 10^{-10}$ M_o yr^{-1}. Red supergiants of intermediate mass (1.4 M_0 < M < 8 M_0) lose a negligible amount of mass during core helium burning and previous evolutionary phase. By estimating the mass of the white dwarf remnant when the outer layers are blown away and by evaluating the time, final mass loss rates of 10^{-5} M_0/yr are found, comparable with the value required by Paczynski (1974) for the formation of planetary nebulae.

8. EVOLUTIONARY IMPLICATIONS

As can be seen from figure 8 there are two regions in the Hertzsprung-Russell diagram where mass loss is sufficiently large to be of possible evolutionary importance.
  a) the region of the massive main-sequence stars (M>20 M_0)
  b) the red-giant and red-supergiant region.
Whether or not the mass loss has important consequences depends on the time the star spends in that region and on the mass-loss rate. These times are shown in table 6, taken from various evolutionary computations.

mass (in M_0)	main sequence lifetime	hydrogen shell burning	helium core burning
15	10^7	450000	1.39×10^6
9	$2 \cdot 10^7$	60000	410000
5	$6.5 \cdot 10^7$	650000	17.47×10^6
3	$2.27 \cdot 10^8$	$5.3 \cdot 10^6$	77×10^6
1.5	$1.57 \cdot 10^9$	$227 \cdot 10^6$	
1	$8.06 \cdot 10^9$	$639 \cdot 10^6$	

Table 6. Lifetimes in various evolutionary stages for moderate mass stars (e.g. cf. Iben 1967; Stothers, 1972).

mass (in M_0)	main sequence lifetime	final mass	average mass loss
50	$4.1 \cdot 10^6$	26.29	$6.5 \cdot 10^{-6}$
40	$4.6 \cdot 10^6$	22.80	$4 \cdot 10^{-6}$
30	$5.4 \cdot 10^6$	19.10	$2 \cdot 10^{-6}$
20	$7.9 \cdot 10^6$	14.00	$0.75 \cdot 10^{-6}$

Table 7. M.S. lifetimes, final mass and the average mass loss in M_0 yr^{-1} for massive stars, losing mass by stellar winds.

a) For stars of moderate mass the mass-loss rate during the main-sequence stage is unimportant. Stars of 1-3 M_0 spen-

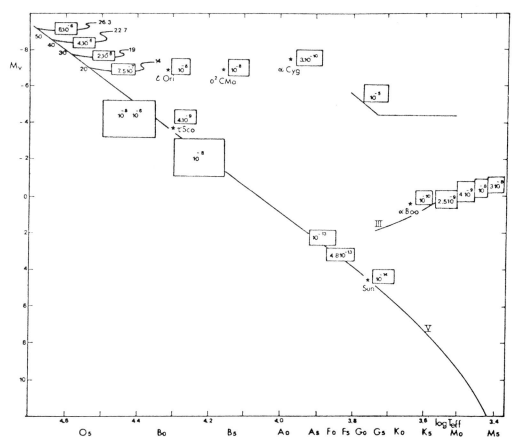

Figure 8. Observed and calculated mass-loss rates for main-sequence stars, giants and supergiants. The $\dot{M}$ values for stars of 50, 40, 30 and 20 $M_\odot$ and their evolution tracks at the upper left, were computed by de Loore et al. (1976) using a simplified Lucy-Solomon model; the numbers at the right end of these tracks give the remaining mass after the stellar wind mass loss. The values for MS Fo and F5 stars are according to de Loore (1968). $\dot{M}$ values at the right end of the giant branch are given by Reimers (1975).

ding between 80.10^6 and 200.10^6 years in the red giant stage, with mass-loss rates between $2.5\ 10^{-9}$ and 3.10^{-8} lose at least 0.2 M_0. No detailed calculations with mass loss included have been performed for single stars in this phase. However, computations of de Loore and De Grève (1975) for a 10 M_0 + 8 M_0 binary system for the second stage of mass transfer, i.e. the remnant of 1.66 M_0 of the first stage starting again mass transfer, show that after a time interval of $3.2\ 10^6$ year a mass of 0.52 M_0 is transferred towards

the companion, and then the mass transfer tends to its end. The average mass-loss rate of $0.16 \ 10^6$ is comparable with the mass-loss rate of these moderate mass stars at the red giant phase, hence we may conclude that they will evolve towards the left part of the HRD and will end as white dwarfs (Rose and Smith, 1970; Paczynski, 1970).
b) Massive stars lose during their main-sequence stage a considerable fraction of their initial mass. The results of evolutionary computations with mass loss included (de Loore et al., 1976) are shown in table 7.
c) When the stars evolve towards the right-hand part of the Hertzsprung-Russell diagram the mass-loss rate, during the first part at the same value as before, decreases slowly with the luminosity. However, as the star moves quickly through this stage, the total amount of mass loss in this phase is quite unimportant (see table 6).

9. MASS LOSS IN T TAURI STARS

T Tauri stars and other young objects show most probably evidence for mass loss. Computations by Larson (1969,1972) show that for protostars of masses $\leq 10 \ M_o$ for core masses of 2-3 M_o, due to heating of the central layers, radiative energy transfer becomes important near the centre of the core in a timescale shorter than the accretion time. The heating of the outer layers, caused by the increase of the core luminosity may increase the radiation pressure and the envelope may be blown away. Model calculations of Appenzeller and Tscharnuter (1974) for a 60 M_o protostar show that $4.5 \ 10^4$ years after the core formation, and when the core mass is 18 M_o, the outer layers expand in such a violent way that the entire envelope attains the escape velocity and a remnant of ~ 17 M_o is left.
More calm processes that involve the stellar core and which may be important in protostars are stellar winds. Outflow of matter occurs already in objects which are very young (10^5 years). This was observed in Herbig-Haro objects by Strom et al. (1974). The objects are still completely hidden in dust clouds. These stars possess deep convection zones and these may be associated with the occurrence of chromospheres and stellar winds. Moreover, there is spectroscopic evidence for enhanced chromospheric activity as mentioned by Strom et al. (1974); in the youngest T Tauri stars the mass-loss rate is apparently so large that it has dominant effects on the stellar spectrum (strong emission lines which can be attributed to dense circumstellar material or extended outflowing matter). As mentioned by Larson (1975) mass loss could play an important dynamical role by dissipating protostellar envelopes and in limiting the growth in mass of stellar cores.

REFERENCES

Ando,H. and Osaki,Y.:1975, Publ.Astron.Soc.Japan 27,581
Appenzeller,I. and Tscharnuter,W.:1974, Astron.Astrophys. 30,423
Bahng,J. and Schwarzschild,M.:1961, Astrophys.J. 134,312
Biermann,L.:1946, Naturwiss. 33,118
Böhm,K.H. and Cassinelli,J.:1971, Astron.Astrophys. 12,21
Böhm,K.H.:1972, "Stellar Chromospheres", NASA Colloquium, p.301
Deubner,F.L.:1976, in "Physique des mouvements dans les atmosphères stellaires", Ed.R.Cayrel and M.Steinberg, p.259
Dupree,A.K.:1976, in "Physique des mouvements dans les atmosphères stellaires", p.439
Evans,R.G., Jordan,C. and Wilson,R.:1975, Monthly Notices Roy.Astron.Soc. 172,585
Ferraro,V.C.A. and Plumpton,C.:1958, Astrophys.J. 127,459
Friedman,H.:1959, Proc.Inst.Res.Eng. 47,272
Fusi-Pecci,F. and Renzini,A.:1975, Mém.Soc.Roy.Sci.Liège, p.383
Gerola,H., Linsky,J.L., Shine,R., Mc Clintock,W., Henry,R.C. and Moos,H.W.:1974, Astrophys.J. 193,L107
de Grève,J.P. and de Loore,C.:1976, Astrophys.Space Sci., in the press
de Groot,M.:1971, "Wolf Rayet and High Temperature Stars", Ed. by M.K.V.Bappu and J.Sahade
Hearn,A.G.:1973, Astron.Astrophys. 23,97
Hearn,A.G.:1975, Astron.Astrophys. 40,355
Hundhausen,A.J.:1972,"Solar Wind and Coronal Expansion", Springer Verlag, Berlin, Heidelberg, New York
Hutchings,J.B.:1968, "Mass Loss from Stars", p.49, Ed.M.Hack
Hutchings,J.B.:1976, Astrophys.J. 203,438
Iben,I.Jr.:1964, Astrophys.J. 40,1631
de Jager,C.:1975, Mitt.Astronom.Gesellschaft 36,15
de Jager,C.:1976, Mém.Soc.Roy.Sci.Liège, 6e série, tome IX
de Jager,C. and Kuperus,M.:1961, Bull.Astron.Inst.Netherlands 16,71
de Jager,C. and de Loore,C.:1971, Astrophys.Space Sci. 11, 284
Kippenhahn,R.:1972, "Stellar Chromospheres", NASA Colloquium, p.266
Kalkofen,W. and Ulmschneider,P.H.:1976, Astron.Astrophys., submitted
Kondo,Y, Morgan,T.H.,Modisette,J.L. and White,D.R.:1975, preprint
Kopp,R.A.:1968, Ph.D.Thesis, Harvard University
Kuan,P. and Kuhi,L.V.:1975, Astrophys.J. 199,148
Kuperus,M.:1965, Recherches Astron.Obs.Utrecht, 17,1
Kuperus,M.:1969, Space Science Rev. 9,713
Kurucz,R.L., Peytremann,E. and Avrett,E.H.:1974,Blanketed Model Atmospheres for Early Type Stars,Smiths.Inst.Cambridge,Mass.

Lamers,H.J.G.L.M. and de Loore,C.:1974 in H.J.Lamers:"Studies on the Structure and Stability of Extended Stellar Atmospheres", Thesis, Utrecht, p.201
Lamers,H.J.G.L.M. and Rogerson,J.B.:1975, in preparation
Lamers,H.J.G.L.M. and de Loore,C.:1976, in "Physique des mouvements dans les atmosphères stellaires",Ed.R.Cayrel and M.Steinberg, p.453
Lamers,H.J.G.L.M. and Morton,D.C.:1976, Astrophys.J., submitted
Larson,R.B.:1975, Mém.Soc.Roy.Sci.Liège, 6e série, tome VIII
Larson,R.B.:1969, Monthly Notices Roy.Astron.Soc. 145,271
Larson,R.B.:1972, Monthly Notices Roy.Astron.Soc. 157,121
Leighton,R.B.:1960, Proc. IAU Symp. 12,321, Nuovo Cinsento Suppl. 22, 1961
Leighton,R.B.,Noyes,R.W. and Simon,G.W.:1962, Astrophys.J. 135,474
Lighthill,M.J.:1952, Proc.Roy.Soc.London, A211,564
de Loore,C.:1967, Mededel.Koninkl.Vlaam.Acad.Wetenschap.29,9
de Loore,C.:1968, Ph.D.Thesis, Astrophys.Institute, Free University Brussels
de Loore,C.:1970, Astrophys. and Space Sc. 6,60
de Loore,C., de Grève,J.P. and Lamers,H.J.G.L.M.:1976, in preparation
Mc Clintock,W.,Henry,R.D.,Moos,H.W. and Linsky,J.L.:1976, Ap.J., in the press
Mewe,R.,Heise,J.,Gronenschild,E.,Brinkman,A.C.,Schrijver,J. and den Boggende,A.J.F.:1975, Nature Phys.Sci. 256,711
Mewe,R.,Heise,J.,Gronenschild,E.,Brinkman,A.C.,Schrijver,J. and den Boggende,A.J.F.:1976, Astrophys. and Space Sci. 42,217
Morton,D.C.:1967, Astrophys.J. 150,535
Mullan,D.J.:1976, Astrophys.J. 209, 171
Nariai,K.:1969, Astrophys. Space Sci. 3,150
Osterbrock,D.E.:1969, Astrophys.J. 134,347
Paczynski,B.:1970, Acta Astronomica 20,47
Paczynski,B.:1974, IAU Symp. 66
Parson,S.B.:1967, Astrophys.J. 150,263
Proudmann,I.:1952, Proc.Roy.Soc. A214,119
Reimers,D.:1975, Mém.Soc.Roy.Sci.Liège, 6e série, tome VIII
Rogerson,J.B. and Lamers,H.J.G.L.M.:1975, Nature 256,190
Rose,W.K. and Smith,R.L.:1970, Astrophys.J. 159,903
Sargent,W.L.W.:1961, Astrophys.J. 107,1
Souffrin,P.:1966, Ann.d'Astrophys. 29,55
Stein,R.F.:1968, Astrophys.J. 154,297
Stothers,S.R.:1972, Astrophys.J. 175,431
Strom,S.E.,Grasdalen,G.L. and Strom,K.M.:1974, Astrophys.J. 191,111
Thomas,R.N.:1973, Astron.Astrophys. 29,297
Ulmschneider,P.H.:1967, Z.Astrophys. 67,193
Ulmschneider,P.H.:1970, Solar Physics 12,403
Ulmschneider,P.H.:1971a, Astron.Astrophys. 12, 297

Ulmschneider, P.H.: 1971b, Astron. Astrophys. 14, 275
Ulmschneider, P.H.: 1974, Solar Physics 39, 327
Ulmschneider, P.H., Schmitz, F., Renzini, A., Cacciari, C., Kalkofen, W. and Kurucz, R.: 1976b, in preparation
Ulmschneider, P.H., Kalkofen, W., Nowak, T. and Bohn, H.U.: 1976a, Astron. Astrophys., in the press
Ulmschneider, P.H. and Kalkofen, W.: 1976, Astron. Astrophys., submitted
Underhill, A.B.: 1969, "Mass Loss from Stars", Ed. M. Hack, Reidel, Dordrecht
Whitaker, W.E.: 1963, Astrophys. J. 137, 914
Williams, I.P.: 1967, Monthly Notices Roy. Astron. Soc. 136, 341
Wright, A.E. and Barlow, M.J.: 1975, Monthly Notices Roy. Astron. Soc. 170, 41

BOUNDARY CONDITION WITH MASS LOSS: THE RADIATIVELY-DRIVEN WIND MODEL

Dimitri Mihalas
High Altitude Observatory
National Center for Atmospheric Research
Box 3000 Boulder, Colorado, USA

A brief summary of the current status of radiatively driven wind models for early-type stars is given. A critique of these models is made both on theoretical and observational grounds, and it is concluded that a pure radiatively driven wind is probably not a realistic approximation for O-star winds. It is argued that probably the wind structure must have an initial high-temperature ("coronal") region through which the trans-sonic flow takes place, followed by radiative accelerations to very high terminal velocities. Full details of the discussion can be found in Stellar Atmospheres, 2nd Edition, by D. Mihalas, to be published by W. H. Freeman and Company, San Francisco, in Fall 1977.

STRATIFICATION OF ELEMENTS IN A QUIET ATMOSPHERE: DIFFUSION PROCESSES

Georges Michaud
Département de Physique, Université de Montréal, Canada.

1. ABUNDANCE ANOMALIES: THE SIGNATURE OF STABILITY

In the absence of turbulence or convection one expects that, in stars, heavy elements would tend to settle gravitationally while light elements would go to the surface. Eddington (1930) however realized that this general tendency could be modified both by the electric field and by differential radiation pressure. In spite of their small mass, electrons do not all float on the surface of stars because an electric field is generated that keeps them from separating from the protons. Instead of settling gravitationally, heavy elements often concentrate on the surface because they absorb relatively much more photons than hydrogen or helium and are dragged to the surface by the radiative flux. Eddington concluded that turbulence was too strong for diffusion to be important in stars since the relation that diffusion predicts between surface abundances and stellar masses did not appear to be realized in most stars.

However, we now know that stars whose outer envelopes are most likely to be stable, thus where diffusion is most important, show surface overabundances of heavy elements and underabundances of helium. These are the Fm, Am, Ap and Bp stars that are slow rotators, and often have strong magnetic fields. Furthermore, the abundance anomalies become larger as the importance of the outer convection zone decreases from Fm to Ap (Smith 1971, 1973, Preston 1974).

In order to derive the diffusion equation (§ 2), I will here start from equilibrium gradients in stellar atmospheres and envelopes. The electric fields and local charge separation in stars will be described. The electric force on protons is half as strong as the gravitational force on protons, and this is true even in convection zones. Surface underabundances appear after 10^4 years (§ 3) but overabundances after as little as ten years.

Comparisons of observed abundance anomalies with diffusion calculations (§ 4) show that even the largest observed anomalies can be explained as can their variation with the effective temperatures of stars; and this without any arbitrary parameter. However such phenomena as isotope anomalies and line asymmetries depend sensively on the structure of the outer atmosphere. Mainly because of our ignorance of the hydrodynamics involved, their explanation requires arbitrary parameters. This may also lead to a better understanding of stellar hydrodynamics.

No attempt is made here to present a complete review. A more complete list of references may be found in Michaud (1975).

2. BASIC PHYSICS

In order to introduce the diffusion equation, we study the equilibrium configurations of gases constituted successively only of hydrogen, only of protons and electrons, and finally of protons, electrons and traces of an element of mass A and charge Z.

2.1 The Equilibrium Configuration of Hydrogen

The reader is probably familiar with the hydrostatic equilibrium of a gas constituted of pure hydrogen (See Fig. 1). Requesting that the

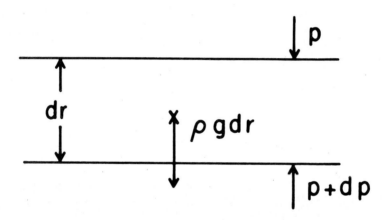

Figure 1. Forces (per cm^2) acting on a slab of material of density ρ, thickness dr, in a gravitational field.

sum of the forces acting on a slab material of thickness dr, density ρ, submitted to a gravitational acceleration g be equal to zero, we have

$$dp = - \rho g \, dr \, . \tag{1}$$

Or replacing ρ by the pressure:

$$\frac{d\ln p}{dr} = - \frac{m_p g}{kT} \, , \tag{2}$$

where m_p is the proton mass, the difference between the hydrogen and the proton mass being negligible here.

2.2 The Electric Field of Stars

When hydrogen is completely ionized we now show that an electric field must appear for equilibrium to be maintained.

At equilibrium, collisions must exchange no momentum, nor anything else, between protons and electrons. Equilibrium is reached when transport phenomena have ceased. In the prevailing electric and gravitational fields one must then have equilibrium separately for electrons and protons. One can write separately for protons and electrons an equation similar to eq. (2).

$$\frac{\partial \ln p_e}{\partial r} = - \frac{m_e g}{kT} - \frac{eE}{kT} , \qquad (3)$$

$$\frac{\partial \ln p_p}{\partial r} = - \frac{m_p g}{kT} + \frac{eE}{kT} . \qquad (4)$$

Since the electron mass is much smaller than the proton mass, the electrons have a tendency to float and an electric field is needed to keep the matter neutral on the whole. To prevent a large charge separation, one must have

$$\frac{\partial \ln p_e}{\partial r} = \frac{\partial \ln p_p}{\partial r} , \qquad (5)$$

or, using equations (3) and (4),

$$eE = \frac{1}{2} (m_p - m_e) g . \qquad (6)$$

In stars, the electric force on protons is equal to half the gravitational force on protons, and the electric force on electrons is much larger than the gravitational force on electrons.

In stellar structure calculations, the electric field is neglected. This is possible since one is generally not interested in the equilibrium configurations of protons and electrons separately. It is trivial to show that equations (3), (4) and (6) lead to the usual hydrostatic equilibrium equation involving the reduced mass $[\mu \equiv (m_p + m_e)/2]$. However the concept of the reduced mass is justified by the presence of the electric field.

We have seen that there must be electric fields in stars, but then the electric field lines must start and end somewhere. There must be some charge separation.

The required charge separation can be estimated at some distance r from the star center, where the gravitational acceleration is g. We will compare the total charge to the total mass of the star within that radius. Let z be the total charge number (total charge within r

$Q(r) = ze$) and a the total mass number within radius r (total mass, $M(r) = am_p$). Eq. (6) requires that:

$$eE = \frac{1}{2} m_p g$$

or

$$\frac{eQ}{r^2} = \frac{e^2 z}{r^2} = \frac{1}{2} m_p \frac{GM}{r^2} = \frac{1}{2} m_p^2 \frac{Ga}{r^2}$$

leading to:

$$\frac{z}{a} = \frac{m_p^2 G}{2e^2} \simeq 10^{-37} \quad . \tag{7}$$

Since the electric interaction is a much stronger interaction than the gravitational interaction, only a small charge is required for the electric force on a proton to equal half the gravitational force on a proton. Eq. (7) implies a charge separation of the same order:

$$\frac{p_e - p_p}{p_e + p_p} \sim 10^{-37} \tag{8}$$

averaged throughout the star. The electron and proton pressure gradients are then equal (eq. [5]) to a very good approximation. (For a more complete discussion, see Milne 1924, Schatzman 1958, Montmerle and Michaud 1976.)

Are stars electrically neutral or where do the electric field lines end? Milne (1924) has shown that, if stars were alone in space and perfectly stable, the field lines would end at $N \sim 10^{-6}$ cm^{-3} where there would be a slight excess of electrons. Since in interstellar space the number density of protons is of order 1 cm^{-3}, the treatment of Milne does not apply. Presumably motions occur and there is a slight electron excess where the number density of protons is of order 1 cm^{-3}. If a star were not neutral it would accrete a few electrons from space to become neutral. Only 10^{-3} electron per cm^2 of stellar surface is needed to cancel the charge of the star. This problem has apparently never been studied in detail. The timescale for the establishment of the electric field will be discussed after the introduction of the diffusion equation.

2.3 The Diffusion Equation

Diffusion is now presented as the first order process transforming a non-equilibrium into an equilibrium configuration (Eddington 1930). Consider a stellar envelope throughout which element A has the abundance $c(A)$

$$c(A) \equiv N(A)/[N(H)+N(A)] \quad . \tag{9}$$

The equilibrium abundance is given by an equation similar to equation (4):

$$\frac{\partial \ln p_{eq}(A)}{\partial r} = \frac{Am_p(g_{rad}-g)}{kT} + \frac{ZeE}{kT} \tag{10}$$

where Z is the charge of the ion (not that of the nucleus), and g_{rad} is the acceleration on element A due to the absorption of photons by that element, i.e. the "radiative acceleration". In the present context diffusion is the transport phenomenon transforming the actual distribution of abundances of element A into the equilibrium distribution. To first order the diffusion velocity is then linearly proportional to the difference between the equilibrium gradient and the actual gradient.

$$w = D \left[\frac{\partial \ln c_{eq}}{\partial r} - \frac{\partial \ln c_{actual}}{\partial r} \right] \qquad (11)$$

where D, the proportionality constant, is called the diffusion coefficient. Using equations (4), (9) and (10), one obtains (if $c(A) \ll 1$)

$$w = D \left[-\frac{\partial \ln c}{\partial r} + \frac{m_p g}{kT}(1-A) + \frac{eE}{kT}(Z-1) + \frac{Am_p g_{rad}}{kT} \right], \qquad (12)$$

where c is used for c_{actual}. If the electric field is determined by the protons only, one uses equation (6) to obtain:

$$w = D \left[-\frac{\partial \ln c}{\partial r} + \frac{m_p g}{2kT}(1-2A+Z) + \frac{Am_p g_{rad}}{kT} \right]. \qquad (13)$$

This diffusion velocity is based on diffusion being a first order process. It is the same as that obtained by Aller and Chapman (1960) from statistical physics considerations except that thermal diffusion is here neglected. Michaud et al. (1976) have shown how thermal diffusion can be taken into account by modifying g. In the atmosphere, thermal diffusion is generally negligible, but not deeper in the envelope, where Z becomes large.

In deriving equation (13) we have assumed $c(A)$ to be small. This is generally an excellent approximation except for helium, whose non-negligible abundance modifies the electric field. Indeed when helium is more than half as abundant as hydrogen, the outward electric force on hydrogen becomes larger than the gravitational pull, and hydrogen is ejected (Montmerle and Michaud 1976) from the region where helium is abundant. The observation of a circumstellar hydrogen shell around σ Ori E (Walborn 1974) is probably related to the electric fields in He-rich stars.

The diffusion coefficient is inversely proportional to the collision probability. It thus depends on the type of interaction between element A and protons, when diffusion occurs in ionized hydrogen. When element A is ionized,

$$D \approx 1.5 \times 10^8 \, T^{5/2}/(NZ^2)$$

where N is the number density of protons. All quantities are in the cgs system (See Aller and Chapman 1960, Chapman and Cowling 1970, Montmerle and Michaud 1976). When element A is not ionized, collisions occur mainly through the polarization of element A, giving

$$D \simeq 3.3 \times 10^4 \, T/(N \, \alpha^{\frac{1}{2}})$$

where α is the polarisability (Ratcl 1975). Values of α for a few elements of interest are given in Table I.

Element	$\alpha(10^{-24}$ cm$^3)$
He	0.20
O	0.77
Si	5.5
Mn	14.
Sr	25.

Table I. Polarizabilities (From Teachout and Pack 1971).

In stellar atmosphere the diffusion coefficient of the neutral elements is about two to three orders of magnitude larger than that of the ionized elements. Even if an element has only 10^{-3} of its atoms neutral, the neutral element is sometimes the dominant contributor to the diffusion velocity, the mobility of elements being so much larger in the neutral state. This in particular can enhance the effect of photons absorbed in the neutral state (See Montmerle and Michaud 1976, § VI). In Fig. 2, we see that, for helium, whereas the radiation force transmitted through the lines is always negligible if one neglects the effect just described,

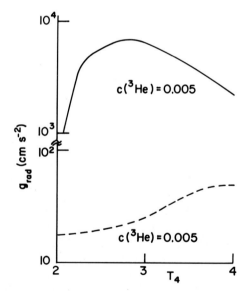

Figure 2. Radiative acceleration transmitted through the lines to ^{3}He, both taking into account (full line) and neglecting (dashed line) the effect of increased mobility in the neutral state. It leads to a two order of magnitude difference. Only when the effect is taken into account can the lines support helium. (Michaud, Praderie and Montmerle, in preparation.)

this force becomes dominant at T = 30,000 °K if one includes the effect of the increased mobility. The problem of timescale for the appearance of the helium isotope anomaly probably disappears once this is taken into account (Vauclair et al. 1974a).

3. DIFFUSION TIMESCALES

Two very different timescales appear: whereas electrons can always separate rapidly enough from protons to create electric fields whatever turbulence there may be, elements cannot necessarily diffuse rapidly enough to lead to abundance anomalies. So all stars have internal electric fields but only a few have abundance anomalies. However if turbulence is small enough, the timescale to create abundance anomalies is much smaller than the stellar lifetime for stars more massive than 1.3 $M_\odot$ and somewhat smaller than the stellar lifetime for stars more massive than the sun. Whether diffusion can modify surface abundances in the sun sensitively depends on hydrodynamical effects which are poorly understood at the present.

3.1 Electric Fields

The timescale for the establishment of the electric field can be estimated using a diffusion equation similar to eq. (13) for the electron-proton mixture. We assume no charge separation at $t = 0$ ($p_{po} = p_{eo}$). Then

$$w_{pe} \simeq D_{pe} [\frac{m_p g}{kT}] . \tag{14}$$

The time for the establisment of the electric field is approximately the time for electrons to move from the point where $p_{eo} = p_{po}$ to the point where

$$\frac{p_{eo} - p_{po}}{p_{eo} + p_{po}} \simeq 10^{-37} .$$

From eq. (3), this is approximately at a distance 10^{-37} of a scale height. Then

$$\Delta t \simeq \frac{\Delta r}{w_{pe}} \simeq \frac{10^{-37} kT}{m_p g\, w_{pe}} \simeq 10^{-25} \text{ sec.} \tag{15}$$

where a value of 10^{10} cm has been used for the scale height, and a number density of 10^{15} has been used in estimating D_{pe}. Even in regions where the density is much larger, the timescale for the "creation" of the field, $\vec{E}$, is very short. As soon as the separation starts (or 10^{-25} seconds later) it stops because the electric field has been created. No turbulence or convection timescale is close to this one. Electric fields are present whenever there is ionized matter in a gravitational field. More sophisticated calculations of the timescale could be made but it hardly seems worth it. Equation (15) is quite accurate enough!

3.2 Timescales for Abundance Anomalies to Appear

There are different diffusion timescales depending on whether an element falls (i.e. settles gravitationally) or is supported by radiation pressure. When an element falls, the downward flux is proportional to the abundance in the convection zone which is assumed completely mixed. The abundance in the atmosphere varies as (See Michaud et al. 1976):

$$c(A)/c_o(A) = \exp(-t/\tau(A)) \quad . \tag{16}$$

Table II, gives the characteristic times for helium diffusion. They were obtained using eq. (13), but taking thermal diffusion into account. If

$M(M_\odot)$	$\tau(He)$ (years)
2.6	4.3×10^4
2.0	1.3×10^5
1.55	1.8×10^6
1.4	4.3×10^6
1.2	1.1×10^8
1.07	1.5×10^9
1 ($\alpha = 1.5$)	2×10^{10}
1 ($\alpha = 1.0$)	5.4×10^9
1 ($\alpha = 0.7$)	5×10^8

Table II. Diffusion Characteristic Times.

the sun is stable below its convection zone, helium would be underabundant by a factor of two in its atmosphere if $\alpha = 1$. (α is the ratio of mixing length to pressure scale height.) However if $\alpha = 1.5$, no anomaly is expected in the solar lifetime and if $\alpha = 0.7$, very large underabundances are possible. For the more massive stars, it is clear that very large anomalies are possible in the stellar lifetimes. In stars of 2.6 $M_\odot$ or more, the gravitational settling timescales for helium are of 10^4 years or so. Helium completely disappears from the surface, except for stars with $T_{eff} \gtrsim 18,000\ °K$ where helium begins to be supported by radiation pressure for $N(He)/N(H) \simeq 10^{-2}$ (See Vauclair et al. 1974a). In stars where helium settles gravitationally, the helium convection zone disappears (Vauclair et al. 1974b).

The timescale for the development of overabundances can be much shorter than the gravitational settling timescale: at a distance r from the center of the star, the radiative acceleration on unabundant elements ($c(A) \lesssim 10^{-9}$) is of the order of

$$g_R \simeq 1.7 \times 10^8 \frac{T_{e4}^4 R^2}{AT_4 r^2} \text{ cm/sec}^2 \tag{17}$$

where R is the stellar radius, T_{e_4} the effective temperature in 10^4 °K, and T_4 the temperature at r in 10^4 °K. This is up to four orders of magnitude larger than the gravitational acceleration, so that the diffusion timescales for elements pushed upwards can be four orders of magnitude smaller than that for those going downwards. Figure 3 shows the time evolution of abundances for some cases of interest. In stars of 3 $M_\odot$ or more, abundance anomalies start appearing after ten years. Note that the important time here is the time for elements to migrate from the envelope to the surface (Michaud et al. 1976). Watson (1971b) and Cowley and Day (1976) have carried out similar though less detailed calculations and obtained similar results.

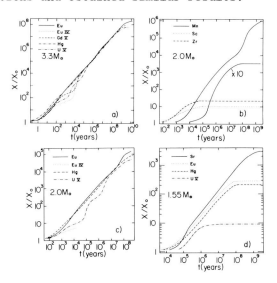

Figure 3. Time evolution of abundances in the atmospheres of main sequence stars. It is here assumed that all elements that get into the atmosphere stay there, and that turbulence is negligible. In general, we thus obtain the maximum overabundances possible.

4. DIFFUSION AND ABUNDANCE ANOMALIES IN F, A AND B STARS

If stars are stable over periods of 10^4 years, diffusion is expected to modify their surface abundances. Since those stars that show abundance anomalies also are the most likely stars to have stable atmospheres or envelopes (Strittmatter and Norris 1971), diffusion appears as a likely explanation of the anomalies.

Indeed the Fm, Am, Ap and Bp stars are, as a group, slow rotators. Meridional circulation is less important in them and can more easily be suppressed by magnetic fields. The hydrogen convection zone is progressively less important (from Fm to Bp) as the effective temperature increases and carries less and less of the energy flux. Indeed, in the Ap and Bp stars magnetic fields are often observed at the surface and are apparently very stable. As the stars rotate, magnetic fields are observed to vary in a periodic way. All well studied magnetic variables can apparently be explained in this way (Preston 1970, 1971); that is, by the oblique rotator model. In the Ap and Bp stars, the magnetic fields apparently succeed in imposing some order to the atmosphere. The Ap and Bp stars show the largest abundance anomalies. In the Fm and Am stars, however, no magnetic field is generally observed, presumably because the surface convection zone carries relatively more energy flux than in the hotter (Ap and Bp) stars. Diffusion probably goes on below the hydrogen convection zone in the Fm and Am stars. The abundance anomalies are then smaller than in the Ap and Bp stars since a larger mass must be contaminated.

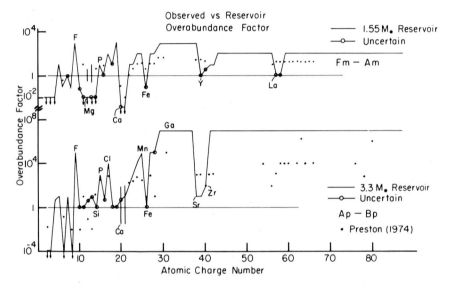

Figure 4. The observed anomalies (fig.7 of Preston, 1974) are compared to the anomalies allowed by the reservoir. No arbitrary parameter is involved. The maximum observed abundance is shown since the reservoir gives us the maximum possible overabundance in a stable envelope. Vertical lines are used when both overabundances and underabundances are observed. Circles indicate calculations in which the radiative and gravitational accelerations are within a factor of three of each other. The result is then uncertain. The only apparent difficulty is with magnesium (Mg).

In Fig.4 the observed abundance anomalies are compared to the abundance anomalies that the diffusion of elements to and from the stable envelope of the stars <u>can</u> lead to. The diffusion equation was here solved for the different elements both in a 1.55 $M_\odot$ and in a 3.3 $M_\odot$ star. The atmosphere of the 3.3 $M_\odot$ star is assumed to be stable but for the

1.55 $M_\odot$ only the zone below the hydrogen convection zone is assumed to be stable. All elements that reach the atmosphere (or the convection zone) are assumed to stay there, that is, they are assumed not to leave the stellar atmosphere via a stellar wind. The radiation forces used were accurate enough for $T_4 \gtrsim 4$ but not at lower temperatures. For the 3.3 $M_\odot$ star the region where $T_4 < 4$ is important and some elements may not be supported in that zone and so may not reach the atmosphere (see Michaud et al. 1976, for details). Turbulence will be discussed in the next paper. It is certainly important for some stars. The calculated abundance anomalies of Fig. 4 are then maximum possible values which should be compared to the envelope of the abundance anomalies observed. On Fig. 4, they are compared to the maximum abundance anomaly of each element from Fig. 7 of Preston (1974). For a few elements both overabundances and underabundances are observed; they are indicated by vertical lines. Stars have been grouped as Am or Ap. Groups 2 and 3 of Preston are both included in "Ap", since the difference between the Mn and the other Ap stars is not expected to be related to the "reservoir" but to the atmosphere (see below and also Michaud 1973). Some of the calculated values are uncertain because the calculated radiative and gravitational accelerations are within a factor of three of each other. The calculated anomaly could then change from an overabundance to an underabundance or vice versa. Even the largest observed abundance anomalies can be explained. Nearly all disagreements appear where the radiative and gravitational accelerations are within a factor of three of each other. The only apparent exception is for Sr, Y, and Zr in the 3.3 $M_\odot$ star and for Mg in the 1.55 $M_\odot$ star. The case of Sr, Y, and Zr is probably related to turbulence

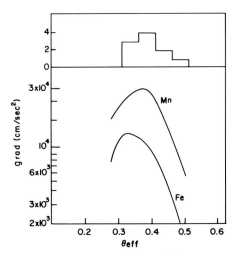

Figure 5. Radiative acceleration transmitted through photoionization to Mn and Fe. In the upper part of the figure is shown the number of Mn stars as a function of effective temperature (Sargent and Searle 1967). Most of the Mn stars have the effective temperature for which the radiative acceleration is the largest.

and will be discussed in the next paper. However, if the abundance of Mg is normal in Am stars, it may turn out to be a serious problem for the diffusion model. Watson (1971a) also obtained that Mg should be underabundant in Am stars. Due to the importance of Mg, Ca, Sr, Ba and Hg in Am and Ap stars, Dr Praderie and myself are conducting a detailed NLTE study of elements which have two electrons more than a closed shell when they are neutral (Be, Mg, Ca, Sr, Ba and Hg) (Praderie 1975).

In the Ap and Bp stars, the atmosphere is apparently stable and the radiation force must be larger than the gravitational force in the line forming region for overabundances to appear. The reverse is true for underabundances. Fig. 5 depicts the radiative force as a function of effective temperature for Mn and Fe. Only the radiation forces transferred through photoionization are shown. The contribution of line absorption is not negligible but is not expected to change the shape of the diagram (Alecian 1976). The number of Mn stars at a given effective temperature is also shown. The largest Mn overabundances appear where the radiation forces are the largest. Similar diagrams can be made for He or O with similar agreement (Michaud 1970).

The great success of the diffusion theory is that without any arbitrary parameter, it can explain the largest abundance anomalies in Fm, Am, Ap, and Bp stars and most of the variations with the effective temperature of the stars.

It has not yet been possible however to reproduce the detailed abundances anomalies of individual stars. The needed calculations require a detailed understanding of the relation between stellar atmospheres and the interstellar matter, and NLTE calculations of radiation forces in the outer atmosphere. Such calculations are underway for strontium and atoms with similar atomic configurations, but many more of them are needed. Turbulence will also be important in this comparison. Because of our lack of understanding of the hydrodynamics of meridional circulation, turbulence, convection and of their relation with magnetic fields some arbitrary parameters will unavoidably appear in this comparison. However, it may be that the abundance anomalies will put important constraints on the hydrodynamics of stars.

5. STELLAR WINDS, LINE ASYMMETRIES AND ISOTOPE ANOMALIES

The Ap-Bp stars are often separated into two groups, the Mn-Hg stars and the others (hereafter Si-Eu stars). In general the Si-Eu stars have magnetic fields, while the Mn-Hg stars do not. A magnetic field strongly influences diffusion in the outer parts of the atmosphere ($\tau < 0.1$ at $\lambda = 5000$ A, which is also the wave length of the optical depths mentioned below) for ionized elements. Whereas ionized elements can leave completely the atmosphere of a star without a magnetic field, they cannot cross magnetic field lines in the Si-Eu stars. Many more elements are expected to be overabundant in the Si-Eu stars than in the Hg-Mn stars. This is observed to be the case (Preston 1974).

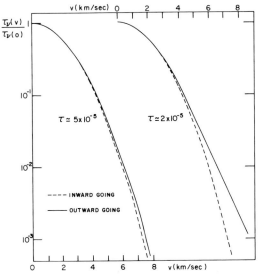

Figure 6. Profile of the optical depth of the line of an element (SrII) in clouds at two different continuum optical depths. The profile is given as a function of the velocity of the absorbing atom. At v = 0, the absorbing atom is at rest with respect to the star. The line is assumed Doppler broadened. Because of the radiative flux, there are more outward than inward going elements. In a cloud at $\tau \simeq 2 \times 10^{-5}$, depending on how saturated the line is (how large is $\tau_\nu(0)$) the maximum of the anisotropy will occur somewhere between 4.5 and 7 km/sec (in the observed line).

In the Mn-Hg stars the only elements expected to be overabundant are those whose radiative force decreases as the element tries to leave the star. The element is then trapped. Such an element is pushed from the envelope and migrates to the upper atmosphere but the radiation force decreases, becomes smaller than the gravitational force and the element stays in the atmosphere. It has been shown that the mercury isotope anomaly could be explained in this model (Michaud et al. 1974). Mercury must stand in a cloud at $\tau \lesssim 10^{-2}$. The other overabundant elements are, similarly, expected to stand in clouds, somewhere between $\tau = 0.5$ and the interstellar matter. Some elements will be higher up than others depending on where in the atmosphere their radiative force becomes smaller than the gravitational force.

One important consequence of this model is for line anisotropies. Using equations (13) and (17), it is easy to verify that at $N \simeq 10^{10}$ cm^{-3}, diffusion velocities of a few kilometers per second become possible. If an element stands in a cloud at that density, anisotropies similarly appear on the line profile. Fig. 6 shows the profile of the optical depth of a Doppler broadened line of an element in such a cloud.

It leads to line anisotropies at 4.5 to 7 km/sec, depending on the strength of the line. Anisotropies at 8 km/sec have been observed by Smith (1976). Whether the difference is unacceptably large is unclear at the moment. This model predicts that the anisotropy should not appear in helium nor in oxygen lines, since those elements are not supported by radiation pressure. Only elements which stand in clouds at $\tau < 10^{-3}$ should show such an effect.

Whereas diffusion explains without arbitrary parameters the "envelope" of the observed abundance anomalies and their variation with effective temperature, the explanation of such features as isotope anomalies and line anisotropies require a more detailed knowledge of the outer atmosphere than is currently possible, and arbitrary parameters must be introduced in the diffusion calculations.

ACKNOWLEDGEMENT

I would like to thank J.-L. Tassoul and R. Racine for a critical reading of the manuscript, A. Chénard for the skilful typing of this manuscript, Y. Charland for carrying out some of the calculations, and R. Martel for the drawings. This research was partially supported by grants from Le Conseil National de Recherches du Canada and Le Ministère de l'Education du Québec.

REFERENCES

ALECIAN G.: 1976, Etude Quantitative des Effets de la Diffusion sur l'Abondance des Eléments dans l'Atmosphère de κ Cancri, Université de Paris VII (thèse de 3e cycle).
ALLER, L.H. and CHAPMAN A.: 1960, Astrophys. J. 132, 461
CHAPMAN S. and COWLING T.G.: 1970, The Mathematical Theory of Non-Uniform Gases, University Press, Cambridge.
COWLEY C.R. and DAY C.A.: 1976, Astrophys. J. 205, 440.
EDDINGTON A.S.: 1930, The Internal Constitution of Stars (reprinted by Dover, New York, 1959) p. 278.
MICHAUD G.: 1970, Astrophys. J. 160, 641.
MICHAUD G.: 1973, Astrophys. Letters 15, 143.
MICHAUD G.: 1975, in W.W. Weiss (ed.), I.A.U. Colloquium Number 32, Physics of Ap Stars. The University Press, Vienna.
MICHAUD G., REEVES H. and CHARLAND Y.: 1974, Astron. Astrophys. 37, 313.
MICHAUD G., CHARLAND Y., VAUCLAIR S. and VAUCLAIR G.: 1976, Astrophys. J. (December).
MILNE, E.A.: 1924, Proc. Cambridge Phil. Soc. 22, 493.
MONTMERLE T. and MICHAUD G.: 1976, Astrophys. J. Suppl. 31, 489.
PRADERIE F.: 1975, in W.W. WEISS (ed.), I.A.U. Colloquium Number 32, Physics of Ap Stars, The University Press, Vienna.
PRESTON G.W.: 1970, in A. Slettebak (ed.), Stellar Rotation, D. Reidel, Dordrecht.

PRESTON G.W.: 1971, Publ. Astron.Soc. Pacific 83, 571.
PRESTON, G.W.: 1974, Ann. Rev. Astron. Astrophys. 12, 257.
RATEL, A.: 1975, Université de Montréal, (Internal Report).
SARGENT W.L.W. and SEARLE L.: 1967, in R.C. Cameron (ed.), The Magnetic and Related Stars, Mono Book Co., Baltimore.
SCHATZMAN E.: 1958, White Dwarfs, North Holland, Amsterdam.
SMITH M.A.: 1971, Astron. Astrophys. 11, 325.
SMITH M.A.: 1973, Astrophys. J. Suppl. 25, 277.
SMITH M.A. and PARSONS S.B.: 1976, Astrophys. J. 205, 430.
STRITTMATTER P.A. and NORRIS J.: 1971, Astron. Astrophys. 15, 239.
TEACHOUT R.R. and PACK R.T.: 1971, Atomic Data 3, 195.
VAUCLAIR S., MICHAUD G., and CHARLAND Y.: 1974a, Astron. Astrophys. 31, 381.
VAUCLAIR G., VAUCLAIR S. and PAMJATNIKH A.: 1974b, Astron. Astrophys. 31, 63
WALBORN N.R.: 1974, Astrophys. J. Letters 191, L95
WATSON W.D.: 1971a, Astron. Astrophys. 13, 263.
WATSON W.D.: 1971b, Nature Phys. Sci. 229, 228.

COMPETITION BETWEEN DIFFUSION PROCESSES AND HYDRODYNAMICAL
INSTABILITIES IN STELLAR ENVELOPES.

by Gérard and Sylvie VAUCLAIR
DAPHE, Observatoire de Meudon, France.

Since the work of Michaud (1970), the abundance anomalies observed in the peculiar Ap and Am stars are increasingly believed to be a consequence of diffusion processes in stellar atmospheres or stellar envelopes. A number of the problems that seemed at first sight insoluble within the framework of diffusion processes have now been solved by it. Diffusion processes can, for example, account for anomalous helium isotopic ratios (Vauclair et al, 1974 (b)) and mercury isotopic ratios (Michaud et al, 1974). Quantitative results on abundance variations due to diffusion processes have been obtained (Michaud et al, 1976 ; Michaud, this conference ; Alecian, 1976). They show that, in general, the relative abundance anomalies obtained from computation are close to the observed ones. It is now well established that the largest abundance anomalies observed in Ap stars (for rare earths) can be interpreted by diffusion processes with a satisfactory time scale, in a completely stable atmosphere. However, the predicted absolute abundance variations often exceed the observed ones, as in the case of Am stars. This suggests that the assumption of stability is not completely valid for the stellar gas : some kind of macroscopic motion, such as a meridional circulation or turbulence or both, must be at work and slow down the diffusion.

It has so far been generally believed that instabilities in stellar atmospheres, such as convection or mass loss, would inhibit diffusion. This led Strittmatter and Norris (1971) to mark in the HR diagram the region where diffusion could take place (i.e. the region where the stellar atmosphere is stable). However, strong macroscopic motions may carry to the surface some of the anomalies created in the deep interior by the diffusion processes. This is the case for the superficial convection zones. It could also be the case for mass loss, as proposed by Osmer and Peterson (1974) to account for helium rich stars : they suggest that a helium overabundance in the stellar interior could be brought up to the surface by mass flow. (The difficulty of this model is that the radiation force is probably as unable to support the helium inside the star as it is at the surface).

However, it is possible that sufficiently slow motions will merely perturb the diffusion without stopping it (time scale not too short compared to the diffusion time scale -cf Schatzman (1969) for the effect of mild turbulent motions).

Let us write the total velocity of elements as :

$$V = V_D + V_M \qquad (1)$$

where V_D is the diffusion velocity, and V_M the macroscopic velocity. In the case of laminar movement, it is clear that if the macroscopic velocity V_M is much larger than the diffusion velocity V_D, there will be no diffusion since a concentration gradient cannot arise. In the case of turbulent motions, V_M is a random velocity, whose form may be found by a simple statistical analysis. Consider a turbulent mixture of two gases 1 and 2 with numerical densities n_1, n_2 (gas 2 = test particles), and a composition gradient along the r axis. The flux of particles 2 due to the turbulent velocity V can be written as :

$$n_2 V_M = - \langle l.V \rangle \cdot \mathrm{grad}\, n_2 \qquad (2)$$

where l is some characteristic mixing length.

We may also write :

$$V_M = -D_T \frac{1}{C} \frac{\partial C}{\partial r} \qquad (3)$$

where $D_T = \langle l.V \rangle$ is the turbulent diffusion coefficient, and C the concentration of particles 2. Taking into account the general expression for the diffusion velocity (see for instance Michaud, 1976, this conference) we find for the total velocity :

$$V = -(D+D_T)\frac{1}{C}\frac{\partial C}{\partial r} + D\left(\frac{m_p g}{2kT}(1-2A+Z) + \frac{Am_p}{kT}g_{rad}\right) \qquad (4)$$

where D is the "microscopic" diffusion coefficient. We see from expression (4) that the effect of turbulence is to increase the backward force produced by the concentration gradient, i.e. to reduce the concentration gradient built up by diffusion. If the turbulent diffusion coefficient D_T is much larger than the diffusion coefficient D, there is no diffusion since the concentration gradients are smoothed out.

Consequently, we may eliminate from the HR diagram the regions where diffusion cannot take place, draw instead a "Possible diffusion Region" or P.D.R. (section 1). The next step will be to study the hydrodynamical stability of the P.D.R. stars and try to guess which of them will in fact be sufficiently stable to undergo diffusion, at which point in their evolution, and for how long (section 2). If V_M is of the same order as V_D or D_T of the same order as D, macroscopic motions may interact with diffusion processes ; this can give rise to new effects which it is interesting to study in detail (section 3).

I. THE "POSSIBLE DIFFUSION REGION".

Figure 1 shows the region in the HR diagram where diffusion processes can have observable effects. The region of mass loss is to the left of the blue boundary. Observations made by the Copernicus satellite show that stars hotter than $\sim B_0$ type lose mass (Lamers, 1976). Smith and Parsons (1975 and 1976) have observed asymmetries in the absorption lines of some Mn stars. They interpret them as an outward radial motion, with a velocity of several Kms.sec^{-1}. They think that this wind could inhibit diffusion. However, Michaud (1976 - this conference), suggests that the observed line asymmetries would be a simple consequence of diffusion processes.

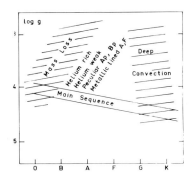

Fig 1.

Fig. 1 : Schematic Possible Diffusion Region.
Diffusion is excluded in the part of the diagram where mass loss or deep convection occur.

Another kind of instability which has not been investigated in detail could be a consequence of the fact that hot stars are rapid rotators. The time scale of the meridional circulation may be shorter than the diffusion time scale.

To the right of the red boundary is the region where the outer convection zones are too deep for diffusion processes to be effective during a stellar lifetime. Michaud et al. (1976) have shown that diffusion below the outer stellar convection zones cannot have observable effects for stars cooler than G. For stars earlier than F2, the diffusion time scales are short compared to the stellar lifetime, and diffusion below the convection zones can lead to large abundance anomalies, as observed in Am and Fm stars. For cooler F stars, the diffusion time scales increase and the expected anomalies are much smaller. This effect could be responsible for the metal abundance dispersion observed in F type stars (e.g. Powell, 1972) as well as light element abundance variations. G type stars probably represent the point at which the diffusion time scales exceed the main sequence lifetime. We do not know enough about convection to be able to fix this limit precisely.

II. THE HYDRODYNAMICAL STABILITY OF PDR STARS.

The earlier limiting regions in the HR diagram were based on the reasonable assumption that strong macroscopic motions cancel the effects of small microscopic motions. However, to ensure that diffusion processes will produce observable abundance anomalies in PDR stars, the diffusion zone must remain stable for a sufficiently long time.

In stars with strong magnetic fields, such as Ap stars, the diffusion zone is stable so long as the magnetic energy density exceeds the kinetic energy of the macroscopic motions :

$$\frac{B^2}{8\pi} > \frac{1}{2} \rho V_M^2 \qquad (5)$$

In A stars, a magnetic field stronger than about a hundred Gauss can inhibit convection.

In the case of non-magnetic stars, the stability conditions, even if satisfied on the main sequence, can be destroyed at some later evolutionary phase : mass loss can increase as the star becomes a giant and/or the convection zone can become deeper. Other phenomena, related to the transport of angular momentum through the star could also modify the stability conditions in the diffusion zone : for example, mixing from the deep interior due to meridional circulation on a time scale comparable to the evolutionary time scale. However these large scale motions have not been studied in detail. One case where the transport of angular momentum has been studied is that of stars which achieve a state of uniform rotation at some phase of their evolution (Vauclair, 1976 a, b). This is of course a specialised case, but one for which it is possible to follow the evolution of the velocity field and its stability. Note that the assumption of uniform rotation is not unreasonable for Am type stars, most of which are components of close binary systems. The model can be briefly summarized as follows : in a rotating star, meridional circulation transports angular momentum. Since angular momentum must be conserved, the meridional circulation induces a differential angular velocity which increases with time. In a homogeneous fluid, characterized by its velocity V, its dimension l and its viscosity ν, turbulence occurs if the Reynolds number exceeds some critical value, currently taken in the range $10^3 - 10^4$. However, since a stellar envelope is stratified in density, this is not a sufficient condition for turbulence to develop. Richardson's criterion must then be applied. In an incompressible fluid, this criterion is simply obtained by comparing the energy stored in the shear $\sim 1/4 \, \rho (\delta V)^2$ with the work done by the inertial forces against gravity $\sim -g\delta\rho\delta z$.

The system is stable if :

$$R_i = \frac{-g(1/\rho)(\partial \rho/\partial z)}{(\partial V/\partial z)^2} > \frac{1}{4} \quad (6)$$

if we take into account the compressibility of the fluid and the radiative losses, we obtain a modified Richardson number which, following Townsend (1958) and Zahn (1974), can be written as :

$$R_i = \frac{\sigma R_e (g/H_p)(\nabla_{ad} - \nabla)}{(dV/dr)^2} \quad (7)$$

where σ is the Prandtl number (ratio of the viscosity to the thermal conductivity), and R_e is the adopted critical Reynolds number. A fluid becomes turbulent when its modified Richardson number is smaller than one. In this model, the differential velocity and its gradient both increase with time. As a consequence, the associated Reynolds number increases while the Richardson number decreases. Both effects converge towards the onset of turbulent motions. In fact, since the Richardson number drops below one after the Reynolds number has become critical, there is a phase of stability. Diffusion can work freely during this period. When the density stratification is no longer able to damp the perturbations, turbulence develops and smooths out the concentration gradients built up during the stability phase. The mixing will not be complete if the stability phase has lasted sufficiently long for the helium to diffuse ; in this case, the molecular weight barrier built up by the diffusion of the helium will resist the mixing for a longer time, since the effect of the density stratification is reinforced by the variation of molecular weight (Mestel, 1965). The Richardson criterion must be changed to take the μ barrier into account :

$$R_i = \frac{-g(1/\mu)(d\mu/dr)}{(dV/dr)^2} \quad (8)$$

The "ordinary" stability phase in the homogeneous fluid t_s and the "μ stability" phase in the μ barrier t_μ have been estimated for a representative Am star of 2 $M_\odot$. A number of values are shown in table 1 for various values of the rotational velocity (Vauclair, 1976 a and b). Before the disruption of the μ barrier, the Am character and the pulsations exclude each other as helium has diffused below the region where it could be partially ionized (Vauclair et al., 1974). When the μ barrier disrupt, the mixing of the envelope is complete. Helium is restored to the appropriate temperature domain, where the κ-mechanism, becoming efficient, forces the star to pulsate. Such a model suggests that Am stars may evolve into δ Scuti stars on a time scale comparable to their main sequence lifetime.

TABLE 1

Stability phases of a rotating 2 $M_\odot$ star

V_0 km sec^{-1}	t_s years	t_μ years
15	3.8×10^6	4.2×10^8
20	1.6×10^6	2.1×10^8
25	8.0×10^5	1.2×10^8

III. THE EFFECTS OF SMALL MACROSCOPIC MOTIONS ON DIFFUSION.

By small macroscopic motions we mean motions whose time scales are comparable to diffusion time scales. As noted above, we distinguish two kinds of motion : "organized" motion and random motion.

a) "organized motion" : e.g. non-turbulent radial mass loss or laminar meridional circulation. Their main effect is to bring fresh matter into the diffusion zone regularly.

1 - in general, diffusion processes become less efficient and the predicted abundance anomalies decrease. This effect has been used by Kobayashi and Osaki (1973) to interpret the Am phenomenon in terms of diffusion processes superimposed on meridional circulation.

2 - in some cases, diffusion leads to the accumulation (or depletion) of certain elements somewhere in the envelope (clouds or holes). A global motion can shift and deform such regions so as to change the predicted abundance anomalies (Vauclair, 1975 a). Figure 2 illustrates such an effect : the abundance profiles for lithium are shown as a function of depth in the envelope of a 2 $M_\odot$ star (after 5×10^5 y) ; one ascending and one descending column of the same size are taken to represent the meridional circulation below the convection zone. The central block in the figure gives the abundance in the convection zone. The left hand block shows the abundance profile in the ascending column ; here, depth increases towards the left. The right hand block shows the abundance profile in the descending column ; here, depth increases towards the right. Going from left to right in the figure, we can get an idea of the mass transport due to meridional circulation. The abundance profile of lithium has bumps - the radiation force is modified by the resonance lines of Li II and Li III. The flux of meridional circulation ($V_c \rho$) is a variable parameter.

3 - the interaction between the global motions and the diffusion processes can also create new clouds and holes, if the total velocity ($V_M + V_D$) changes sign. This has been proposed as an explanation for

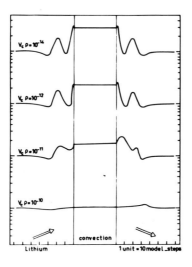

Fig. 2

Fig. 2 : Role of the meridional circulation on the abundance profiles for lithium in a 2 $M_\odot$ envelope.

helium anomalies in helium rich and helium variable stars (Vauclair, 1975 b). This effect is illustrated on figure 3. If somewhere in the atmosphere, the downward diffusion flux of helium is balanced by an opposite global mass loss flux, the helium will accumulate at this level because the diffusion velocity decreases with increasing depth

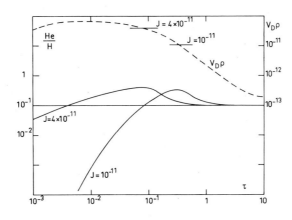

Fig. 3

Fig. 3 : Role of a non turbulent radial mass loss on the helium abundance profile in the atmosphere of a 20,000°K main sequence model. (after Vauclair, 1975 b).

in the star. Above this level, the downward diffusion flux exceeds the upward global flux : helium sinks by diffusion. Below this level, the upward global flux exceeds the downward helium diffusion flux : helium is transported upwards by the global flow. If the accumulation level falls in the line forming region, we shall have a helium rich star with somewhat special helium line profiles, since the helium abundance is not constant throughout the atmosphere. On the other hand, if the accumulation level is deeper than the line forming region, the star will appear to be helium weak. If mass loss is constrained along magnetic field lines, this process could produce helium rich patches on the stellar surface and account for helium variable stars.

b) "random motion" : we have already noted that turbulent motions partially mix the region where diffusion processes are at work. As a consequence, the concentration gradient built up by diffusion is smoothed out by the turbulent mixing. The important parameter in this problem is the ratio of the turbulent diffusion coefficient D_T to the particle diffusion coefficient D (see eq. 4), so that the effects of turbulence vary from atom to atom. One interesting possibility is that for a given D_T, the downward helium diffusion could be drastically slowed down while the upward diffusion of some other elements could continue normally. An A star with such a turbulent diffusion coefficient would appear as a pulsating star with abundance anomalies. As such stars are not observed on the main sequence but only among giant stars (Kurtz, 1976), this constrains to some extent the value of the turbulence and its time scale.

Another interesting consequence is that "clouds" and "holes" are less pronounced and broader. We show on figure 4 some results on strontium abundance variations in a 2.6 $M_\odot$ star (in which the radiation force has been computed according to Michaud et al, 1976). Figure 4a shows profiles of the strontium abundance versus depth through the envelope, for certain time intervals. Turbulence has not been included. The computations were made for the bottom of the hydrogen convection zone in which the abundance is maintained homogeneous. Since the radiation force on strontium is less than gravity at the bottom of the convection zone (where it is mainly Sr III, a noble configuration), the strontium sinks. Below, at a point where the fractional mass exceeds 10^{-10}, the radiation force is larger than gravity and the Sr is pushed upwards. There is an accumulation of Sr in the layers where the radiation force balances gravity. The accumulation peak increases with time while the convection zone progressively loses its strontium. Turbulence is then introduced into the computations. As an example, the turbulent diffusion coefficient is chosen to vary as ρ^{-1} (figure 4b). The value given for D_T in the figure is that fixed at the bottom of the convection zone. For $D_T = 10^5$, the situation is similar to that previously described for $D_T = 0$, except that the profile is smoother. In figure 4c, where the results for $D_T = 10^6$ are presented, the result is qualitatively different : the peak is broadened so much by the turbulence that it enters the convection zone, and produces an overabundance of strontium.

DIFFUSION PROCESSES AND HYDRODYNAMICAL INSTABILITIES

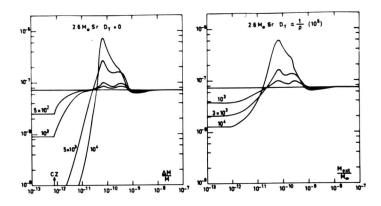

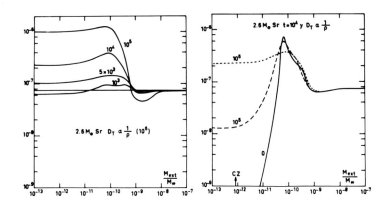

Fig. 4

Fig. 4 : Influence of turbulence on the diffusion of strontium in a 2.6 $M_\odot$ star. The diagrams show the abundance profiles (Sr/H) as a function of depth (represented by the external mass fraction). Fig. 4a (up left) : without turbulence. The profiles are shown after 5×10^2, 10^3, 5×10^3 and 10^4 y. Sr sinks in the convection zone. Fig. 4b (up right) : the turbulent diffusion coefficient varies as ρ^{-1}. Its value at the bottom of the convection zone is 10^5. The profiles are shown after 10^3, 3×10^3 and 10^4 y. Fig. 4c (down left) : same as fig. 4b, with $D_T = 10^6$ at the bottom of the convection zone. Here the strontium abundance increases with time in the convection zone. The overabundance is of one order of magnitude after 10^5 y. Fig. 4d (down right) : the abundance profiles are shown for the same epoch (10^4 y) with the three different values of D_T.

The situation is summarized in figure 4d, where strontium abundance profiles are given for the same epoch (10^4 y) but for different values of the turbulent diffusion coefficient. It is clear from the figure that an underabundance may be transformed into an overabundance just by increasing the turbulence. This effect could have drastic consequences on the predicted abundance anomalies and could help solve some of the difficulties of the diffusion theory raised by Michaud (this conference).

To conclude, this kind of computation applied to abundance anomalies should give us some information about the hydrodynamics of stellar envelopes, about which little is known. The competition betwween diffusion and mixing allows us to study the hydrodynamics of the deep stellar interior : in practice, one needs good spectral observations as well as a sound theoretical analysis of line formation in these kind of stars. Finally, the radiation forces need to be computed accurately, as well as the time dependent abundance variations. Of course, as many elements as possible should be included in the analysis in order to minimize the uncertainty on the results.

ACKNOWLEDGEMENTS-REMERCIEMENTS

C'est un plaisir pour nous de remercier Ludwik Celnikier qui a passé un bon moment à corriger l'anglais de cet article, et Catherine Leguay qui a passé un mauvais moment à le taper.

REFERENCES

Alecian, G. 1976, Thèse de 3è Cycle, Université de Paris VII
Kobayashi, M., Osaki, Y. 1973, Publ. Astron. Soc. Japan $\underline{25}$, 495.
Kurtz, D.W. 1976, Astrophys.J., in press.
Lamers, H.J.G.L.M. 1976, IAU General Assembly
Mestel, L. 1965, in Stellar Structure, ed. L.H. Aller and D.B. McLaughlin, University of Chicago Press, Chicago, p. 465.
Michaud, G. 1970, Astrophys.J. $\underline{160}$, 641.
Michaud, G., Charland. Y., Vauclair, S., Vauclair, G. 1976, Astrophys.J. in press.
Michaud, G., Reeves, H., Charland, Y. 1974, Astron. Astrophys. $\underline{37}$, 313.
Osmer, P.S., Peterson, D.M. 1974, Astrophys.J. $\underline{187}$, 117.
Powell, A.L.T. 1972, Monthly Notices Roy. Astron. Soc. $\underline{155}$, 483.
Schatzman, E. 1969, Astron. Astrophys. $\underline{3}$, 331.
Smith, M.A., Parsons, S.B. 1975, Astrophys.J. Suppl. $\underline{29}$, 341.
Smith, M.A., Parsons, S.B. 1976, Astrophys.J. $\underline{205}$, 430.
Strittmatter, P.A., Norris, J. 1971, Astron. Astrophys. $\underline{15}$, 239.
Townsend, A.A. 1958, J. Fluid Mech. $\underline{4}$, 361.
Vauclair, G. 1976 a, Astron. Astrophys. $\underline{50}$, 435.
Vauclair, G. 1976 b, to be published.
Vauclair, G., Vauclair, S., Pamjatnikh, A. 1974 a, Astron. Astrophys. $\underline{31}$, 63.
Vauclair, S. 1975 a, in IAU Colloquium N° 32 : Physics of Ap stars.
Vauclair, S. 1975 b, Astron. Astrophys. $\underline{45}$, 233.
Vauclair, S., Michaud, G., Charland, Y. 1974 b, Astron. Astrophys. $\underline{31}$, 381.
Zahn, J.P. 1974, in IAU Symp. N° 59 : Stellar Instabilities and Evolution, Ledoux et al., Ed. Reidel Publishing Company, p. 185.

MIXING BETWEEN BURNED CORE MATERIAL AND SURFACE LAYERS

Martin Schwarzschild
Princeton University Observatory

Dr. Pagel has given us a fine summary of the observations relevant to the topic I am to cover. Drs. Michaud and Vauclair have described in detail the potential effects of diffusive processes which might well explain, at least in part, the apparent abundance anomalies observed in the peculiar A stars and the magnetic A stars. In the paper following mine, Dr. Boesgaard will present the observational evidence regarding the abundance of the light elements lithium, beryllium and boron, and relate these observations to the theory of stellar envelopes. With these important topics expertly covered I shall restrict myself to the single question: How do some reasonably common types of stars, such as the carbon stars and the S stars, manage to show off at their surfaces, for us to observe, a fair sampling of the products of the nuclear burning going on deep in their cores?

In search for the answer to this question theoreticians have naturally first investigated the classical process of direct convective mixing. Could this simple process under special circumstances act all the way from the observable surface down to the burning layers? To the best of my knowledge no model constructed under our present standard assumptions has been found with such deep direct convective mixing -in spite of extensive searches, particularly regarding the shell flash evolution phases. This generally negative result does not apply to the particular case of the carbon isotope ratio. Models have been found with convective envelopes reaching down to temperatures not high enough for the carbon cycle to provide an energy source, but high enough to transmute C^{12} to C^{13}. Such enrichment of C^{13} at the bottom of the convective envelope can show up in diluted but observationally significant degree at the surface. This special item, however, does not change the fact that as yet no model seems to have been found in which an active burning shell is connected to the observable surface by direct convective mixing.

Classical convective mixing might nevertheless bring the products of nuclear burning to the surface through a process other than that of direct convective mixing, a process we might refer to as delayed convective mixing. Dr. Iben (1975) has recently found a very clean case of such delayed convective mixing. He has followed a medium mass star through its second red giant phase during which helium shell flash burning occurs. After every flash for a little while a convective zone stretches outwards from the helium shell and deposits some of the flash products quite a way out in the star. Subsequently, when the heat wave caused by the flash works its way outwards the inner convective zone dies, but, for just a little while, the convective envelope extends far deeper into the star than normally, indeed deep enough to just reach into the region in which some burning products had been deposited previously. This temporarily extended convective envelope will then transport some burning products to the surface. Thus in a two-phase, delayed process nuclear burning products are, in fact, brought to the photosphere. In this specific case the effectiveness of delayed convective mixing is enhanced by the fact that the shell flash occurs cyclically so that the entire process repeats over and over again. This form of delayed convective mixing does not, however, appear to work for low mass stars (such as all the stars of the disk population or of Population II) according to several independent numerical investigations. One possible exception to this last statement might be represented by F G Sagittae, which appears to be a low mass star. It ejected a planetary nebula some time ago and may more recently have suffered one more shell flash sending it on a wild, fast loop through the Hertzsprung-Russell Diagram. In such an unusual state, characterized by a minute mass in the envelope, convection may be able to transport burning products to the surface in a magnificently observable way. I am not aware, however, of any way of interpreting this peculiar sequence of events which occurs extremely late in the life of a low mass star, as in any way connected with the processes which produce the common carbon and S stars.

A second case of delayed convective mixing has been presented at an earlier meeting during this Assembly by Dr. Lamb. He found that during an advanced evolution phase of a massive star the convective envelope reaches the layers in which, at a much earlier phase, strong hydrogen burning had occurred. Thus, here again, the products of nuclear burning at an early epoch are later brought to the surface by classical convection -however diluted.

The two cases I have cited for effective delayed convective mixing refer to medium mass or high mass stars. I am not aware of any similar theoretical success in this direction for low mass stars. In contrast, the observations clearly include low mass carbon and S stars. Altogether, therefore, I feel led to the conclusion that thus far we have not found the observationally required mixing mechanism in terms of classical convection. There remains, of course, the question whether present estimates of overshooting, i.e., the extension of turbulent motions from a convective zone into an adjacent stable zone, are reasonably correct. If the estimates derived by Shaviv and Salpeter (1973) are approximately right the effects of overshooting seem to be negligible in our present context. If, however, the extent of overshooting should be larger than these estimates by an order of magnitude -a possibility emphasized by Dr. Roxburgh in our discussions- the conclusions I have just described might well be altered. As for myself, however, I must admit that I have not yet understood a physical process that might lead to the required extent of overshooting.

Obviously, convection is not the only mechanism that can transport material from the stellar interior to the surface. Permit me to add some speculative comments on three phenomena which potentially could cause other transport mechanisms: rotation, magnetic fields, and steady mass loss. Rotation causes meridional circulation. The speed of this circulation might be sufficient for our present purposes, particularly if we permit ourselves the assumption of fast rotating stellar cores. Dr. Mestel, however, has shown that the rotation-driven meridional circulation generally prefers to divide into separate circulation cells rather than transgress a region in which the mean molecular weight changes. Since nuclear burning by its nature generally occurs in regions with substantial molecular weight gradients, rotation does not seem to be too likely a candidate to solve our problem through the meridional circulation it causes. On the other hand, differential rotation will generally cause turbulent motions within stars through sheer-flow instabilities. This turbulence provides a form of macroscopic diffusion which in turn might mix material from the inside with that of the surface layers. Dr. Schatzman has described to us his recent estimates of this process under the assumption that the sheer-flow instability of the differential rotation exceeds the neutrality limit everywhere only slightly. Under this assumption he finds macroscopic diffuse mixing potentially important for the transport of lithium, beryllium and boron inwards to destructively high temperatures. He estimates, however, that this process would not likely be sufficiently effective to transport nuclear burning products from the deep interior outwards.

My only comment regarding magnetic fields is that we obviously cannot at this time exclude internal magnetic fields as the agents or catalyst by which the products of nuclear burning get transported to the surface. On the other hand, I am not aware of a concrete process by which magnetic fields could accomplish our desired aim -clearly a statement referring to lack of knowledge rather than lack of physical possibilities.

Finally, regarding steady mass loss, we have listened to a vivid description by Dr. Mihalas of the characteristics and problems of stellar atmospheres and chromospheres which are dominated by a steady outflow of mass through them. Such atmospheres are obviously a fascinating topic in themselves and highly relevant for red giants, i.e., the type of stars which most concerns us here. If I understand Dr. Mihalas' developments correctly, however, atmospheres with stationary mass flow differ from static atmospheres very basically at low optical depths but not substantially at the higher optical depth of the photosphere -at least not when the mass loss rates are no larger than those indicated by the observations. Since conditions in the photosphere provide the boundary conditions for the stellar interior we may conclude that even substantial steady mass loss, though of great potential influence on the eventual evolution of the star, is not likely to alter our interior models noticeably and thus will probably not help us in the particular problem we are considering here.

In summary then, I would feel that for stars of high and medium mass we have at least some concrete models in which the products of nuclear burning are transported out to the surface with an efficiency quite likely high enough to fulfill the observational requirements. In contrast, for low mass stars which may well include the majority of those stars showing nuclear burning products at their surfaces, we do not yet seem to have found a concrete process to explain these observations. What new observations then may plausibly help most to drive theoretical investigations in the right direction? My estimate at this time would be that an observational concentration on low mass stars with abundance anomalies likely caused by nuclear burning might be both the most practical and the most effective approach. Finally, I suspect that the most vital guidance for the theoretical resolution of this fascinating problem may come from observations devised to tell us whether the relevant abundance anomalies occur exclusively after helium ignition, i.e., during the helium core burning phase (horizontal branch) or the helium shell burning phase (the second red-giant phase or asymptopic branch), or can already occur in the hydrogen shell burning phase, i.e., during the subgiant and the first giant phases.

REFERENCES

Iben. Jr. I.: 1975, Astrophys. J. 196, 525
Shaviv, G. and Salpeter, E.E.: 1973, Astrophys. J. 184, 191

DECAY OF LIGHT ELEMENTS IN STELLAR ENVELOPES

Ann Merchant Boesgaard
Institute for Astronomy, University of Hawaii

I should like to confine this discussion of the decay of the light elements to Li, Be, and B since the observations of D and He^3 in stars are not relevant to the decay in stellar envelopes. The light elements are trace constituents best observed in the resonance lines. Only Li I and Be II have resonance lines which can be studied from ground-based observations. Therefore we know the most about abundances in stars where these ions can be observed and how those abundances are affected by stellar evolution. I will first discuss the cosmic or initial abundances of Li, Be, and B to establish the "zero-point" from which the decay occurs. Then observations and interpretations relevant to the decay and to stellar evolution will be considered.

1. COSMIC OR INITIAL ABUNDANCES

Zappala (1972) determined the value for "initial" Li from a combination of observations of the Li content in T Tauri stars, young clusters (Hyades, Pleiades, NGC 2264), meteorites. His result, in the ratio of the number of atoms, is $Li/H = 10^{-9}$ within a range of ± a factor of 2.

In a careful study of the solar Be abundance Chmielewski, Müller, and Brault (1975) find $Be/H = 1.4 \times 10^{-11}$ from high-resolution, center-and-limb observations and a non-LTE analysis. I have recently redetermined the Be abundance in 33 F and G dwarfs (Boesgaard 1976a). The observed Be II ($\lambda 3131$) equivalent widths were compared with line strengths predicted from Carbon and Gingerich (1969) model atmospheres; values of effective temperature and gravity appropriate for each star were found from the Strömgren narrow-band photometric indices, $H\alpha$, c_1, b-y, as mentioned by Cayrel in his talk this morning. Figure 1 shows the Be/H values as a function of temperature. For 27 stars (including the sun) the average value Be/H is 1.3×10^{-11}; the range of values is ± a factor of 2. (I will discuss the 6 stars which appear depleted in Be in the next section about the decay of light elements.) The values

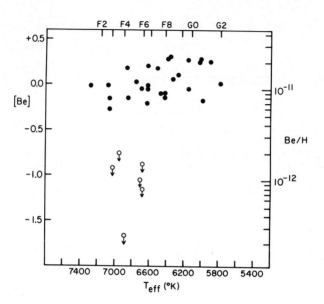

Figure 1. Be in F and G dwarfs.

of Be/H found in meteorites, the sun, and six Hyades dwarfs agree with the field-star average.

The resonance lines of B I are near 2500 Å and of B II at 1362 Å. New Copernicus observations at 0.05 Å resolution of B II have been made in Vega and Sirius by Praderie, Boesgaard, Milliard, and Pitois (1977). Figure 2 shows a sample of these spectra; the B II line is blended with V III in Vega while only the V III feature is present in Sirius. The analysis, including non-LTE effects, results in a ratio of B/H of 1.5×10^{-10}. Kohl, Parkinson, and Withbroe (1976) have made center-and-limb rocket observations of the B I resonance lines in the solar photosphere at a resolution of 0.03 Å. From spectrum synthesis they derive B/H = 4×10^{-10}. The results for the sun and Vega are in approximate agreement, but B is depleted in Sirius.

In round numbers, the cosmic or initial abundances are: Li/H = 10^{-9}, Be/H = 10^{-11}, B/H = 10^{-10}.

2. LIGHT ELEMENT DECAY AND STELLAR EVOLUTION

The light elements are fragile and are destroyed by proton fusion at temperatures of $2-4 \times 10^6$ °K. The most fragile is Li, which forms 2 helium nuclei after fusion with a proton. Thus the light elements exist only in a thin outer region of the star where the temperature is $< 3 \times 10^6$ °K. For example, at the end of the main-sequence lifetime of a 1 solar-mass star, Li will remain only in the outer 2.5% (by mass) of the star, Be in the outer 4.8% and B in the outer 18%.

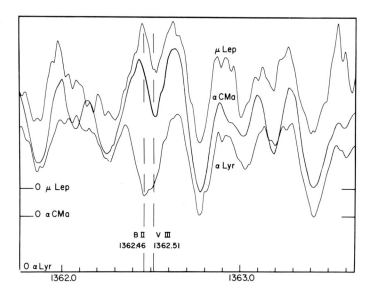

Figure 2. <u>Copernicus</u> observations for the B II resonance line.

To understand the destruction of light elements, it is necessary to know where the base of the convection zone is relative to the layer where the temperature is hot enough for destruction. If the temperature at the bottom of the convection zone is greater than the critical temperature for proton fusion, then Li, Be, and B will be depleted on the stellar surface.

During pre-main sequence evolution, low mass stars are fully-convective and according to calculations by Bodenheimer (1965) virtually all the Li will be destroyed in pre-main-sequence and main-sequence evolution in K and M dwarfs. Observations of the Li content in young stars by various workers (Zappala, 1972; Danziger, 1967; Catchpole, 1971) show that post-T Tauri/pre-main-sequence stars have 2-50 times less Li than the T Tauri stars.

For main-sequence stars Herbig (1965) has shown that the Li content is a function of both stellar mass and age: there is a range in Li at a given main-sequence spectral type and there is a decrease in the maximum Li content for stars of later spectral types. The correlation with age is in the sense that younger main-sequence stars have more Li. However, stellar and solar models indicate that the convective zones in F and G dwarfs are not deep enough to burn Li. Several ideas have been proposed to explain the observed Li depletion. The effect of a deeper convective zone can be achieved by convective overshoot, by rotational braking leading to turbulence below the convective zone, by diffusion of Li. Straus, Blake, and Schramm (1976) have recently re-discussed convective overshoot; they find that the amount of overshoot possible is greater for larger mass stars, i.e. F dwarfs could have more overshoot than G dwarfs. This is necessary to explain (1) why some F stars show no Li and (2) why there is a gradual, rather than rapid, decrease of Li with stellar mass. At an IAU colloquium on stellar ages, S. Vauclair (1972) showed that diffusion of Li below the convective zone could cause a slow depletion of Li in agreement with the observations.

Now there is an additional complication in the F and G dwarfs: Figure 1 shows a group of Be-deficient stars which are markedly separate from the group of stars with normal Be content. The six Be-deficient stars are all Li-deficient and all hotter than 6600 °K, but have no other similarities in age, metallicity, duplicity, position in the galaxy, etc. (see Boesgaard 1976a). The challenge is to find a mechanism(s) which can result in one-third of the hotter stars being deficient in both Be and Li, two-thirds of the hotter stars having Be but some with and some without Li, while all the cooler stars show Be. Clearly observations of both Li and Be in main-sequence stars are indicators of some aspects of internal stellar structure.

There is a second major mechanism of decay of Li, Be, and B which occurs in post-main-sequence evolution. The idea of dilution of the surface content of Li was first suggested by Iben (1965) to explain the Li content in the two $3M_\odot$ components of Capella. The stars leave the main-sequence with a thin outer shell which contains Li and virtually no outer convection zone; as they evolve to the red giant region of the HR diagram, the convective envelope deepens and spreads the Li throughout the convection zone. This results in a decrease of the surface Li even though the Li is not destroyed since the temperature at the bottom of the convective envelope is not high enough for nuclear burning of Li (or Be or B).

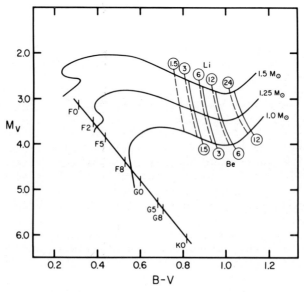

Figure 3. Curves of equal Li (---) and Be (——) dilution.

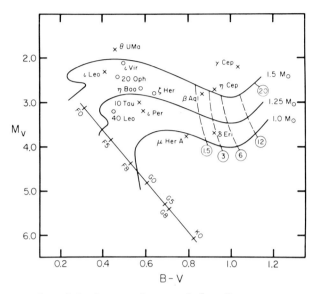

Figure 4. Subgiants observed for Be.

Figure 3 shows an HR diagram with curves of equal dilution of Li and of Be for stars in the 1-1.5 $M_\odot$ range (Boesgaard 1976b). Herbig and Wolff (1966) examined the Li content in a number of F and G subgiants, but unfortunately only one of their stars in the Li dilution region showed enough Li to give more than an upper limit on the Li abundance. In addition, the large range of Li values in main-sequence stars makes the interpretation of the Li content in evolved stars complicated. Boesgaard and Chesley (1976) have looked at the Be content in subgiants. Their results are shown in Figure 4. In the post-main sequence/pre-Be-dilution region there are stars with the main-sequence Be abundance (X in Figure 4) and some stars which are Be-deficient (O in Figure 4), the descendents of the main-sequence Be-deficient stars. Those stars in the Be-dilution region show less Be and less by the predicted amounts. This work appears to confirm that dilution of light elements does take place during post-main-sequence evolution and indicates some of the details of stellar structure and convective mixing.

Figure 5 shows an HR diagram for the Hyades where observations of Be in both dwarfs and giants have been made by Boesgaard, Heacox, and Conti (1977) and of Li in the dwarfs by Zappala (1972) and in the giants by Bonsack (1959). (The stars in which Be was observed are indicated by open circles; the points A, B, and C correspond to positions where the calculations of Li and Be dilution were made.) Relative to the dwarfs, Be is deficient in the giants by at least a factor of 30 and Li is deficient by about 180 times. Dilution alone can explain a deficiency of 25x for Be and only 60x for Li. An additional effect is needed to account for the larger observed deficiencies. The most probable explanation is mass loss. Both Li and Be are confined to a thin

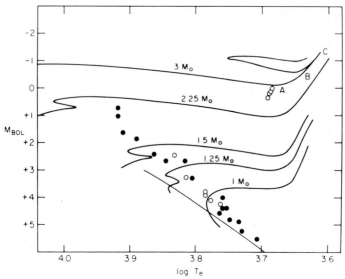

Figure 5. HR diagram for the Hyades.

outer shell. If 1% of the outer mass is lost it will increase the effect of the Li dilution by a factor of 2 (or 120x) and of the Be dilution by 20 percent. If the mass is lost on the main-sequence, the rate required is about 10^{-12} $M_\odot$/yr and if it is lost during post-main sequence evolution, it would be at the rate of about $10^{-9} - 10^{-10}$ $M_\odot$/yr. Such slow mass loss rates are not inconsistent with observations which show no evidence of circumstellar material.

Weak support for this slow mass loss is found in the observations of the Li content of field giant stars. Figure 6 shows the Li abundance as a function of surface temperature or spectral type for F, G, K, and M giants. (The arrowheads indicate upper limits.) There is a huge range in the Li content from cosmic Li/H of 10^{-9} to 10^{-13}. The stars represent a large range of masses and the Li contents reflect pre-main-sequence and main-sequence depletion, post-main-sequence dilution (with typical amounts being factors of 30-60) and mass loss effects. We see F giants with the cosmic Li abundance and F giants with little Li due to main-sequence depletion. We see M giants with the cosmic Li diluted by 60x and M giants showing cosmic Li down by a factor of 10^4 due to depletion by nuclear burning, dilution, and mass loss.

Decay of the light elements is caused by 1) nuclear destruction--including convective depletion, diffusion, convective overshoot, turbulence from rotational braking, meridional circulation and 2) dilution and 3) mass loss effects. The observations can be well understood in terms of these effects and give strong support for theoretical ideas on stellar structure and evolution.

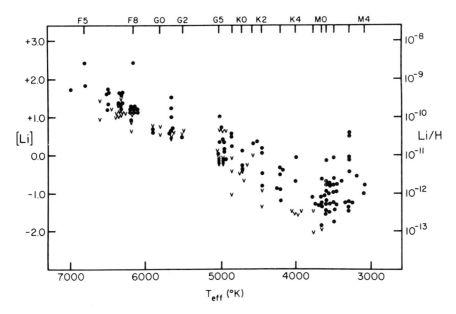

Figure 6. Li in giant stars.

REFERENCES

Bodenheimer, P.: 1965, Astrophys. J. 142, 451.
Boesgaard, A.M.: 1976a, Astrophys. J. 210, in press.
Boesgaard, A.M.: 1976b, Publ. Astron. Soc. Pacific 88, 353.
Boesgaard, A.M. and Chesley, S.E.: 1976, Astrophys. J. 210, in press.
Boesgaard, A.M., Heacox, W.D., and Conti, P.S.: 1977, Astrophys. J. in press.
Bonsack, W.K.: 1959, Astrophys. J. 130, 843.
Carbon, D. F. and Gingerich, O.C.: 1969, in Proceedings of the Third Harvard-Smithsonian Conference on Stellar Atmospheres, ed. O. Gingerich (Cambridge: MIT Press), p. 377.
Catchpole, R.M.: 1971, Monthly Notices Roy. Astron. Soc. 154, 15^p.
Chmielewski, Y., Muller, E.A., and Brault, J.W.: 1975, Astron. Astrophys. 42, 37.
Danziger, I.J.: 1967, Astrophys. J. 150, 733.
Herbig, G.H.: 1965, Astrophys. J. 141, 588.
Herbig, G.H. and Wolff, R. J.: 1966, Annales d' Astrophys. 29, 593.
Iben, I., Jr.: 1965, Astrophys. J. 142, 1447.
Kohl, J.L., Parkinson, W.H., and Withbroe, G.L.: 1976, Astrophys. J. Letters, in press.
Praderie, F., Boesgaard, A.M., Milliard, B., and Pitois, M.L.: 1977, Astrophys. J., in press.
Straus, J.M., Blake, J. B. and Schramm, D. N.: 1976, Astrophys. J. 204, 481
Vauclair, S.: 1972, L'Age des Etoiles, I.A.U. Colloquium No. 17, G. Cayrel de Strobel and A.M. Deplace, eds. (Paris: Meudon Obs.), p. 38-1.
Zappala, R. R.: 1972, Astrophys. J. 172, 57.

Figures 1, 3, and 6 are reproduced through the courtesy of the *Publications of the Astronomical Society of the Pacific*.

Figures 4 and 5 are reprinted with permission of *The Astrophysical Journal*, published by the University of Chicago Press for the American Astronomical Society. Figure 4 is from a paper by A.M. Boesgaard and S.E. Chesley appearing in the Dec. 1, 1976 issue, and Figure 5 is from a paper by A.M. Boesgaard, W.D. Heacox, and P.S. Conti appearing in the May 15, 1977 issue. These figures are copyrighted by the American Astronomical Society. All rights reserved.

CONCLUSION

by

R. CAYREL

Observatoire de Paris-Meudon

This joint discussion has amply demonstrated that stellar atmospheres are not passive and dumb bodies just sitting there at the surface of a star.

First of all the surface "feel" fundamental parameters of the star as a whole, specifically the ratio L/R^2 and m/R^2. As these fundamental parameters are affected by stellar evolution, the surface layers reflect to a large extent the degree of evolution of the star. This is the main basis for photometric and spectral classification systems. One could even claim that the observation of the stellar atmosphere would univocally determine the state of evolution of the star if i) it was always possible to determine the atmospheric abundances of the key elements to internal structure and ii) if the chemical composition of the atmosphere was always representative of the chemical composition of the whole star. Unfortunately a serious problem occurs for i) with the abundance of helium which element has no photospheric absorption line for all spectral types later than B and also with ii) in a more limited portion of the HR diagram, mainly in late B and A slow rotators. But whereas point i) is a mere problem of contingency, point ii) is a basic problem, greatly overlooked in the past, and to which one fourth of the joint discussion was devoted.

The most elementary type of motions, very clearly described in Dr Michaud's paper, arises from the fact that the individual particles are driven by forces depending upon their mass, charge and radiative cross section, personal to each species, resulting in a trend for each species to stratify with its own scale height. This type of motion, currently known as diffusion, has the capital effect of destroying the initial chemical composition of the atmosphere, well mixed with the envelope during the Hayashi phase of stellar contraction. If these diffusion motions are not counteracted by other types of motions, then the chemical composition of the atmosphere is expected to evolve with time, with a "diffusion" time scale. That is now believed to be the correct explanation of several peculiarities occuring in late B and A stars.

However, most of the time, these diffusive processes are disturbed by several types of macroscopic motions. One is the existence in stellar atmospheres of random motions due to convective instability (or other instabilities) occuring almost everywhere in the HR diagram. These random motions tend to destroy the diffusive separation of the elements but do not prevent selective losses of elements at the boundaries of a more or less homogeneous " reservoir" well mixed by random motions. Then come some organized motions as mass loss and meridional circulation. If meridional circulation is still poorly known, great progress has been made in the empirical knowledge (thanks to observations from space) and understanding of mass loss. It turns out that high luminosity stars have mass loss rates such that the matter in the atmosphere is turned over in a characteristic time of a minute or so. Such a rate of mass loss is large enough to alter the mass of the star within the nuclear time scale and then needs to be taken into account in the computation of evolutionary tracks.

Elsewhere in the HR diagram mass loss rates seem to be several orders of magnitude smaller but are still existing and may be significant when one is concerned with diffusion processes and loss of angular momentum.

Stellar atmospheres are indeed a very complex kind of upper boundary condition for internal structure, due to the fact that the decrease in density by ten orders of magnitudes occuring in the atmosphere-chromosphere-corona regions leads to supersonic flows, extremely high temperatures, and mass loss.

Perhaps simpler in principle, but not yet satisfactorily accounted for by the theory, is the chemical composition alteration of the surface layers which could be caused by mixing of burned products in the core with more superficial layers. This still remains an extremely attractive subject of study, as a way of having direct evidence of nuclear events which are occurred in the interior of the star. The best chance of explaining a mixing of this sort seems to be the saw-tooth alternance occuring on the asymptotic branch, the bottom of the convective zone being possibly able to reach at times regions where nuclear activity has taken place before.

More advanced is the interpretation of the abundances of the light elements Li, Be ,B in stellar atmospheres. There is little doubt that these elements are destroyed in stars having a deep convective zone.

The fine interpretation of the abundances of these elements establishes a direct connection between observables at the surface of the star and things happening deep below at temperatures of millions of degrees.

Summarizing, this joint discussion has made clear how stellar atmospheres can be used not as an object of study per se , but in order to achieve a more complete understanding of the star as a whole and of its evolution.

JOINT DISCUSSION NO. 6

THE SMALL SCALE STRUCTURE OF SOLAR MAGNETIC FIELDS

(Edited by F. L. Deubner)

Organizing Committee

F. L. Deubner (Chairman), J. O. Stenflo.

CONTENTS

6. THE SMALL SCALE STRUCTURE OF SOLAR MAGNETIC FIELDS

J. HARVEY / Observations of Small-Scale Photospheric Magnetic Fields 223

N.O. WEISS / Small Scale Solar Magnetic Fields: Theory 241

E. WIEHR / Some Comments on the Measurement of Small Scale Strong Magnetic Fields on the Sun 251

E.N. FRAZIER / Line Profiles of Faculae and Pores 255

G.A. CHAPMAN / Facular Models, the K-Line, and Magnetic Fields 261

H.C. SPRUIT / Small Magnetostatic Flux Tubes 265

C.J. DURRANT / Flows in Magnetic Flux Tubes 267

DISCUSSION 271

H.U. SCHMIDT / Concluding Remarks 273

OBSERVATIONS OF SMALL-SCALE PHOTOSPHERIC MAGNETIC FIELDS

J. Harvey
Kitt Peak National Observatory*, Tucson, Arizona U.S.A.

1. INTRODUCTION

If the Sun is observed like a star, without spatial resolution, its magnetic field seldom exceeds 1 Gauss. But with high spatial resolution the field is seen to be largely concentrated into kG structures. Observations of the structure and dynamics of solar magnetic fields can therefore provide a guide to the nature of magnetic fields of other stars which cannot be resolved. Solar activity and the structure of the chromosphere and inner corona are intimately linked with magnetism and a complete understanding of these features often depends on magnetic field details. There are unsolved physical problems involving solar magnetic fields which have challenged many physicists. For example, confinement of small-scale fields in kG structures is a problem of current interest (Parker, 1976; Piddington, 1976; Spruit, 1976). Solar observers are no less challenged since the Sun presents us with a complicated magnetic field having a range of scales from global to less than the scale of our best observations as illustrated in Figures 1, 2, and 3. This paper is a survey of observational techniques and results at the small-scale end of the spectrum of sizes in the solar photosphere. This topic has been frequently reviewed (e.g. Athay, 1976; Beckers, 1976; Deubner, 1975; Howard, 1972; Mullan, 1974; Severny, 1972; Stenflo, 1975) so that recent work is emphasized here.

Many techniques are available with which to observe or infer solar magnetic field properties ranging from direct *in situ* measurements such as those made by the *Helios* spacecraft to purely theoretical inferences. Discussions of these techniques have been published recently (Beckers, 1971, 1976; Schröter, 1973; Staude, 1974; Vrabec, 1974). This survey is limited to observations which depend directly or indirectly on the Zeeman effect in photospheric spectrum lines since this is the most powerful technique presently available. Stenflo (1971) reviewed how solar spectrum lines are altered by the Zeeman effect. The observations

---

*Operated by the Association of Universities for Research in Astronomy, Inc. under contract with the National Science Foundation.

are not easy to interpret because we cannot resolve the small-scale inhomogenieties directly and properties must be inferred by relying more or less on models. This procedure is laden with traps for the unwary and the emphasis here is on the observational results rather than the models used to interpret them.

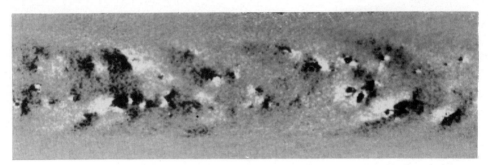

Figure 1. A synoptic map of measurements of the line-of-sight component of the photospheric magnetic field covering one solar rotation in March 1970. North is at the top and east is to the left. (KPNO magnetogram)

In reference to Figures 1, 2, and 3 it is convenient to consider the small-scale part of the complicated pattern as composed of a) sunspot fields, b) active region fields, c) network fields, and d) inner network fields. Sunspots deserve a separate discussion such as that presented by Beckers (1975) and are not discussed further here. Outside

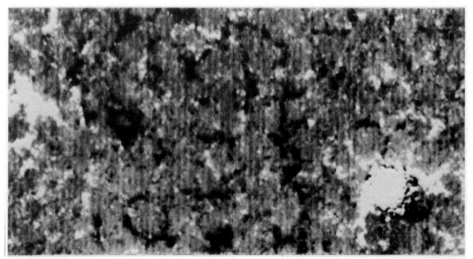

Figure 2. Line-of-sight component of the quiet photospheric magnetic field near the disk center on 9 September 1974. An old sunspot is at the lower right and strong field concentrations can be seen in a network pattern. Small inner network fields are found all over the area of 200" x 400". (KPNO magnetogram)

of sunspots, the magnetic field appears to be organized into elements with remarkably similar properties; the main distinction between active region fields and network fields is apparently the number of 'basic' elements per unit area. In areas with a large number of elements there is a tendency for the elements to cluster together to form larger cohesive structures so that a spectrum of size scales is observed, but even in large active regions the 'basic' small-scale structure is clearly identifiable. For that reason we treat the small-scale structure of active region and network fields as basically identical. Little is known about inner network fields but they appear qualitatively different from the network fields. The following discussion divides observational results according to whether one or more spectrum lines were used. Within each section the most direct results are discussed first.

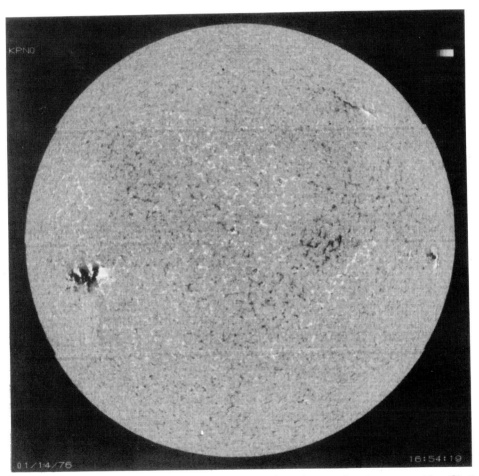

Figure 3. Line-of-sight component of the photospheric magnetic field on 14 January 1976. Faint remnants of old active regions form a large pattern near disk center while young active regions are seen near the equator in the south and a new cycle active region is located in the northwest. (KPNO magnetogram)

2. OBSERVATIONS WITH SINGLE SPECTRAL LINES

2.1 Direct observations of line profiles

When Zeeman splitting is small relative to observed line widths, the Zeeman effect introduces broadening, line profile changes, and (usually) changes in line equivalent width in unpolarized light. Measurement of these quantities offers the possibility of learning the average magnetic field strength (weighted as mentioned below) in volumes which include network and non-network regions. The problem with this technique is that many other mechanisms in addition to the Zeeman effect can cause line broadening and changes in line profile and equivalent width.

The Zeeman effect introduces polarization in line profiles and few, if any, other mechanisms can be imagined which cause closely similar polarization effects. Thus, most measurements of non-sunspot magnetic fields have made use of the polarization properties of the Zeeman effect. The outer (σ) components become oppositely circularly polarized in a field along the line of sight. If the Zeeman splitting is large then the difference between two opposite circularly polarized spectra clearly reveals the σ components and a measure of their separation yields the average of the line-of-sight component of the magnetic field in the observed volume weighted by any variation of the relative strength of the spectrum line as a function of line-of-sight field strength. Systematic velocity fields associated with the magnetic field will shift both the σ components relative to the position of the unpolarized spectrum line and velocity gradients will cause asymmetric σ components. In the visible spectrum the Zeeman splitting in non-sunspot field is not large enough to directly detect the σ components. By using an infrared line at 15648 Å which exhibits about 3 times the Zeeman splitting of visible lines, Harvey and Hall (1975) were able to resolve the σ components in circularly polarized spectra in non-sunspot fields (Figure 4). Splittings corresponding to average line-of-sight field strengths *in the magnetic elements*, $<B_\ell>$, between 1200 and 1700 G were measured; peak field strengths in excess of 2 kG were not excluded by the observations. A systematic red shift of the σ components corresponding to 2.2 ± 0.7 km s^{-1} was observed. The weakness of the σ components suggested that the magnetic elements filled less than 10% of the resolution element ($\sim 2"$) unless the spectrum line systematically weakens in the magnetic field region.

In the visible spectrum the σ components are blended with the unpolarized spectrum line and with each other and a less direct technique is required to extract average field strengths within the magnetic elements. Seares (1913) derived simple expressions for line profiles expected in polarized and unpolarized light which are valid in the case of a homogeneous field and optically thin lines. He further showed that in the case of small Zeeman splitting the wavelength displacement of the line center or center of gravity of the circularly polarized line profile is proportional to the line-of-sight component of the field.

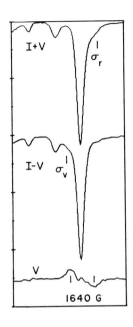

Figure 4. The 15648 Å line observed in opposite circular polarizations in a network field element. The difference spectrum at the bottom shows separation of the σ components corresponding to a mean line-of-sight field strength of 1640 G. (Harvey and Hall, 1975)

These results have been used to interpret observations even when the Zeeman splitting is not small and the spectrum line is not optically thin. Semel (1971) showed that the displacement of the center of gravity of circularly polarized line profiles is not very sensitive to violations of the Seares assumptions. Thus, the Seares results can be used to interpret observations of moderately strong spectrum lines and Zeeman splitting to yield estimates of the strength of the line-of-sight component of the field *averaged over the resolution element* and weighted by correlated changes of line strength with field strength. The symbol (B_ℓ) will be used below to denote this quantity. Values of (B_ℓ) are sensitive to spatial resolution and have steadily increased with improved observations. Sheeley (1967) found values of (B_ℓ) of 350 G. Steshenko (1967) reported values ranging up to 1400 G based on visual measurements which can take advantage of the best moments of seeing. The most recent study of this sort (Simon and Zirker, 1974) achieved a spatial resolution not likely to be soon exceeded with ground-based observations and they reported 100 G < (B_ℓ) < 1500 G. They also found a good association between magnetic regions and relative downward motion. They believe their observations resolve the magnetic structures and find sizes > 1".5 but it is possible that they resolved only clumps of smaller elements and not the smallest elements themselves.

Title and Andelin (1971) initially employed the same technique above and found 100 G < (B_ℓ) < 500 G. Tarbell and Title (1975) reanalyzed the data using a Fourier transform technique (Title and Tarbell, 1975) to extract $<B_\ell>$ values typically around 1500 G in regions 2-3" in size with one case of 1950 G reported. Numerical experiments by Heasley (1976) suggest that the Fourier transform method of extracting $<B_\ell>$ values is not always reliable so results must be treated with caution until this new method is more fully developed.

Beckers and Schröter (1968) investigated network fields (which they called magnetic knots) using circularly polarized spectra and deduced values of (B_ℓ) in the range 250-400 G for structures 2-3" in size. They proceeded beyond earlier investigations by using the Unno (1956) theory for the formation of spectrum lines in magnetic fields to match their observed profiles. After correction for dilution of the observed spectrum by stray light from non-magnetic regions they inferred values 600 G < $<B_\ell>$ < 1400 G in regions with sizes of about 1".3. They also reported a tendency for the magnetic structures to be associated with relative downward motion and dark intergranular regions.

2.2 Indirect determinations of line profiles

Spectroheliograms and filtergrams with good spectral resolution taken in circularly-polarized light with different parts of the profile of a Zeeman-sensitive spectrum line can be used to infer the profile of the σ components in magnetic regions relative to the mean line profile in non-magnetic regions. Using this technique, Giovanelli and Ramsay (1971), Sheeley (1971) and Schoolman and Ramsey (1976) all report a redward systematic displacement of the σ components corresponding to a downflow of about 0.5 km s^{-1}. This value should be independent of spatial resolution and represents a weighted mean value in the magnetic region in a volume where the core of the 6103 Å CaI line is formed. Unfortunately, determination of the amount of splitting of the σ components by measurement of the amount of polarization of the magnetic elements in the line wings is resolution dependent and only values of (B_ℓ) can be estimated.

2.3 Measurements in line wings

The Seares expressions provide a foundation for determinations of (B_ℓ) by measurement of the shift of the center-of-gravity of oppositely circularly polarized spectra. The magnitude of the shift can be inferred from the intensity difference between the opposite circular polarizations measured in the wing of a suitably selected spectrum line. This is the principle of operation of most solar magnetographs. Unfortunately calibration of this technique requires the assumption that the line profile in the magnetic region is unchanged in shape, strength and average wavelength compared with surrounding non-magnetic regions. Although these requirements are generally violated the procedure outlined is widely used to interpret longitudinal magnetograph observations. This violation leads to the expectation that most such determinations of (B_ℓ) are underestimates of true values. Photographic (Sheeley, 1966) and photoelectric (Livingston and Harvey, unpublished) observations with spatial resolution approaching 1" typically yield (B_ℓ) values of a few hundred Gauss with peaks of 700-800 G in network fields. Lynch (1974) expressed his measurements of network field clumps near the disk center in terms of net flux and found typical values of 2×10^{19} Mx.

Magnetograph observations in the best seeing conditions with high sensitivity (Figure 2) reveal inner network magnetic fields (Livingston and Harvey, 1975). The inner network fields exhibit a granular pattern of mixed polarities with a scale of about 2" and net flux values within resolved elements of about 5×10^{16} Mx. Inner network fields are transient with a time scale of the order of 30 min as shown by 2 frames from a movie in Figure 5. Smithson (1975) also detected inner network fields and suggested that the true field strength in these features is less than in network fields. The small-scale mixing of opposite polarities in the inner network fields makes their detection very sensitive to spatial resolution. A similar dependence of measured (B_ℓ) values in network fields on spatial resolution was found by Stenflo (1966).

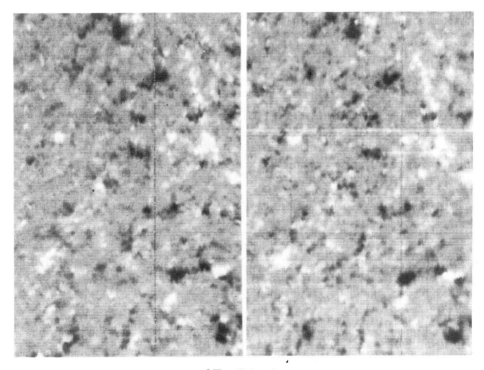

$\Delta T = 70$ min

Figure 5. Two frames from a movie of the quiet magnetic field near disk center showing time changes in the network and inner network field in a period of 70 min. (Livingston and Harvey, 1975)

3. OBSERVATIONS WITH MULTIPLE SPECTRAL LINES

As noted above, single spectral lines are affected by so many variables in addition to the magnetic field that observations are often difficult to interpret. An old solution to this problem is to use sets of spectrum lines selected to emphasize sensitivity to some desirable parameter and suppress sensitivity to other variables. This approach has been widely used in magnetic field studies and is quite powerful. One danger of this seductive approach is a tendency to forget that observations are spatial averages which can effectively hide some kinds of fine structure even from powerful line ratio techniques.

3.1 Line profile effects

Unno (1959) used pairs of lines from selected multiplets in an attempt to determine if a systematic broadening with increasing Zeeman sensitivity could be detected. No positive results were obtained and an upper limit of 300 G for the average field strength was established. Howard and Bhatnagar (1969) found a difference of 20 ± 15 G between the magnetic field strength of granular and intergranular regions from the correlation of line width differences with Zeeman sensitivity on a high quality spectrogram. This value, however, depends on how well the intergranular lanes were resolved and this is not known.

Chapman and Sheeley (1968) studied the variation of the central intensities of several spectral lines as functions of Zeeman and temperature sensitivity and concluded both effects produced line weakenings in network elements. Chapman (1976) observed changes in several complete line profiles in network features and matched the profiles with a model atmosphere having magnetic elements with values of <B> of about 1620 G at the level of formation of weak spectrum lines.

3.2 Use of one line as a reference

Simultaneous observations of (B_ℓ) using two spectrum lines were pioneered by Soviet astronomers (e.g. Severny, 1966). These observations were interpreted in terms of height variations of the field. Harvey and Livingston (1969) found simultaneous observations of (B_ℓ) with the 5250 and 5233 Å lines to give a large but nearly constant discrepancy (Figure 6). Following a suggestion by Chapman and Sheeley (1968), they assumed that a high sensitivity of the 5250 Å line to temperature and a presumed increase in temperature in network field elements was responsible for the discrepancy. But interpretation of discrepancies simply in terms of height or temperature effects alone is incorrect. As Stenflo (1968) pointed out, there are many potential sources of discrepancies in (B_ℓ) measurements and all potential sources must be observationally proven to be unimportant for a particular spectrum line before any single source can be safely ignored.

Using the same magnetograph, Frazier (1970) confirmed the observational results of Harvey and Livingston (1969) and further showed that velocity measurements made with the 5233 and 5250 Å lines also show a discrepancy in network field elements in the sense that the 5233 Å line showed a systematic downdraft but the 5250 Å line did not. Frazier (1974) later proposed that the velocity discrepancy was due to large Zeeman splitting of the 5250 Å line in network elements which caused a loss of sensitivity to network velocities (Zeeman saturation). The field strengths required fell in the range 1300-2600 G depending on model assumptions.

Howard and Stenflo (1972) used a different instrument and non-simultaneous observations and found that the discrepancy between measures of (B_ℓ) with the 5233 and 5250 Å lines decreased at low values of (B_ℓ).

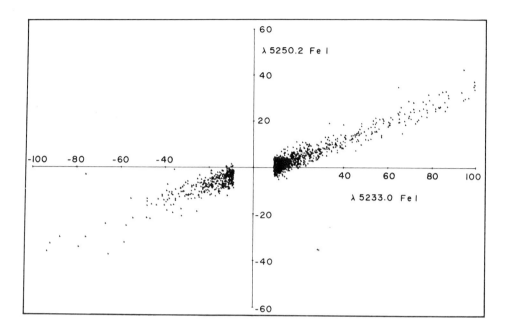

Figure 6. Simultaneous measurements of (B_ℓ) in Gauss with the 5233 and 5250 Å spectrum lines. A systematic discrepancy is obvious. (Harvey and Livingston, 1969)

This was interpreted as due to time changes in the pattern of field distribution with a scale of hours. Using a model proposed by Stenflo (1971), the observed discrepancy was interpreted as due to a systematic reduction of the 5250 Å line signal in magnetic elements with large field strength. This model allowed an estimate to be made that with a resolution of 17" more than 90% of the flux observed with the 5233 Å line was subject to the discrepancy. They also discovered that the discrepancy shows a significant center-to-limb variation (Figure 7) which was interpreted as due to a rapid decrease of field strength with increasing height.

Further studies of the 5250-5233 Å discrepancy by Frazier and Stenflo (1972) used higher spatial resolution observations taken mainly in active regions. The center-to-limb variation was confirmed and it was found that a for positive values of $(B_\ell)_{5233}$ there was a negative bias in $(B_\ell)_{5250}$ of a few Gauss (and *vice versa* for negative values of $(B_\ell)_{5233}$). The effect may be seen in Figure 6. Using a two-component model this effect was explained as due to weak, opposite polarity fields systematically associated with the strong network elements. The bias has not been confirmed in other observations and we now suspect a computer program error affecting Kitt Peak observations in 1968 and 1969 is the cause of the bias.

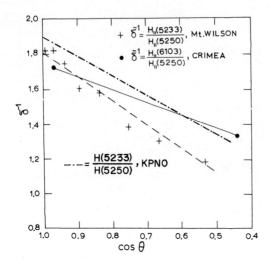

Figure 7. Center-to-limb variation of mean discrepancies in measures of (B_ℓ) with pairs of spectrum lines at the Mt. Wilson (dashed), the Crimean (solid) and the Kitt Peak (dash-dot) observatories. (Howard and Stenflo, 1972; Frazier and Stenflo, 1972; Gopasyuk et al., 1973)

Gopasyuk et al. (1973) studied observations made with the Crimean Observatory double magnetograph using many different pairs of spectrum lines. They found intrinsic scatter in values of the discrepancy made with a given pair of lines at a specific disk position. They confirmed the center-to-limb variation (Figure 7) and showed that the average value of the discrepancy is a strong function of the Zeeman sensitivity of the spectrum lines involved (Figure 8) in the sense that larger values of (B_ℓ) were observed with less Zeeman-sensitive spectrum lines.

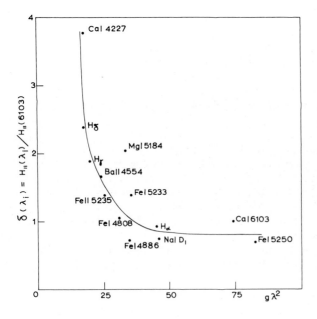

Figure 8. Variation of mean discrepancies in (B_ℓ) measurements with different pairs of spectrum lines as a function of Zeeman sensitivity. (Gopasyuk et al., 1973)

In an unpublished study, Harvey (1972) used higher sensitivity than had previously been available to confirm that values of $(B_\ell)_{5233}/(B_\ell)_{5250}$ show larger scatter at given values of $(B_\ell)_{5233}$ than instrumental sources can explain (Figure 9). On the other hand, it was found that the number distribution of $(B_\ell)_{5233}$, corrected for noise, could be fit very well by a sum of the $(B_\ell)_{5250}$ distribution plus the $(B_\ell)_{5250}$ distribution scaled and widened by a constant to get the best fit. This confirmed that about 93% of the magnetic flux measured with the 5233 Å line is subject to some effect causing a discrepancy in (B_ℓ) measurements with the 5250 Å line at a spatial resolution of about 3".

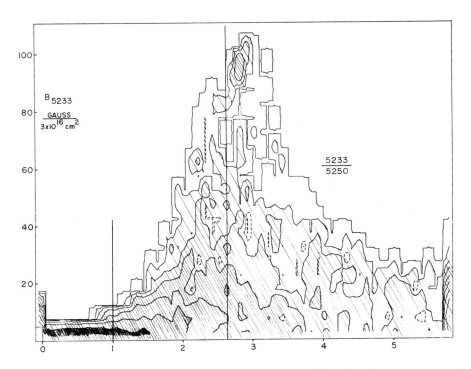

Figure 9. Isopleths increasing by factors of 2 showing a range of values of the ratio of $(B_\ell)_{5233}/(B_\ell)_{5250}$ at given values of $(B_\ell)_{5233}$. At large values of $(B_\ell)_{5233}$ a mean discrepancy of 2.6 is found but this approaches 1 as $(B_\ell)_{5233}$ decreases. Values of $(B_\ell)_{5233} > 80$ G are instrumentally saturated. (KPNO observations)

Livingston *et al.* (unpublished) are critically examining the nature of (B_ℓ) discrepancies using several pairs of lines. One of their results has considerable significance because none of the proposed sources of errors in (B_ℓ) measurements predicts that the 5233 Å and similar lines should give large errors. Using the 5253 Å line as a reference, Livingston measured (B_ℓ) using different parts of the 5233 Å line and found a variation of more than a factor of two in the sense that larger values of (B_ℓ) are measured near the core of the 5233 Å line than in its wings. The same behavior is observed in the 8688 Å line. An explanation of this curious effect is not obvious but the direct implication is that the line profile weakens considerably more in the wings than in the core in network field elements. This might be due simply to temperature increases in the network or to a decrease in optical depth within the magnetic field element relative to the surroundings with increasing geometric depth. Livingston suggests that the heating hypothesis is only likely to be consistent with observations of line profile changes in faculae (Stellmacher and Wiehr, 1971) if the heated area is too small to be resolved in facular line profile observations at low heights. An optically-thin magnetic structure at low height implies a depressed tube-like structure similar to sunspots (without darkening) and one might expect a strong variation of observable parameters as a function of distance from disk center in vertical tubes. In fact, another of Livingston's observations is a strong variation of the (B_ℓ) discrepancy in the range $1.0 > \mu > 0.95$ using several different line pairs. Further, the ratios of (B_ℓ) values measured with different parts of the 5233 Å line vary most at $\mu \simeq 1.0$ and show considerably less variation at values of $\mu \leq 0.8$. Observations with the 8688 Å line yield similar results.

3.3 Use of two lines of the same multiplet

In order to reduce the effects of all variables except magnetic field from (B_ℓ) observations, Stenflo (1973) made simultaneous observations using two lines nearly identical in every aspect except Zeeman sensitivity. Observations at the same place in the line wings should give identical values of (B_ℓ) unless one line suffers more Zeeman saturation than the other. Stenflo observed the ratio of the values of (B_ℓ), which he called k, to vary with distance from line center and to approach a value of 1 only in the extreme wings. Stenflo interpreted his observations using a two-component model and various vertical field strength profiles together with the Unno (1956) theory. Values of damping and Doppler broadening were assumed and the line strength, field strength, and relative velocity within the magnetic region were left as free parameters. Line strength and velocities less than 1.2 km s^{-1} were found to have little effect on k. Doppler shifts larger than 1.9 km s^{-1} could be excluded. Only Zeeman splitting had a significant effect on k. Depending on the choice of field strength profile, peak values of B_ℓ ranging between 1670 and 2300 G and mean values $<B_\ell>$ between 620 and 835 G best fit the observed values of k. A model having a homogeneous magnetic field component and a non-magnetic component gives a value of $<B_\ell> = 1100$ G (Stenflo, 1975). A mean downdraft velocity of about 0.5 km s^{-1} could be inferred from the observations. Assuming a flux of 2.8×10^{17} Mx in a basic magnetic element Stenflo deduced that the size of the elements was in the range 100-300 km. Stenflo (1975) later used these observations together with facular observations to derive a complete model atmosphere for the magnetic elements.

Wiehr (1976) used 3 lines of the same multiplet (6302, 6334, and 6408 Å) with Zeeman splitting factors from 1 to 2.5 to study strong network features with 2" resolution. Measured values of (B_ℓ) fell between 50 and 500 G. He tentatively confirmed the existence of a more or less unique field strength for network fields with $<B_\ell>$ in the range 1200 to 1700 G and he inferred from varying amounts of measured flux that network elements are less than 400 km in size, faculae range from 500 to 900 km and pores are larger than 1100 km.

Livingston et al. are studying measurements of k using the same lines as Stenflo and have found real variations in k at a constant distance from line center. This might be due in part to varying velocity fields but it has been interpreted as due to differing amounts of Zeeman saturation. Using a two-component Unno model with the assumption of a uniform field inside the magnetic component, the result is a range of values of $<B_\ell>$ from less than 500 G to 1700 G with a median of 1100 G. Livingston discovered that k values are dependent on μ, changing rather rapidly from $\mu = 1.0$ to $\mu = 0.9$ and more gradually to small values of μ. This result is tentatively identified with a drop in field strength with increasing height. This type of observation was also attempted on inner network fields but so far, without success. Svalgaard (1976) reports that K = 1.0 for observations of the Sun as a star which implies that a significant amount of flux exists at field strengths less than 500 G.

4. OTHER OBSERVATIONS

If the size and structure of a magnetic element could be measured then a determination of flux would allow the field strength to be inferred rather directly. We must be careful here to distinguish between tiny "basic" elements and clumps of such elements. Partial solar eclipses offer the possibility of measuring the time for a magnetic feature to be covered and thus its size. (An attempt to do this during an unfavorable eclipse at Kitt Peak by timing the uncovering of features failed for the simple reason that one never knows where a feature is until it is uncovered). At the Crimean Observatory, the last two transits of Mercury were used to determine the size of radio emission features (Efanov et al., 1974) and magnetic, velocity and brightness features (Severny, 1976). No magnetic or radio structures smaller than 1".5 were seen. Speckle interferometry is a technique which involves rapid recording of an image with a very tiny scanning aperture. Provided the scanning is rapid enough, information at spatial scales as small as the diffraction limit of the telescope can be determined. I have attempted magnetic speckle observations at Kitt Peak so far without success. The problem is one of poor signal-to-noise ratio owing to the small scanning aperture required.

It is well known that brightenings observed in weak to moderately strong spectrum lines (Chapman and Sheeley, 1968; Chapman, 1972) and in the continuous spectrum near disk center (Liu, 1974; Skumanich et al., 1975) are closely associated with magnetic fields. Very high resolution continuum photographs near disk center (Dunn and Zirker, 1973;

Mehltretter, 1974) show that the dimensions of local brightenings (filigree) associated with magnetic fields are as small as 100-200 km. If the brightenings strictly indicate the locus of the network field then there is good agreement between the filigree dimensions and the dimensions of field elements which can be inferred from various measurements discussed earlier. At present it is not clear if the filigree and the magnetic field are different manifestations of strictly the same structure. Simon and Zirker (1974) concluded that the field is more diffuse than the filigree while Beckers (1976), using observations like Figure 10, argues that the nearly perfect agreement he finds between bright features and areas showing circular polarization in the wing of a Mg line can be extrapolated slightly downward to the filigree structures. A rapid variation of flux tube diameter is required in the first 400 km above the photosphere in a recent model by Spruit (1976) in order to match center-to-limb variations in facular contrast.

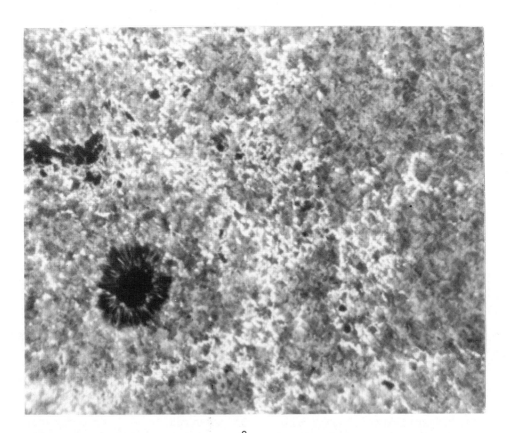

Figure 10. Filtergram taken 0.4 Å in the red wing of the Mg I line at 5183 Å near disk center on 8 May 1973. The bright network elements are cospatial with features showing large Zeeman effect. Scale is about 1" mm^{-1}. (Beckers, 1976)

Since just determining the structure and basic properties of small-scale magnetic fields is a very difficult observational problem, the determination of time variations is even harder and little is known with confidence. One serious problem is that in the presence of telescope polarization, instrumental crosstalk is possible between measured magnetic field and other parameters such as Doppler shift and brightness (Wiehr, 1971; Jäger, 1972). Motion pictures of network and inner network fields reveal horizontal motions with velocities rarely faster than 1 km s^{-1} (Smithson, 1973; Livingston and Harvey, 1975). Larger magnetic structures tend to be more stable. Stenflo (1975) has suggested that magnetic structures decay at a constant rate of -10^{15} Mx s^{-1} regardless of size but the observational evidence for this is rather indirect. Conflicting statements concerning oscillations of network field elements have been made and new observations are probably required to clarify this matter.

5. SUMMARY

It is likely that our picture of small-scale photospheric magnetic fields will continue to change rapidly in the future as it has in the past as new observations are made and analyzed. Therefore, any definitive conclusions are out of order. At the present, observers might be able to agree with most of the following statements regarding photospheric magnetic fields outside of sunspots:

1. There is no evidence for an unresolved "microturbulent" magnetic field.
2. Most of the magnetic flux observed with a resolution approaching 1" is concentrated into small elements but there is increasing evidence for the existence of an unknown amount of flux in the form of fields less than 500 G in strength.
3. The small elements tend to cluster into larger structures which can act cohesively so that a broad spectrum of sizes of magnetic structures is observed.
4. Magnetic flux tends to cluster in a network pattern which coincides with the boundaries of supergranule cells.
5. The main difference between network and active region magnetic fields is the number density of small flux elements.
6. The size of magnetic elements increases with increasing height.
7. Polarities are mixed on a small scale so that estimates of total magnetic flux are lower limits.
8. The fields are probably mainly vertical but observational evidence is very weak.
9. The field strength in magnetic elements decreases rapidly with increasing height.
10. At the height of formation of most photospheric lines at the disk center in network elements, a range of $<B_\ell>$ values from less than 500 G up to 1900 G is found with a high sample frequency at around 1100 G.
11. Peak values of B_ℓ within network magnetic elements may exceed 2 kG.

12. Systematic downdrafts are associated with magnetic fields with mean values in the magnetic elements of 0.5 km s^{-1} at the height corresponding to the core of the 6103 Å line and 2.2 km s^{-1} at the lower height corresponding to the wings of the 15648 Å line.

It now seems clear that rather complicated models of the sort developed by Stenflo (1975), Chapman (1976) and Spruit (1976) are required to quantitatively interpret magnetic field observations. The reason is inadequate spatial resolution of the tiny magnetic elements and systematic association of the elements with differential velocities, line weakenings, temperature changes and gradients in these quantities.

6. ACKNOWLEDGMENTS

I am grateful to many colleagues for providing results in advance of publication and particularly to W. C. Livingston for a continuing association which assisted greatly in the preparation of this paper.

7. REFERENCES

Athay, R. G.: 1976, *The Solar Chromosphere and Corona: Quiet Sun*, D. Reidel, Dordrecht, Ch. 4.
Beckers, J. M.: 1971, in R. Howard (ed.), 'Solar Magnetic Fields', *IAU Symp.* **43**, 3.
Beckers, J. M.: 1975, Environmental Research Papers **499**, Air Force Cambridge Research Laboratories.
Beckers, J. M.: 1976, to be published in the Proceedings of the AGU International Symposium on Solar Terrestrial Physics (Boulder, Colorado).
Beckers, J. M. and Schröter, E. H.: 1968, *Solar Phys.* **4**, 142.
Chapman, G. A.: 1972, *Solar Phys.* **26**, 299.
Chapman, G. A.: 1976, submitted to *Astrophys. J.*
Chapman, G. A. and Sheeley, N. R.: 1968, *Solar Phys.* **5**, 442.
Deubner, F.-L.: 1975, *Oss. e Mem. Oss. Arcetri*, no. 105, 39.
Dunn, R. B. and Zirker, J. B.: 1973, *Solar Phys.* **33**, 281.
Efanov, V. A., Moiseev, I. G., and Severny, A. B.: 1974, *Nature* **249**, 330 (also *Izv. Krymsk. Astrofiz. Obs.* **53**, 121).
Frazier, E. N.: 1970, *Solar Phys.* **14**, 89.
Frazier, E. N.: 1974, *Solar Phys.* **38**, 69.
Frazier, E. N. and Stenflo, J. O.: 1972, *Solar Phys.* **27**, 330.
Giovanelli, R. G. and Ramsay, J. V.: 1971, in R. Howard (ed.) 'Solar Magnetic Fields', *IAU Symp.* **43**, 293.
Gopasyuk, S. I., Kotov, V. A., Severny, A. B. and Tsap, T. T.: 1973, *Solar Phys.* **31**, 307.
Harvey, J. and Hall, D.: 1975, *Bull. Am. Astron. Soc.* **7**, 459.
Harvey, J. and Livingston, W.: 1969, *Solar Phys.* **10**, 283.
Heasley, J.: 1976, personal communication.
Howard, R.: 1972, in C. P. Sonnett, P. J. Coleman and J. M. Wilcox (eds.) 'Solar Wind', *NASA SP-308*, 3.

Howard, R. and Bhatnagar, A.: 1969, *Solar Phys.* 10, 245.
Howard, R. and Stenflo, J. O.: 1972, *Solar Phys.* 22, 402.
Jäger, F. W.: 1972, *Solar Phys.* 27, 481.
Liu, S.-Y.: 1974, *Solar Phys.* 39, 297.
Livingston, W. C. and Harvey, J.: 1975, *Bull. Am. Astron. Soc.* 7, 346.
Lynch, D. K.: 1974, *Bull. Am. Astron. Soc.* 6, 191.
Mehltretter, J. P.: 1974, *Solar Phys.* 38, 43.
Mullan, D. J.: 1974, *J. Franklin Inst.* 298, 341.
Parker, E. N.: 1976, *Astrophys. J.* 204, 259.
Piddington, J. H.: 1976, *Astrophys. Space Sci.* 40, 73.
Schoolman, S. A. and Ramsey, H. E.: 1976, *Solar Phys.*, in press.
Schröter, E. H.: 1973, *Mitt. Astron. Ges.* 32, 55.
Seares, F. H.: 1913, *Astrophys. J.* 38, 99.
Semel, M.: 1971, in R. Howard (ed.), 'Solar Magnetic Fields', *IAU Symp.* 43, 37.
Severny, A. B.: 1966, *Soviet Astron. - A. J.* 10, 367.
Severny, A. B.: 1972, in E. R. Dyer (ed.) 'Solar Terrestrial Physics/1970', *Astrophys. Sp. Science Library* 29, 38.
Severny, A. B.: 1976, personal communication.
Sheeley, N. R.: 1966, *Astrophys. J.* 144, 723.
Sheeley, N. R.: 1967, *Solar Phys.* 1, 171.
Sheeley, N. R.: 1971, in R. Howard (ed.) 'Solar Magnetic Fields', *IAU Symp.* 43, 310.
Simon, G. W. and Zirker, J. B.: 1974, *Solar Phys.* 35, 331.
Skumanich, A., Smythe, C., and Frazier, E. N.: 1975, *Astrophys. J.* 200, 747.
Smithson, R. C.: 1973, *Solar Phys.* 29, 365.
Smithson, R. C.: 1975, *Bull. Am. Astron. Soc.* 7, 346.
Spruit, H. C.: 1976, preprint.
Staude, J.: 1974, *Z. Meteorologie* 24, 214.
Stellmacher, G. and Wiehr, E.: 1971, *Solar Phys.* 18, 220.
Stenflo, J. O.: 1966, *Ark. Astron.* 4, 173.
Stenflo, J. O.: 1968, *Acta Univ. Lund.* II, no. 2.
Stenflo, J. O.: 1971, in R. Howard (ed.), 'Solar Magnetic Fields', *IAU Symp.* 43, 101.
Stenflo, J. O.: 1973, *Solar Phys.* 32, 41.
Stenflo, J. O.: 1975, in V. Bumba and J. Kleczek (eds.), 'Basic Mechanisms of Solar Activity', *IAU Symp.* 71, in press.
Stenflo, J. O.: 1975, *Solar Phys.* 42, 79.
Steshenko, N. V.: 1967, *Izv. Krymsk. Astrofiz. Obs.* 37, 21.
Svalgaard, L.: 1976, personal communication.
Tarbell, T. D. and Title, A. M.: 1975, *Bull. Am. Astron. Soc.* 7, 459.
Title, A. M. and Andelin, J. P.: 1971, in R. Howard (ed.) 'Solar Magnetic Fields', *IAU Symp.* 43, 298.
Title, A. M. and Tarbell, T. D.: 1975, *Solar Phys.* 41, 255.
Unno, W.: 1956, *Pub. Astron. Soc. Japan* 8, 108.
Unno, W.: 1959, *Astrophys. J.* 129, 375.
Vrabec, D.: 1974, in R. G. Athay (ed.), 'Chromospheric Fine Structure', *IAU Symp.* 56, 224.
Wiehr, E.: 1971, *Solar Phys.* 18, 226.
Wiehr, E.: 1976, personal communication.

SMALL SCALE SOLAR MAGNETIC FIELDS: THEORY

N. O. WEISS
Department of Applied Mathematics and Theoretical Physics
University of Cambridge

1. INTRODUCTION

One of the most exciting developments in solar physics over the past eight years has been the success of ground based observers in resolving features with a scale smaller than the solar granulation. In particular, they have demonstrated the existence of intense magnetic fields, with strengths of up to about 1600G. Harvey (1976) has just given an excellent summary of these results.

In solar physics, theory generally follows observations. Intergranular magnetic fields had indeed been expected but their magnitude came as a surprise. Some problems have been discussed in previous reviews (Schmidt, 1968, 1974; Weiss, 1969; Parker, 1976d; Stenflo, 1976) and the new observations have stimulated a flurry of theoretical papers. This review will be limited to the principal problems raised by these filamentary magnetic fields. I shall discuss the interaction of magnetic fields with convection in the sun and attempt to answer such questions as: what is the nature of the equilibrium in a flux tube? how are the fields contained? what determines their stability? how are such strong fields formed and maintained? and what limits the maximum field strength?

We also need to know what field strengths are possible beneath the surface of the sun, for magnetic fields are important probes for investigating its interior. In the photosphere the magnetic pressure in a flux tube is nearly equal to the gas pressure outside. If this balance persisted deep in the convective zone, fields of 10^7G might be formed - and dynamo theories of the solar cycle would have to be altered to accommodate them.

2. MAGNETIC FLUX TUBES AND CONVECTION

In a compressible fluid, convection is dominated by rising and expanding plumes. The numerical experiments of Graham (1975) show broad upwellings and narrow sinking regions. In a cell with fluid rising at its centre, the horizontal velocity is directed outwards over most of the depth. This picture is confirmed by observations: in a granule there is a broad central region with hot rising gas, surrounded by a narrower ring of rapidly sinking gas at the periphery (Kirk and Livingston, 1968; Deubner, 1976).

The kinematic transport of weak magnetic fields is well understood. Flux is rapidly swept aside and concentrated at the edges of convection cells and particularly at corners where several cells meet (Parker, 1963; Clark, 1965, 1966; Weiss, 1966; Clark and Johnson, 1967). The effect of radial inflow in three dimensional convection has been strikingly demonstrated by Galloway (1976). At the same time, the field within the cell is distorted by the motion and eventually expelled from a persistent eddy. In the sun, however, the lifetime of turbulent eddies is too short for this process to be completed (Weiss, 1966).

Flux concentration is limited by the Lorentz force. The magnetic field separates into ropes, where large scale convection is suppressed, while convection proceeds in the field-free region in between (Weiss, 1964). The formation of ropes is confirmed by dynamical calculations for two-dimensional (Peckover and Weiss, 1972, 1976; Weiss, 1975) and axisymmetric (Galloway, 1976) models. Pre-existing flux ropes must influence the pattern of motion itself so as to maintain the separation of vigorous convection from the almost stagnant flux tubes.

Opinions differ as to the structure of magnetic fields in the sun. On the one hand, turbulent dynamos seem to require strong small scale fields everywhere (e.g. Krause, 1976). On the other, Piddington (1974, 1975a,b,c, 1976a,b,c,d), in a spate of papers has castigated most other authors as supporters of "diffuse field theories". Those of us who have argued for flux ropes may not recognize his version of our views. Observations show that there is an intimate relationship between small scale magnetic fields and convection (e.g. Dunn and Zirker, 1973; Mehltretter, 1974). On a supergranular scale, the network fields display similar behaviour. Most of the flux seems to remain at cell boundaries, while the inner network fields (Harvey, 1976) involve only a relatively small flux. Apparently magnetic fields in the solar convective zone are normally concentrated into ropes. It does not follow, however, that these ropes are formed independently of convection, nor is it necessary to introduce twisted strands of a primeval field in order to provide models that are compatible with observations. Indeed, the only viable theories for explaining the structure of small scale magnetic fields rely on the interaction of those fields with convection.

Traditionally, it has been supposed that the field strength, B, in a flux rope cannot exceed the equipartition field, B_e. The equipartition field has an energy density equal to the kinetic energy density of the motion,

$$B_e^2/8\pi = \tfrac{1}{2}\rho U^2$$

where ρ is the density and U is the maximum speed of convection. This appealing, and uniquely simple, dimensional argument is incorrect. In the photosphere, B_e is less than 600G, yet fields of 1500G have been observed. The justification for the equipartition limit is that pressure fluctuations in a convecting fluid should be of order $\tfrac{1}{2}\rho U^2$ and, if the density and temperature in the flux rope are similar to those outside, the magnetic pressure cannot be greater than these pressure fluctuations. But if the flux tube is evacuated the internal pressure falls and the equipartition limit is irrelevant. A slight pressure excess is sufficient to squeeze the flux tube and to induce a downward flow of gas. Since the density in the sun increases rapidly with depth, the displaced matter can easily be accommodated.

3. STATIC EQUILIBRIUM IN A FLUX TUBE

Observations show that the magnetic field in a pore or sunspot drops to zero at the boundary in a distance too small to be resolved. The current sheet at the boundary has a very small but finite thickness owing to the finite electrical conductivity of the gas. It is reasonable, therefore, to adopt an equilibrium model in which the field is discontinuous at the boundary. Continuity of normal stress then requires that

$$P_i + B^2/8\pi = p$$

at the boundary, where P_i is the gas pressure within the flux tube and p is the pressure outside. Hence the magnetic field at the boundary cannot be greater than the value

$$B_p = (8\pi p)^{\frac{1}{2}}$$

that balances the external pressure when $P_i = 0$.

It has been suggested (e.g. Stenflo, 1975) that a twisted force-free field might have a mean longitudinal component $\langle B_z \rangle$ greater than B_p, contained by an aximuthal field at the boundary. The field on the axis may indeed be larger than B_p but Parker (1976a) has proved that $\langle B_z \rangle < B_p$ for cylindrical force-free fields. There are two effects that might allow the observed field to be greater. If the boundary of the flux tube is inclined at an angle θ to the vertical then the central field is greater than that at the boundary. With a monopole field, for instance,

$$\langle B_z \rangle = B_p \sec^2\tfrac{1}{2}\theta \approx B_p(1+\tfrac{1}{8}\theta^2).$$

For small flux ropes this correction is negligibly small. Secondly, the Wilson depression in the flux rope makes it possible to observe fields at a greater geometrical depth (Spruit, 1976). The local pressure scale height in the external gas is about 200Km; if the level with optical depth unity is depressed by 100Km then the central field may rise by 30%. The influence of these effects on the average field is relatively slight and so $\langle B_z \rangle$ cannot appreciably exceed B_p. In the photosphere $B_p \approx 1600G$. Direct measurements of Zeeman splitting in an infrared line (Harvey, 1976) give values of $\langle B_z \rangle$ that are close to this. Reports of higher average fields are not likely to be verified.

Within the flux tube, large scale convection is suppressed by the magnetic field and the heat transport is reduced, so that the gas is cooled (e.g. Cowling, 1976a). Nonradiative transport in a strong magnetic field is highly anisotropic and the jump in temperature at the boundary of a flux tube can therefore be sustained. In small flux ropes the lateral radiative flux becomes significant (Zwaan, 1967; Spruit, 1976). Parker (1974c,d, 1975a, 1976a,d) has suggested that sunspots are refrigerated owing to the efficient transformation of energy into Alfvén waves. However, the amount of energy emerging from sunspots into the corona is comparatively small, and it is improbable that hydromagnetic waves can be generated so efficiently (Cowling, 1976a,b). Magnetic inhibition of convection remains the most likely explanation of cooling in sunspots and small flux ropes.

The simplest models of small flux tubes have assumed an axisymmetric vacuum field (Simon and Weiss, 1974; Meyer et al., 1976). More sophisticated magnetohydrostatic models, allowing for the internal pressure, have been put forward by Chapman (1974), Stenflo (1975), Spruit (1976) and Wilson (1976). Spruit has constructed a family of models with fluxes increasing from 3×10^{17}mx to 10^{19}mx, and surface radii from 80Km to 500Km. By a height of 400Km the radius has doubled: adjacent tubes must therefore merge and this effect may explain the more diffuse fields found by Simon and Zirker (1974). Spruit suggests that the appearance of bright faculae near the limb is caused by emission from the walls of flux tubes, seen in projection. Chapman, on the other hand, explained the disappearance of faculae at the centre of the disc by a temperature stratification with a cool layer overlaid by a hot region, heated presumably by hydromagnetic waves.

So far, only static models have been mentioned. The observations reported by Harvey (1976) indicate the presence of downward velocities in the photosphere, though it is not clear whether these are permanent or transient effects (see Durrant, 1976). Steady motion along the field lines has been discussed by Ribes and Unno (1976) and by Parker (1976b,c). To conserve mass there must also be upward motions, and the small scale fields are probably the sites of spicules with rapid upward surges (Parker 1974a,b; Unno et al., 1974).

4. PRODUCTION AND MAINTENANCE OF INTENSE MAGNETIC FIELDS

The manifest failure of the equipartition argument has stimulated various attempts to replace it. It has even been suggested

(Sreenivasan, 1973; Stenflo, 1975) that force-free fields can spontaneously be amplified by a Beltrami flow; of course, the energy must be supplied from somewhere and work is done by the external gas in compressing the flux tube. Provided that the total magnetic energy in the flux tube remains small compared with the kinetic energy of a granule, there is no difficulty in supplying enough energy to form a strong field within the lifetime of a granule. Parker (1974a,b) has considered mechanisms of hydraulic concentration by turbulent pumping, kneading and massaging of the flux tube (see the discussion by Durrant, 1976). The resulting increase in the field strength is only the appropriate equipartition field, which is still too low.

In sunspots, which are much larger than individual granules, inhibition of convection leads to cooling and a collapse to a final state with a strong magnetic field, as originally suggested by Biermann (see Cowling, 1976a). This thermal mechanism cannot be applied to flux ropes that are much smaller than the local scale of convection, for the field is swept aside before any thermal instability can grow (Galloway et al., 1976). So we have to discover what limits the concentration of flux by convection once the dynamical effects of the magnetic field have become important.

The idealized problem of laminar convection in a Boussinesq fluid with an imposed magnetic field has been fairly thoroughly investigated. Since pressure does not enter the equation of state, the pressure within the flux tube can always be reduced to ensure a magnetohydrostatic balance. In fact, pressure can be altogether eliminated from the governing equations, so the equipartition argument (which depends on balancing contributions to the total pressure) is obviously irrelevant. Busse (1975) pointed out that the peak field depended on the relative rates of viscous and ohmic dissipation, and could be made arbitrarily large by choosing a sufficiently high value for the ratio of the viscous to the magnetic diffusivity. This has been confirmed for fully nonlinear convection by several series of systematic numerical experiments (Peckover and Weiss 1971, 1976; Weiss 1975; Galloway, 1976) in which peak fields distinctly greater than the equipartition field have been produced. (In the most extreme case, with the diffusivity ratio equal to 10, the peak field $B^* \approx 6B_e$, though the particular numerical value is of no significance.)

In these computations runs were made with the thermal boundary conditions kept fixed while B_0, the average magnetic field over a whole convection cell, was varied. When B_0 is small, flux concentration is purely kinematic and B^* is limited by diffusion. B^* remains proportional to B_0 throughout the kinematic regime but, as B_0 is increased, dynamical effects eventually become important. Thereafter, in the dynamic regime, flux concentration is limited by the Lorentz force and the rate of working of the buoyancy force is balanced by ohmic dissipation. The peak field B^* reaches its maximum value, B_m, at the transition from the kinematic to the dynamic regime, when ohmic dissipation in the flux rope becomes comparable with viscous dissipation in the convective cell.

For solar convection a similar criterion should apply
(Galloway et al., 1976). Consider a flux rope of radius δ concentrated
between granular convection cells with a radius d (where d $\gg \delta$). The
rate of dissipation of kinetic energy in a granule is approximately
$\frac{1}{2}\rho U^2 d^3/\tau$, where the lifetime τ of a convective eddy is approximately
the turnover time d/U. The rate of ohmic dissipation in the flux rope
is approximately $(\eta B^{*2}/\delta^2) \cdot \delta^2 d$, where η is the magnetic diffusivity,
and is therefore independent of δ. The two rates are equal when

$$B^* \approx B_m \sim (Ud/\eta)^{\frac{1}{2}} B_e.$$

Hence fields much greater than B_e can be maintained if the magnetic
Reynolds number $(Ud/\eta) \gg 1$. However, the value of η depends not on
the laminar diffusivity but on the effective diffusivity provided by
small scale oscillatory convection within the flux rope. This
effective diffusivity can be estimated by taking a velocity of 1km s^{-1}
and a horizontal scale of 100Km, corresponding to microturbulent
velocities in sunspots (Beckers, 1976); then $\eta \approx 10^{11} cm^2 s^{-1}$. Hence,
for intergranular fields $B_m \sim 15 B_e \sim 5000G$. Thus it is possible to
produce photospheric magnetic fields that are much greater than the
equipartition limit. Since B_m exceeds the critical field B_p that
balances the external pressure, the limit is in practice set by B_p.
Depending on the flux contained, small ropes should have average fields
of up to 1600G.

Thermal effects are only important near the photosphere. Deeper
down, the magnetic pressure is much smaller than the gas pressure.
Arguments similar to those above suggest that the peak field in flux
ropes between supergranules should not be greater than about 10000G
and even in the deep convective zone the field is not likely to reach
significantly higher strengths (Galloway et al., 1976). Certainly the
limit B_p cannot be approached except at the upper boundary of the
convective zone.

5. STABILITY OF FLUX ROPES

In all simple magnetohydrostatic models the radius of the flux
tube increases with height, owing to the vertical pressure gradient
outside. Hence the field fans out at the boundary and is concave
towards the external plasma. Parker (1975b) and Piddington (1975a)
have argued that the configuration should therefore be intrinsically
unstable, and liable to interchange (or flute) instabilities. A more
careful treatment, based on the energy principle of Bernstein et al.
(1958), shows that the field is stabilized locally by buoyancy effects
(Meyer et al., 1976), provided that the flux is greater than about
10^{19}mx. Thus there is no need to invoke twisted magnetic fields
(which have not been observed) to stabilize a sunspot. Small flux
tubes are weakly unstable, though the instability probably grows too
slowly for it to be significant. To hold the flux together, a deeper
collar is still required and this may be provided by the interaction
with external convection. Without such a collar, the total magnetic
energy can be reduced by splitting up the flux tube (Wilson, 1976b).

Observations show that sunspots and pores have lifetimes much
longer than the timescale for dynamical instability (the time taken by
an Alfvén wave to travel across the flux rope). Obviously they are
stable. Small scale fields also survive for longer than the dynamical
timescale, but their behaviour is dominated by convection in the
photospheric granules that surround them. Individual points and
crinkles in the filigree are buffeted and jostled by granules, and
their characteristic lifetime (about 10 minutes) is similar to the
lifetime of a granule (Dunn and Zirker, 1973). Magnetic flux is
shunted to and fro in the intergranular lanes and concentrated particularly at junctions. The field strength attained depends on the
amount of flux brought together. Once the convection pattern alters,
the flux disperses owing to the effective diffusion caused by small
scale motion (cf. Meyer et al., 1974); for rope 100Km in radius, the
diffusive timescale is about a minute.

6. CONCLUSION

The jump in gas pressure at the boundary of a flux rope, like the
cooling of sunspots, is a shallow phenomenon, confined to the region
where the magnetic pressure can balance the external pressure. The
formation of a sunspot, which is much larger than individual granules,
begins when a flux rope protrudes into the photosphere from below or
when smaller flux tubes are assembled by supergranular convection.
Magnetic inhibition of convection reduces the supply of energy to the
photosphere and the temperature falls, so reducing the pressure. The
spot then collapses until a stable equilibrium is reached.

Filamentary magnetic fields follow a different scenario. Their
scale is determined locally by granular convection; in particular, the
inner network fields can only be concentrated by small scale convection.
Once the field is strong enough to impede heat transport, the temperature falls within the flux rope and the pressure difference $(p-p_i)$ is
sufficient to squeeze the tube and to drive gas, mainly downwards,
along the field lines. Deeper down, where the density is much greater
and the magnetic pressure is much less than the external pressure,
the density in the flux tube is actually increased and the cooling
correspondingly enhanced in order to maintain a hydrostatic balance.
Evacuation of the flux tube at the photosphere continues until the
magnetic pressure there is strong enough to balance the external
pressure. The resulting configuration is then maintained by a slow
flow of gas across the field, driven by convection in adjacent granules.

A more detailed treatment of small scale magnetic fields is still
needed. So far, we lack a decent theory of convection in a magnetic
field, and the energy transport in a magnetic field can only be
calculated by making rather arbitrary assumptions (cf. Deinzer, 1965;
Spruit, 1976). With a better theory of convection, we would be able
to construct improved equilibrium models and then to study their
stability.

Nevertheless, we do now have a consistent qualitative understanding of photospheric magnetic fields. They are rapidly concentrated into ropes, whose appearance depends on a single parameter: the magnetic flux contained in them. For fluxes less than about 10^{17} mx the field is concentrated kinematically; the peak field strength is proportional to the flux and is limited by the effective (turbulent) diffusivity (Galloway et al., 1976). Thus small filamentary fields with strengths of several hundred to a thousand gauss must be common. For greater fluxes, the field reaches the limiting value B_p that balances the external pressure. So for fluxes above 10^{17} mx we expect to have fields of about 1500G. As the flux rises, the area of the tube increases with it until, for fluxes greater than about 10^{19} mx, the radius becomes comparable with that of a granule. Cooling is then more effective and a dark pore is formed. For yet higher fluxes, the field has to spread out in order to achieve a magnetostatic equilibrium. At the boundary the field becomes increasingly inclined, while the field strength at the centre becomes significantly greater than 1500G. Eventually, the field is almost horizontal at the edge of the pore. For fluxes greater than about 3×10^{20} mx, convection finally penetrates into the magnetic field, forming the filamentary penumbra characteristic of a sunspot.

REFERENCES

Beckers, J. M.: 1976, Astrophys. J., 203, 739.

Bernstein, I. B., Frieman, E. A., Kruskal, M. D. and Kulsrud, R. M.: 1958, Proc. Roy. Soc. A, 244, 17.

Busse, F. H.: 1975, J. Fluid Mech. 71, 193.

Chapman, G.A.: 1974, Astrophys. J., 191, 255.

Clark, A.: 1965, Phys. Fl., 7, 1455.

Clark, A.: 1966, Phys. Fl., 9, 485.

Clark, A. and Johnson, A. C.: 1967, Solar Phys., 2, 433.

Cowling, T. G.: 1976a, Magnetohydrodynamics, Hilger, Bristol.

Cowling, T. G.: 1976b, Mon. Not. R. Astr. Soc., in press.

Deinzer, W.: 1965, Astrophys. J., 141, 548.

Deubner, F. L.: 1976, Astron. Astrophys., 47, 47.

Dunn, R. B. and Zirker, J.B.: 1973, Solar Phys., 33, 281.

Durrant, C. J.: 1976, these proceedings.

Galloway, D. J.: 1976, Ph.D. dissertation, University of Cambridge.

Galloway, D. J., Proctor, M. R. E. and Weiss, N. O.: 1976, to be published.

Graham, E.: 1975, J. Fluid Mech., $\underline{70}$, 689.

Harvey, J.: 1976, these proceedings.

Kirk, J. G. and Livingston, W.: 1968, Solar Phys., $\underline{3}$, 510.

Krause, F.: 1976, in V. Bumba and J. Kleczek (eds), Basic mechanisms of solar activity, Reidel, Dordrecht.

Mehltretter, J. P.: 1974, Solar Phys., $\underline{38}$, 43.

Meyer, F., Schmidt, H. U. and Weiss, N. O.: 1976, to be published.

Meyer, F., Schmidt, H. U., Weiss, N. O. and Wilson, P. R.: 1974, Mon. Not. R. Astr. Soc., $\underline{169}$, 35.

Parker, E. N.: 1963, Astrophys. J., $\underline{138}$, 552.

Parker, E. N.: 1974a, Astrophys. J., $\underline{189}$, 563.

Parker, E. N.: 1974b, Astrophys. J., $\underline{190}$, 429.

Parker, E. N.: 1974c, Solar Phys., $\underline{36}$, 249.

Parker, E. N.: 1974d, Solar Phys., $\underline{37}$, 127.

Parker, E. N.: 1975a, Solar Phys., $\underline{40}$, 275.

Parker, E. N.: 1975b, Solar Phys., $\underline{40}$, 291.

Parker, E. N.: 1976a, Astrophys. J., $\underline{204}$, 259.

Parker, E. N.: 1976b, Astrophys. J., (in press).

Parker, E. N.: 1976c, Astrophys. J., (in press).

Parker, E. N.: 1976d, in V. Bumba and J. Kleczek (eds), Basic mechanisms of solar activity, Reidel, Dordrecht.

Peckover, R. S. and Weiss, N. O.: 1972, Computer Phys. Comm., $\underline{4}$, 339.

Peckover, R. S. and Weiss, N. O.: 1976, to be published.

Piddington, J. H.: 1974, in R. G. Athay (ed.), Chromospheric fine structure, p. 269, Reidel, Dordrecht.

Piddington, J. H.: 1975a, Astrophys. Space Sci., 34, 347.

Piddington, J. H.: 1975b, Astrophys. Space Sci., 35, 269.

Piddington, J. H.: 1975c, Astrophys. Space Sci., 38, 157.

Piddington, J. H.: 1976a, Astrophys. Space Sci., 40, 73.

Piddington, J. H.: 1976b, Astrophys. Space Sci., 41, 79.

Piddington, J. H.: 1976c, Astrophys. Space Sci. (in press).

Piddington, J. H.: 1976d, in V. Bumba and J. Kleczek (eds), Basic mechanisms of solar activity, Reidel, Dordrecht.

Ribes, E. and Unno, W.: 1976 Astron. Astrophys., in press.

Schmidt, H. U.: 1968, in K. O. Kiepenheuer (ed.), Structure and development of solar active regions, p. 95, Reidel, Dordrecht.

Schmidt, H. U.: 1974, in R. G. Athay (ed.), Chromospheric fine structure, p. 35, Reidel Dordrecht.

Simon, G. W. and Weiss, N. O.: 1970, Solar Phys., 13, 85.

Simon, G. W. and Zirker, J. B.: 1974, Solar Phys., 35, 331.

Spruit, H. C.: 1976, to be published.

Sreenivasan, S. R.: 1973, Physica, 67, 330.

Stenflo, J. O.: 1975, Solar Phys., 42, 79.

Stenflo, J.O.: 1976, in V. Bumba and J. Kleczek (eds), Basic mechanisms of solar activity, p. 69, Reidel, Dordrecht.

Unno, W., Ribes, E. and Appenzeller, I.: 1974, Solar Phys., 35, 287.

Weiss, N. O.: 1964, Phil. Trans. Roy. Soc. A. 256, 99.

Weiss, N. O.: 1966, Proc. Roy. Soc. A, 128, 225.

Weiss, N. O.: 1969, in D. G. Wentzel and D. E. Tidman (eds), Plasma instabilities in astrophysics, p. 153, Gordon and Breach, New York.

Weiss, N. O.: 1975, Adv. Chem. Phys., 32, 101.

Wilson, P. R.: 1976a, to be published.

Wilson, P. R.: 1976b, to be published.

Zwaan, C.: 1967, Solar Phys., 1, 478.

SOME COMMENTS ON THE MEASUREMENT OF SMALL SCALE STRONG MAGNETIC FIELDS ON THE SUN

E. Wiehr
Universitäts-Sternwarte Göttingen, GFR

Since Dr. Harvey has already mentioned the main results of my observations, I want to restrict myself to two topics:
i) a brief description of the improvements of my method as compared to that used by Stenflo (1973),
ii) some puzzeling aspects of my results which might be discussed in this meeting.
The main differences between the two methods of simultaneous Zeeman polarization measurements used by Stenflo (1973) and by myself are based on suggestions by Schröter (1973):
a) use of three instead of only two differently split lines;
b) use of less temperature sensitive lines;
c) use of model atmospheres instead of a Milne-Eddington model for the calculation of the 'calibration curves';
d) observation of selected Ca^+K features in order to avoid the statistical method with the scatter-plot diagram,

Items a) and b) lead to the construction of a three-channel magnetograph for the Fe-lines 6302.5, 6336.8 and 6408 (Fig.1) at the Locarno station of the Göttingen observatory. According to Stellmacher and Wiehr (1970) these 3.6 eV lines remain unchanged going from photosphere to umbra. However, they show a reduced central depth going from photosphere to faculae (Stellmacher and Wiehr, 1971). Because of the rather equal strength (W_λ = 103, 138 and 125 mÅ, resp., according to the 'Liège atlas') this 'rest intensity effect' is almost equal for all three lines as has been verified by calculation with the 'Zeeman lines computer program' by Wittmann (1974).
Concerning item c), it has already stated by Stenflo (1975) that the approximation by a M.E. atmosphere does not alter significantly the 'calibration curves' $P_{circ}(B_{long})$ Fig.2

The investigation of individual network points (item d) resulted in some puzzeling facts which I would like to bring to your attention:

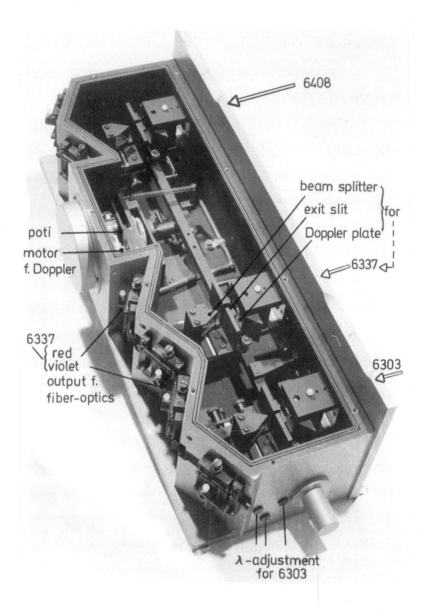

Fig.1: Three-channel magnetograph output adapted for the plate holder of the Locarno spectrograph.

As reported by Dr. Harvey, my observation of various bright
Ca^+K features mainly confirm Stenflo's (1973) results for
quiet network and furthermore indicate their validity also
for enhanced network, faculae and (probably) pores. However,
a certain number among each of these features yield no
'true' field strength since either the apparent P_{circ}-
ratios 6303 : 6337 : 6408 exceed the possible upper limit
2.5 : 2.0 : 1.0 (which represents the case $B_{long} \rightarrow 0$;
c.f. Fig.2) or the P_{circ}-ratios 6303:6337 , 6303:6408
and 6337:6408 yield three different values for B_{true}.
These cases of 'not fitting polarization' cover about
10 - 20% of the Ca^+K features observed. Possibly they
are equivalent to those points in Stenflo's scatter-
plot diagram which largely differ from the average slope.
Similar difficulties were reported by Semel (1976) from
an extended study of a large number of Zeeman lines.

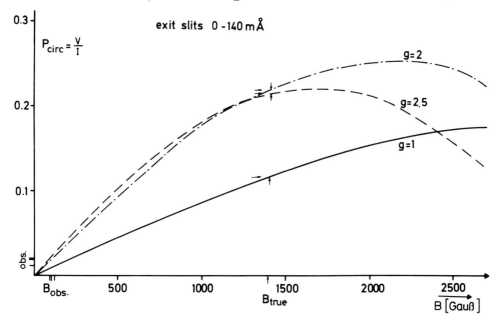

Fig.2: Calibration curves for Fe 6303, 6337, 6408
calculated for the apparent line profiles
and the facula model by Stellmacher/Wiehr(1973)

Another point I would like to mention concerns the basic
idea behind the whole measuring method used by Stenflo
and by myself: the independence of the P_{circ}-ratios on
seeing. This pre-condition seems to be severely violated
for a considerable number of Ca^+K points where with
increasing image quality and hence larger apparent field
strength the P_{circ}-ratios vary in such a way that the
finally deduced 'true' field strength decreases (as compared
to bad seeing).

These effects have to be investigated in detail before considering the (preliminary) concept of a 'limided range of true field strengths' for all solar magnetic fields outside sunspots.

References:
Schröter,E.H. 1973, priv.comm.
Stellmacher,G., Wiehr,E. 1970,Astron.Astrophys.$\underline{7}$, 432
Stellmacher,G., Wiehr,E. 1971, Solar Phys.$\underline{18}$, 220
Stellmacher,G., Wiehr,E. 1973,Astron.Astrophys.$\underline{29}$, 13
Stenflo,J.O. 1973,Solar Phys.$\underline{32}$, 41
Stenflo,J.O. 1975,Solar Phys.$\underline{42}$, 79
Wittmann,A. 1974,Solar Phys.$\underline{35}$, 11

LINE PROFILES OF FACULAE AND PORES

Edward N. Frazier
The Aerospace Corporation

1. INTRODUCTION

Historically, attempts to model the temperature structure of faculae have generally suffered from a rather basic contradiction. Models which were based on center to limb measurements of the continuum contrast of faculae disagree with models that are based on measurements of line profiles in faculae. The "continuum" models predict line weakenings which are of larger amplitude than what is observed, and the "line profile" models predict a continuum contrast that is less than what is observed. Chapman (1976) discusses this problem in some detail. It is the purpose of this paper to show that there is a fundamental reason for this historical contradiction between line profile measurements and continuum contrast measurements: The line profile and the continuum contrast of a given facular are both a function (the two functions are different) of the size of that facula. The first indication of this fact was given by Frazier (1971). Figure 1 shows the contrast of faculae in the core of the line Fe Iλ 525.0 nm, and in the continuum, as a function of the observed magnetic flux. One can see immediately that the contrast in each channel depends on Φ in a much different manner. Therefore, one can conclude that the shape of the entire line profile will vary as a function of Φ. On the basis of Figure 1, we must expect that this variation of the line profile will be continuous from infinitesimally small faculae up through very large faculae, and indeed, all the way up to pores.

We now wish to show that, for any given spectral line, there exists an entire family of facular and pore line profiles. We will show examples of this family of line profiles for two spectral lines; Fe Iλ 525.0 nm and Fe Iλ 524.7 nm. It will then become evident that such an entire family must be observed in order to provide an adequate set of data on the temperature structure of faculae and pores. From the data that is available so far, it appears that this family can be characterized by a single parameter. The best parameter to use for this purpose is the amount of magnetic flux, Φ, contained within a facula. At first thought, it might seem that the magnetic field strength, B, is the best

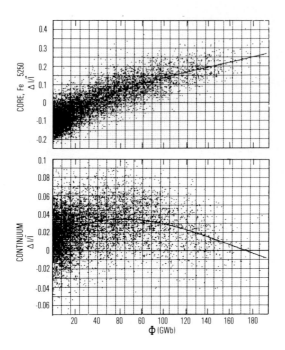

Figure 1. Facular contrast in the core of a temperature sensitive line (Fe λ 525.0 nm) and in the continuum as a function of observed magnetic flux. In this figure only, the contrast is defined as $(I_f - I)/I$, where I is the average brightness of the region scanned. Adapted from Frazier (1971).

parameter, but it has been shown by Frazier and Stenflo (1972) that the magnetic field strength (the entire distribution of field strength) is the same for all faculae. Therefore, "strong" faculae differ from "weak" faculae by virtue of the total magnetic flux they contain, not the field strength. The area of a facula changes in accordance with the flux. From this point of view then, the terms, "strong" or "weak" faculae should be replaced by "large" or "small" faculae. The radius of a facula would be an equivalent parameter to use instead of the magnetic flux. However, the magnetic flux can be easily measured by a magnetograph, and the radius cannot be easily measured, because it is usually less than the resolution element of the telescope.

Therefore, the task is to observe a family of line profiles (including the continuum), with each profile identified by its corresponding magnetic flux. Operationally speaking, there are two different ways in which one can obtain this data: One can either take the spectrum (plus a magnetic measurement) of one facula after another until a large number of faculae have been observed, or else one can scan a large area (i.e., many faculae) at a set position within a line profile ($\Delta\lambda$), then change the $\Delta\lambda$, repeat the scan, and so on until many $\Delta\lambda$'s have been observed. Both techniques yield a brightness measurement as a function of $\Delta\lambda$ and Φ. Since the faculae and pores are being identified only by their magnetic flux, and not by e.g. their location, or their age, the two different techniques produce equivalent data. The second technique is, however, much easier, because it is easily adapted to the operation of a magnetograph. The present observations use this second technique.

2. THE OBSERVATIONS

All the observations were made with the Kitt Peak multi-channel magnetograph. The same pair of lines were used in this study as were used earlier by Stenflo (1973); Fe Iλ 525.022 nm and Fe I 524.706 nm. Four different exit slit geometries were chosen to sample the line profile. The exit slits were always 2.5 pm wide, and were centered successively at $\Delta\lambda$ = 2.25, 4.75, 7.25 and 9.75 pm from the line center. Four separate raster scans were made of each region that was observed. The only instrumental parameter that was changed between the four scans was the position of the exit slits. The magnetograph was calibrated every time the exit slit positions were changed. The scanning aperture was always 2.4 × 2.4 arc sec, and the size of the raster scans were usually 160 × 160 arc sec.

The observations were made during the period of May 4, 1974 through May 11, 1974. For each absorption line, 4 quantites were recorded; the longitudinal component of the magnetic field and of the velocity, the brightness in the wing, and the brightness in the core. Additionally, the continuum brightness was recorded. Both the magnetic channels and the velocity channels were calibrated by standard techniques, using a drift scan across the disk of the sun. In the subsequent data reduction, the magnetic channels were calibrated in units of giga Webers (GWb). The brightness channels were not calibrated in terms of the disk center continuum intensity at the time the observations were made. So all the brightness measurements were simply transformed to contrast units, $\Delta I/I_{ph}$, where I_{ph} is the brightness in neighboring non-magnetic regions.

3. STATISTICAL CONTRAST PROFILES

For each available brightness channel of each raster scan, a scatter plot was made of the observed contrast as a function of the observed magnetic flux. Such scatter plots are similar to that shown in Figure 1. These points were then averaged over 10 GWb intervals of the observed magnetic flux. This results in a series of tables which contain the observed contrast as a function of $\Delta\lambda$ and Φ.

One correction must be made to these tables. The observed flux at a given $\Delta\lambda$, say $\Delta\lambda = 2.25$ pm, is less than the observed flux at another $\Delta\lambda$, say $\Delta\lambda = 7.25$ pm because, when the magnetograph exit slits are set at the narrower $\Delta\lambda$, the instrument suffers from more Zeeman saturation. It would be best to evaluate the total Zeeman saturation for each exit slit setting ($\Delta\lambda$) and to correct the observed magnetic flux at that $\Delta\lambda$ to the true magnetic flux. However, this would require the use of a specific facular model to calculate the distribution of field strength of faculae, and it is desired to keep this data model-independent. There is a way to correct for the differential Zeeman saturation between the various $\Delta\lambda$'s in a completely empirical manner. This allows us to place all of the data on a common scale of magnetic flux which is internally self-consistent, but which may be different from the true magnetic flux due to Zeeman saturation. This empirical flux correction procedure is based on the fact that the same continuum contrast of faculae is always measured for each different exit slit setting. Each scan then produces one curve of continuum contrast as a function of observed magnetic flux, (for example, Figure 1b) and all such curves should be identical to each other. The differential Zeeman saturation is revealed by a change in the horizontal scale for different $\Delta\lambda$'s. It is a simple matter then to renormalize these flux scales to the flux scale from a chosen standard $\Delta\lambda$. $\Delta\lambda = 4.75$ pm was the chosen standard.

With this correction, one has the desired family of line profiles. These profiles are somewhat unorthodox in that the observed points in any given profile do not come from a single feature. Instead, each point is an average measurement over many features, all containing the same amount of magnetic flux. For this reason, these profiles might best be termed "statistical contrast profiles". The relatively low spectral resolution (2.5 pm) is the result of using a magnetograph to measure the profiles instead of a spectrometer. This is a disadvantage which one must contend with in return for the advantage of a simultaneous magnetic flux measurement at every point.

One of the regions scanned near the center of the disk had a significantly higher proportion of large faculae and pores, so the results from that region are presented in Figure 3. Large faculae were observed in sufficient number up to a flux level of about 150 GWb.

4. CONCLUSION

The single principal conclusion of this paper is obvious from Figure 2. The shape of a line profile, and therefore the $T(T)$ relation, of faculae changes siginficantly and continuously from small faculae to pores. This conclusion carries with it two very important implications. The first implication is that it should be possible to calculate a generalized model of a "magnetic feature" which explicitly contains the magnetic flux as the fundamental parameter. This model would then be capable of reproducing the appearance of both faculae and pores just by changing the amount of flux.

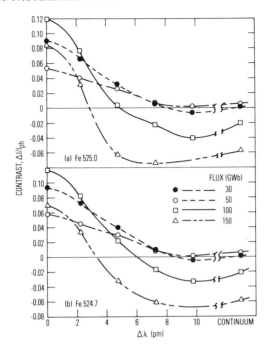

Figure 2. Observed statistical contrast of faculae and pores. These observations were made of an active region at $\mu = .91$. The contrast profiles are labelled by the observed facular magnetic flux. (a) Fe Iλ 525.022 nm. (b) Fe Iλ 524.706 nm.

The second implication is that past observations of line profiles of faculae have suffered from observations selection. In making facular observations, it has generally been necessary to perform some kind of a seach procedure first (e.g. search for the brightest Hα plage, search for the highest intensity in the core of the desired line, etc.). The very fact that a search process is used means that one member of this famicly of profiles has been selected preferentially. If different observations have used different search techniques, the selection effects will also be different, and it will be very difficult to compare the observations with each other, or with any given model. For example, continuum contrast observations tend to select those faculae which are bright in the continuum (small Φ), whereas line profile observations tend to select those faculae which are bright in the cores of lines (large Φ). Therefore models based on these two different types of observations refer to different types of faculae, and they should not be expected to agree with each other.

ACKNOWLEDGEMENTS

The data reported in this paper was obtained at Kitt Peak National Observatory. The author wishes to express his appreciation to Dr. J. Harvey and Mr. B. Gillespie for their invaluable assistance in operating the McMath telescope and the multi-channel magnetograph. Extensive programming assistance was given by Mrs. T. Becker. This project was supported by The Aerospace Corporation company-sponsored research program.

REFERENCES

Chapman, G. A.: 1976, Astro. Phys. J., in press.
Frazier, E. N.: 1971, Solar Phys. $\underline{21}$, 42.
Frazier, E. N. and Stenflo, J. O.: 1972, Solar Phys. $\underline{27}$, 330.
Simon, G. W. and Weiss, N. O.: 1971, Solar Phys. $\underline{13}$, 85.
Stenflo, J. O.: 1973, Solar Phys. $\underline{32}$, 41.

FACULAR MODELS, THE K-LINE, AND MAGNETIC FIELDS

G.A. CHAPMAN
Space Sciences Laboratory, The Aerospace Corporation

Photospheric faculae are now believed to be closely associated with the small-scale solar magnetic field. In order to obtain reliable observations of solar magnetic fields, one needs to have a good description of faculae, which is presently lacking. The problem in obtaining good observational data is severe because faculae are usually not spatially resolved, particularly so in the case of spectroscopic observations. Proper use of spectroscopic observations also requires some knowledge of solar velocity fields and atomic physics in the case of a non LTE analysis. Many of these problems can be avoided by making use of the wings of the Ca II K-line. The wing of this line is unaffected by magnetic and velocity effects. The formation of the line has become increasingly well understood and most of the wing (with the exception of the inner 1 - 2 Å) is formed in LTE. The line is so strong that its formation spans the whole depth of the photosphere.

For some time we have been making calculations of the K-line wing for a variety of facular models. These calculations have assumed LTE for the source function and the ionization state. We have determined the damping constant by matching the profile calculated for the HSRA model (Gingerich et al., 1971) with the observations of White and Suemoto (1968). This procedure is quite similar to that followed by Shine and Linsky (1974) in their study of facular models. Figure 1 shows line profiles obtained by these techniques with the observations of White and Suemoto indicated. The calculated continuum intensity for the HSRA is 2.16×10^{-5} erg s^{-1} cm^{-2} Hz^{-1} ster^{-1}. The absorption coefficient is assumed to be quadratic of the form (Aller, 1963)

$$\frac{a(\nu)}{a_o} = \frac{\delta' \Delta\nu}{\pi^{1/2} (\nu - \nu_o)^2} \qquad (1)$$

where ν_o is the line center frequency, $\Delta\nu$ is thermal and turbulent broadening, and $\delta' = (8.6 \times 10^7 + S)/2\pi$. The quantity S is given by

$$S = \pi \left(\frac{3\pi^2}{4}\right)^{2/5} N V^{3/5} C_6^{2/5} \qquad (2)$$

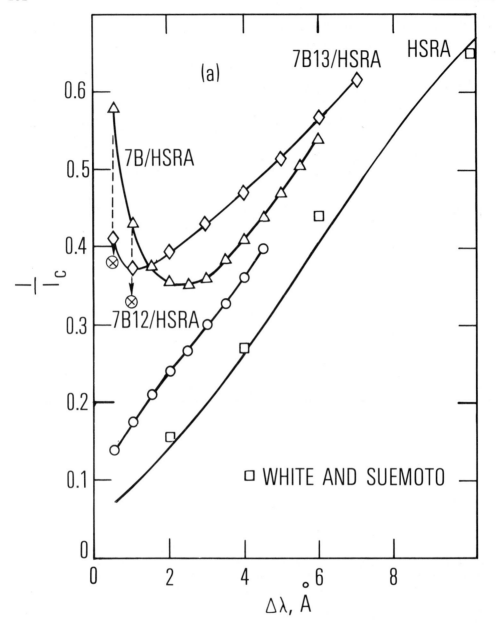

Figure 1. Profiles of the Ca II K-line for various facular models. The circled x's show the effects of partial redistribution for facular model 7B/HSRA.

where N and V are the Cgs number density and velocity of the perturbing atoms. For an assumed Ca/H abundance ratio of 2.4×10^{-6}, we obtained $C_6 = 2.5 \times 10^{-32}$.

The ideal observational data would be spatially resolved spectra or ultra-narrow band filtergrams at clean windows in the K-line wing. Such observations have not been obtained to the authors knowledge. The best available data are the broad-band filtergram contrast measurements of Mehltretter (1974). These observations were obtained with an interference filter having a FWHM of 16 Å centered on the K-line. In order to compare calculated and observed contrasts we have convolved the profiles of Figure 1 (assumed symmetric about $\Delta\lambda = 0$) with a Lorentzian filter transmission of 16 Å FWHM. Each normalized profile has been rescaled by the appropriate continuum intensity. The resultant filtergram contrast, $\Delta F/F$, is given in Table 1 where

$$F = \int_{-\infty}^{\infty} T(\lambda) I(\lambda) d\lambda \tag{3}$$

with $\quad T(\lambda) = \left[1 + (\Delta\lambda/\Delta\lambda_{HWHM})^2\right]^{-1}. \tag{4}$

Table 1

K-line filtergram contrasts

facular model	7B13/HSRA	7B12/HSRA	Stenflo(1975)	observed
contrast	0.324	0.261	0.173	0.65±0.31
contrast with $I_{K_{232}} = I_c$	0.382	0.325	0.238	

The effects of partial redistribution on K-line profiles of facular models, discussed by Heasley et al. (1976), do not significantly alter these results.

Faculae are hotter than their surroundings in the upper photosphere because the associated magnetic field is involved with mechanical energy dissipation. The actual run of temperature and density with depth in the facula is related to the behavior of the magnetic field with depth as clearly shown by Dicke (1970). Thus we have a potentially powerful tool for testing facular models by (a) obtaining the depth variation of $\delta\rho$ and δT, assuming hydrostatic equilibrium from observations in the K-line, (b) calculating a self-consistent magnetic field from the run of $\delta\rho$ and δT, and (c) comparing the computed magnetic field with magnetic fields observed simultaneously with the K-line. We are presently analyzing such data obtained at Kitt Peak National Observatory with the help of Gillespie, Livingston and Lynch. A more thorough analysis will appear in the near future (Chapman and Lynch, 1976).

References

Aller, L. H. 1963 Astrophysics : The Atmospheres of the Sun and Stars, 2nd ed. Ronald Press, New York.

Chapman, G.A. 1976 Ap. J. Suppl., in press.

Chapman, G.A., and Lynch, D.K. 1976, in preparation.

Dicke, R.H. 1970 Ap. J. 159, 25.

Gingerich, O., Noyes, R.W., Kalkofen, W., and Cuny, Y. 1971 Solar Phys. 18, 347.

Heasley, J.N., Kneer, F., and Chapman, G.A. 1976, in preparation.

Mehltretter, J.P. 1974 Solar Phys. 38, 43.

Shine, R.A., and Linsky, J. 1974 Solar Phys. 37, 145.

Stenflo, J.O. 1975 Solar Phys. 42, 79.

White, O.R., and Suemoto, Z. 1968 Solar Phys. 3, 523.

SMALL MAGNETOSTATIC FLUX TUBES

H.C. Spruit
The Astronomical Institute
Zonnenburg 2
Utrecht, The Netherlands.

In an attempt to interpret the observed properties of small scale magnetic fields at the solar surface, a set of models has been calculated based on the assumption of a magnetostatic equilibrium. The basic assumptions made are:

i. The observed magnetic elements are magnetostatic flux tubes.
ii. The efficiency of convective heat transport inside the tube is reduced with respect to that in the normal convection zone; the *horizontal* convective heat transport in the tube is suppressed completely by the magnetic field.
iii. Close to the tube, horizontal convective heat transport is reduced due to the proximity of the magnetic field.

In addition to these assumptions, a number of approximations has been made in the calculations:

1. Energy transport (both by radiation and by convection) is described by a diffusion process with anisotropic, inhomogeneous coefficients. The diffusion coefficients are fixed in advance, taking into account some of the properties (notably the Wilson depression) of the expected solution.
2. The tube is axially symmetric, perpendicular to the solar surface.
3. The magnetic field of the tube is a potential field, bounded by a current carrying outer surface.
4. The heat flux inside the tube, flowing from the deeper layers to the surface, is specified; it is estimated by comparison with the flux of large tubes (sunspots). (This is necessary, because the mechanism of heat transport in a magnetic field is not known sufficiently to derive the appropriate diffusion coefficients theoretically).

The consequences of the magnetostatic character of the tubes shown by these models, are:

a. The tube shows a Wilson depression, like a sunspot does. The value (z_w) depends on the magnetic field strength (at a specified level), and (indirectly) on the radius of the tube. For a given field strength, z_w increases with the radius (R) of the tube.

Some representative cases are:
- Facular point (R = 50 km, B = 1700 G): z_w = 90 km
- Pore (R = 500 km, B = 2000 G): z_w = 260 km

b. Because the interior of the tube is more transparent than the normal convection zone, heat flows laterally into the tube. The depth range over which this influx is important is somewhat larger than z_w. The amount of influx depends on the details of the assumed convective heat flow in the convector close to the tube, on the Wilson depression and on the radius of the tube. If the ratio $2R/z_w$ is of the order of unity, the influx can be comparable in magnitude to the normal solar flux through an area corresponding to the cross section of the tube

c. Through the lateral influx effect, small tubes channel an additional heat flux through the solar surface, i.e. the lateral influx is not a completely local effect; it also taps the heat reservoir of the whole convection zone. Representative values of the additional heat flux F_e caused by the tube (averaged over its cross section) are:
- Facular point: $F_e \approx 0.6$
- Pore: $F_e \approx 0.15$
- (spot: $F_e \approx 0.0$)

d. Geometrical (projection) effects dominate the CLV of the tubes seen in the continuum. Seen near the solar limb, the walls of the tubes can contribute strongly to the contrast of the observed features. The temperature of the wall is not large (5900-6200 K), but near the limb the wall looks bright compared to the ambient photosphere because its radiation has a different angular dependence. The models suggest that this effect may explain the strong CLV of faculae in the continuum.

e. In spectral lines which are formed at a height higher than $h \approx z_w$ above the continuum, the wall effect (see above: d) is much smaller, because the temperature difference between the tube and its surroundings decreases rapidly with height. The CLV in moderately strong lines, therefore, is expected to be much weaker than in the continuum. This fact may indicate an explanation of the observed behaviour of facular contrast in lines.

FLOWS IN MAGNETIC FLUX TUBES

C.J. Durrant
Fraunhofer-Institut, Freiburg

Harvey has reported in this discussion observations of mean downflows in magnetic elements that increase in speed with depth. It is therefore pertinent to ask what role such flows may play in the production and maintainance of the flux concentrations.

EQUILIBRIUM MODELS

The first thing to be said is that such flows cannot directly influence the horizontal momentum balance, i.e. their dynamic pressure is negligible.

A steady internal flow such as that proposed by Ribes and Unno (1975) for chromospheric concentration would require supersonic flows with very large curvature at photospheric levels. There is no evidence for transverse flows of such magnitude.

The rejection of such possibilities implies the rejection of the mechanism proposed by Stenflo (1973) inspired by the general argument of Sreenivasan (1973). This envisages the production of a magnetic field with energy density greater than the equipartition value by evolving a force-free field as some sort of 'response' to a particular velocity field, a so-called hydromagnetic Beltrami flow, that maintains the force-free character of the field. However, by definition, the local velocity field can have nothing to do with the increase of magnetic energy as the concentration occurs. The crucial point to note is that, in an isolated fluid mass, the only field that is force-free everywhere is the trivial case in which it is identically zero. Any increase in magnetic energy in a force-free region is due entirely to the work done at the boundary and the energy transport across the boundary in response to work done in the external, non-force-free, region.

In the absence of dissipation, the rate of change of magnetic energy M in a force-free region V, bounded by the surface S is

$$\frac{\partial M}{\partial t} = \frac{\mu}{4\pi} \int_V \underline{v} \times \underline{H} \cdot \nabla \times \underline{H}\, dV - \frac{\mu}{4\pi} \int_S \underline{H} \times (\underline{v} \times \underline{H}) \cdot d\underline{S}$$

$$= -\frac{\mu}{4\pi} \int_S H^2 \underline{v} \cdot d\underline{S} + \frac{\mu}{4\pi} \int_S (\underline{H} \cdot \underline{v}) \underline{H} \cdot d\underline{S}$$

                flux of magnetic   rate of working
                energy across boundary   at boundary

Concentration arises through changing surface stresses. Naturally there exist pathological internal flow fields that maintain the force-free nature of the field as it evolves. Perhaps the simplest example is that of a squeezed uniform field

$$\underline{B} = (0,\ 0,\ v_o B_o t)$$
$$\underline{v} = (v_o x,\ 0,\ 0)$$

The magnetic energy and flux within fixed boundaries at $x = \pm X$ continually increase but the configuration remains force-free within the boundaries. Work is done in pushing the field across the boundary.

    However, in general, the internal flow is governed by an energy equation not by the maintainance of a force-free state. If the magnetic energy density is greater than the flow and thermal energy densities, any magnetic stresses will tend to be transferred by Alfven waves to the boundaries and an almost-force-free state will arise. On the other hand, there is no guarantee that the flow will even approach the Beltrami-type since relatively small departures from a force-free magnetic field configuration will cause large changes of pressure.

    Thus we conclude the only sufficiently general and uniform force field that could maintain a magnetic concentration is the external thermal gas pressure. This allows balancing with

$$\frac{B^2}{8\pi} \sim P_{go} \lesssim P_{gi}$$

This reduced internal pressure further implies that in the observable layers the field must be almost force-free except at the boundary of the tube. The field lines of such a thin flux tube will be almost straight in the low photosphere.

    However, flows cannot be ruled out if the density in the tube is small enough and the tube narrow enough for mass balance in the atmosphere to be no problem. In this case the stratification need not even be close to hydrostatic along the field lines.

It may be argued that pores and sunspots show no great departures from these conditions so that in the absence of physical inconsistency we should first examine the possibilities of the simplest static model, an approach which has been successfully pursued by Spruit and described in this discussion.

STABILITY OF MODELS

Flows might not be necessary in order to construct equilibrium models but there is an implicit suggestion that they might be necessary for stability. For example, linear stability analysis shows that an infinite, twisted tube embedded in a conducting, incompressible medium is unstable in the absence of flow, but that it can be stabilized if the flow energy exceeds the magnetic energy (Chandrasekhar, 1961).

The equipartition solution in which flow along field lines has an energy density equal to the magnetic energy density is always stable. Parker (1976a) has shown that this case has the interesting property that a slow systematic variation of the pressure around a particular thin flux tube causes the internal conditions to evolve close to successive equipartition solutions. Pressure changes communicated to the tube interior can cause the flow to accelerate, the field to increase and the tube to shrink (in order to conserve flux). Presumably any final concentration thus produced is stable.

No such simple solution is possible in a stratified medium. The demands of both flux and mass conservation are incompatible with it. Instead, Parker (1976b) shows that in a one-dimensional approximation to a very thin flux tube, only one type of steady flow can exist that stretches from the top to the bottom of the atmosphere. In this flow, a neck occurs at the point where the scale height gradient is equal to $-1/2$ and only there are the flow and magnetic energies equipartitioned. In the solar atmosphere this occurs at the top of the convection zone. But this result seems of doubtful relevance to the present discussion for the following reasons:

1. Existence theorems based on first-order terms of expansions should be regarded with suspicion especially when such terms cannot be small everywhere.

2. Parker does not demonstrate that the configuration is stable nor that it can evolve naturally from mild to strong concentration.

3. The process breaks down when the flows approach the sound speed and compressibility effects must be taken into account. This

limits the degree of pressure reduction since

$$\frac{B^2}{8\pi} \sim \tfrac{1}{2}\rho v^2 < \tfrac{1}{2}\rho c_s^2 \simeq p_{gi} \quad \text{and} \quad \frac{B^2}{8\pi} + p_{gi} \simeq p_{go}$$

and hence $\quad p_{gi} > \tfrac{1}{2} p_{go}, \quad \dfrac{B^2}{8\pi} < \tfrac{1}{2} p_{go}$

whereas observations imply a greater reduction which can be achieved by limiting the energy input. The less efficient mechanism is unlikely to stabilize the more efficient one.

These dynamic considerations are only essential so long as one accepts the premise that simple hydrostatic solutions are unstable. The recent analyses of Meyer et al. (1977) and Wilson, reported here, show that axial cooling of sunspots and, by extension, smaller flux tubes, is sufficient to guarantee stability. There is then no compelling theoretical requirement for steady flows in flux tubes.

On the other hand, it might be argued that strong, steady flows are unlikely. If material flows out of both ends of a tube, either into subsurface regions or into the corona, the tube would be rapidly evacuated and then what? Giovanelli (1976) has suggested that the ionization in the temperature minimum regions is low enough to allow a leakage of neutral material across the lines of force and into the flux tube, but it is doubtful whether such a mechanism can supply flows of the quoted speeds.

We must therefore conclude that our present understanding of the physics of magnetic flux elements does not rely on the presence of downdrafts within the elements. They should be regarded as a transient detail of our model and not an intrinsic feature.

References

Chandrasekhar, S.: 1961, Hydrodynamic and Hydromagnetic Stability, Oxford Univ. Press, Oxford, p. 551 ff.

Giovanelli, R.G.: 1976, Private communication

Meyer, F., Schmidt, H.U., Weiss, N.O.: 1977, submitted to Monthly Notices Roy. Astron. Soc.

Parker, E.N.: 1976a, in preparation

Parker, E.N.: 1976b, in preparation

Ribes, E., and Unno, W.: 1975, Proc. First European Solar Meeting, Feb. 1975, Firenze, p. 52

Sreenivasan, S.R.: 1973, Physica 67, 323, 330

Stenflo, J.O.: 1973, Solar Phys. 32, 41.

DISCUSSION

A part of the discussion was devoted to further clarification of some instrumental and observational aspects. It became clear that the assumption of the existence of strongly concentrated fields in the photosphere does not remove entirely the difficulties of interpreting magnetograph signals. According to Semel it appears at present impossible to derive consistent models of this magnetic field when taking a large number of spectral lines measured simultaneously into account. It was also debated by Severny and Wiehr, whether the calibration curves used in the line ratio procedure should be derived from measured or rather from computed line profiles.

The question for the minimum observed size of magnetic structures raised by Newkirk remained unanswered. Recent attempts to derive this quantity directly by observation yielded no information beyond the 1" limit. Spectra taken in the 6302.5 line at the Sac Peak vacuum tower by Koutchmy show some magnetic features in the 1 kgauss range definitely larger than the filigree structures observed simultaneously near the photosphere. The Crimean observers plainly question the existence of any magnetic structures in the photosphere smaller than 1.5" on the grounds of measurements obtained during a transit of Mercury, which was used as an occulting disc; these observations failed to show any sudden jumps in the field strength averaged across the $(1")^2$ aperture. They also find that the short magnetic diffusion time ($\sim 10^3$ s) is at variance with the existence of structures of the order of 100 km diameter (Severny).

The difficulties encountered in deriving facular models, consistent with the center to limb variation of the facular contrast in various lines as well as in the continuum was pointed out by Wiehr.

Finally the question of velocity fields associated with the small scale flux concentrations was discussed again. Downward velocities measured in the chromosphere are typically 0.5 km s^{-1} accelerating to ~ 2 km s^{-1} in photospheric metallic lines. It is unlikely that the peak velocity of the downdrafts occurs within the flux tube because this would immediately evacuate the chromospheric structure extending above the flux element. A motion picture shown by Giovanelli, based on a series of circular polarisation filtergrams taken in the CaI 6103 line clearly demonstrated average downward motions in the polarized structures as well as superimposed vertical oscillations with a period of ~ 5 min.

In discussing the role of convection for the generation and confinement of the magnetic field, Frisch stressed the importance of intermittency effects in turbulence, while Souffrin pointed out that we should expect a hierarchy of sizes for the magnetic structures.

Syrovatsky remarked that the Boussinesq approximation is not useful in the case of strong magnetic fields, and drew the attention to the importance of acoustic MHD waves.

Several models to explain the small-scale structure of the fluxtubes were proposed:
Lynch has used an approach developed by Dicke to derive a grid of fluxtube models corresponding to various positive (faculae, network) or negative (sunspots) brightness contrasts in the continuum. The models will be used later for more detailed fits with recent observations.

Wilson considered twisted and untwisted magnetostatic fluxtubes. In the isothermal case the untwisted tube is unstable to the exchange instability and may split up into smaller tubes. However, when the temperature and granulation features are taken into account the fragmentation does not continue indefinitely, but the total energy is minimized for a flux tube of finite size. With sufficient cooling the overall energy can be minimized in a single structure rather than in a subdivided configuration.

Nordlund proposed a mechanism, which explains the cooling inside the fluxtube as a result of the almost adiabatic motion of the downdrafts. In his model, a net negative buoyancy along the field lines is produced, which sustains the downdrafts.

CONCLUDING REMARKS

H. U. SCHMIDT
Max-Planck-Institut für Physik und Astrophysik
München

Recent interest in the subject of this discussion gained momentum by an observation which did not concern magnetic fields directly. The filigree which Dr. Richard Dunn on Sacramento Peak found 2 Angstrom off the center of H_α is a bright and crisp structure in the photosphere with a width of 1/5 arcsec. It was described in proper detail by Dunn and Zirker (1973). Even in the printed pictures in their paper one clearly sees one step beyond the solar granulation. The filigree is certainly related to the small scale structure of the photospheric magnetic field, but it is not yet clear whether the flux elements are exactly cospatial and have the same small dimensions. Simon and Zirker (1974) concluded from spectra that the field structure is wider than the filigree. On the other hand Harvey (1976) in his excellent review of the observations has also presented the arguments of several authors who conclude that the sizes of the flux elements are as small as those of the filigree. This discrepancy certainly needs further study before such even more delicate questions as the spatial extent of the downdraft inside and around the flux elements can be reliably answered from observations. The theoretical interpretation of the downdraft depends on this answer as different sources of the mass flow are involved: the overlying atmosphere and the convergent massflow of the surrounding convection. The latter stays partly outside the flux element, partly diffuses into it with an efficiency that might be enhanced by convection of still smaller scale.

Can the population of discrete flux elements be described as a family of axisymmetric configurations of meridional fieldlines with the flux as the sole parameter? Though this question has to be anwered by continued observations, there are at present no obvious other parameters which do or should interfere efficiently with such a description and we have heared e.g. Dr. Weiss clearly denying the need to invoke a twist in the theoretical description of any flux element during the discussion after his presentation. But does this family of flux elements have members all the way down to a radius wth optical depth unity in the horizontal direction or where else does it cease? This question is obviously related to the evolution of these elements. There must be a number of observable phenomena which constitute or are closely related to the steps of this evolution. The elements may originally appear in the photosphere either by vertical transport of compact fluxropes from below or by convective concentration of previously dispersed flux. Weiss (1976) in his review gave a beautiful quantitative description of the latter process which turns out to be extremely powerful. He derives a concentration of order square root of the magnetic Reynolds number way beyond the equilibrium between ram pressure and magnetic pressure. The only limitation he finds is set by the gas pressure of the surrounding photosphere. After its formation the element may migrate over the solar surface in a random walk forced by convection, it may coagulate with others of equal polarity into a larger element, it may coagulate with opposite polarity and disappear from the photosphere by reconnection and retraction up and down, and it may become unstable and fall apart into smaller elements or transmute temporarily into totally dispersed flux, e.g. at the lower end of the range of fluxes. But it may not retract below the photosphere together with all the connected atmospheric flux including the corresponding opposite photospheric flux element since there is neither sufficient energy nor sufficient coherent downward motion in the observable flow pattern above the solar photosphere. There seems to be no alternative to reconnection very near to the photosphere as the final fate of any photospheric flux. On the other hand it is an open question how often and how long any compact flux changes into a totally dispersed state.

There are also several different possibilities to define a lifetime of a flux element. The time it takes for an element to migrate far enough to escape observable correlation with its origin may be smaller than the time it takes to fall apart or to disperse or to coagulate with equal polarity and those times may be still much smaller than the time it takes to disappaer completely from the photosphere by reconnection and it may take another length of time til any flux element is connected into interplanetary space by the solar wind. Such types of processes and their timescales must be deduced from observation in order to construct a consistent picture of the evolution of photospheric flux elements.

In all these necessary investigations observations of high accuracy are needed and such accuracy can only be proven by tests. Some such tests may be provided from theory. The sourcefree character of the magnetic field may be used for such a test if the net flux is integrated over an observed area as a function of time. This net flux will not exactly vanish initially but it should change with time only by migration of elements across the boundaries and by emerging flux which appears on the boundaries themselves. Another example is the limitation of the photospheric field at the surface of an flux element by the photospheric gas pressure.

In closing let me emphasize my enthusiasm for the new observations we have seen. They have shown some exciting new land beyond the arcsecond. We ought to conquer it. And if I look at some of the new tools which are now forged in the numerical treatment of convection and radiative transport at intermediate optical depth, I think we will.

References

Dunn, R. B. and Zirker, J. B.: 1973, Solar Phys., 33, 281.

Harvey, J.: 1976, these proceedings.

Simon, G. W. and Zirker, J. B.: 1974, Solar Phys., 35, 331.

Weiss, N. O.: 1976, these proceedings.

JOINT DISCUSSIONS NO.7

THE IMPACT OF ULTRAVIOLET OBSERVATIONS ON SPECTRAL CLASSIFICATION

(Edited by L. Houziaux)

Organizing Committee

L. Houziaux (Chairman), W. P. Bidelman, A. D. Code, C. Jaschek, L. Nandy.

CONTENTS

7. THE IMPACT OF ULTRAVIOLET OBSERVATIONS ON SPECTRAL CLASSIFICATION

L. HOUZIAUX / Introduction 281

C. JASCHEK / Present Status of Spectral Classification in the Conventional Wavelength Range with Emphasis Upon Early-Type Stars 283

K. NANDY / Spectral Classification of Early-Type Stars from the Low Dispersion Ultraviolet Spectra 289

A. CUCCHIARO, M. JASCHEK, and C. JASCHEK / Spectral Classification of B and A Stars from the Line Features of S2/68 Spectra 303

J.M. VREUX and J.P. SWINGS / Behaviors of B Star Continua and Absorption Features Determined from the TD1 S2/68 "Ultraviolet Bright Star Spectrophotometric Catalogue" and from Copernicus Spectra 305

DISCUSSION 307

R.J. VAN DUINEN and P.R. WESSELIUS / Spectral Classification Using ANS Photometric Data 311

K.G. HENIZE, S.B. PARSONS, J.D. WRAY, and G.F. BENEDICT / Spectral Classification with Objective-Prism Spectra from Skylab 315

DISCUSSION 323

A.D. CODE / An Atlas of Ultraviolet Stellar Spectra 325

J. ROUNTREE LESH / A New Temperature Scale for B Stars Based on OAO-2 Data 339

M. LAMPTON, B. MARGON, and S. BOWYER / Extreme Ultraviolet Observations of White Dwarfs 341

R. FARAGGIANA, H.J.G.L.M. LAMERS, and M. BURGER / The Near Ultraviolet Spectrum of Fe III as a Classification Criterion 349

T.P. SNOW, Jr. and E.B. JENKINS / A Catalogue of 0.2 Å Resolution Far-Ultraviolet Stellar Spectra Measured with Copernicus 353

W.P. BIDELMAN / Spectral Classification from Copernicus Data 355

M. HACK, J.B. HUTCHINGS, Y. KONDO, and G.E. McCLUSKEY / The Ultraviolet Spectrum of β Lyrae 361

Y. KONDO / The Mg II Features Near 2800 Å and Spectral Classification 363

DISCUSSION 365

L. HOUZIAUX / Concluding Remarks 367

INTRODUCTION

The idea of organizing a joint discussion on the impact of ultraviolet observations on spectral classification arose during the IAU symposium n°72 held at Lausanne, in the honour of W.W. Morgan. It became evident at that time that the wealth of spectroscopic and spectrometric data already obtained from various orbiting telescopes should shed some new light on the problem of spectral classification. I whish to thank here the members of the Organizing Committee who planned this Joint Discussion : W.P.Bidelman, A.D.Code, C.Jaschek, and K.Nandy. I also thank the authors of invited reports and contributed papers, as well as those who participated in the discussions.

L. HOUZIAUX
Chairman of the Organizing Committee

PRESENT STATUS OF SPECTRAL CLASSIFICATION IN THE CONVENTIONAL
WAVELENGTH RANGE WITH EMPHASIS UPON EARLY-TYPE STARS

C. Jaschek
Observatoire de Strasbourg

A spectral classification system is a morphological device for
arranging in order a finite series of observations of individual
objects. One has thus three elements: an "order", i.e. an abstract
discontinuous scheme, a series of objects and a certain type of observation (spectrogram) that is performed on all objects of the sample.
The "order" is in principle arbitrary; one could think for instance of
an order based on richness of lines, or on the presence of hydrogen
lines, etc. However such an idealization is clearly simplistic, because
it would imply that the classifier is completely unaware of theory -
but he is not, if for no other reason than simply that he would never
get his PhD. Thus one should add the constraint that the scheme one
adopts be physically sound.

The advantages of spectral classification are easy to point out.

a) One gets a quick description of the object which provides a set of
parameters upon which further analysis can be based.

b) One can quickly single out objects having unusual characteristics.

c) One gets a luminosity class and thus the possibility of obtaining a
distance.

It is probably convenient to quote some figures which bear upon
the "quickness" of the method. Let me start by pointing out that from
a purely instrumental viewpoint, with a 100 cm telescope one gets a 7^m5
star at 120 Å/mm dispersion in about 10 minutes. Since longer exposures
become rapidly prohibitive and also since very few big telescopes are
allocated for spectral classification, one can set the practical limit
of spectral classification, on slit spectrograms, except for special
cases, at m = 9 or m = 10. One can certainly go a step beyond with
objective prism techniques. Here the limit is imposed by the overlapping
of the images and it seems that with 120 Å/mm one cannot go much beyond
the 12th magnitude for an all-sky survey.

Up to the tenth magnitude, there are 3.3×10^5 stars. At this moment there exist MK types for about 45 000 stars (~ 14% of the figure quoted), with a rapid growth in the southern hemisphere. There exist also unidimensional types (Harvard or similar) for about 5×10^5 objects. On the other extreme, high dispersion analyses exist only for about 10^3 stars (Morel et al., 1976).

With regard to photoelectric photometry, which has often been praised as "the" solution, one has at the present time about 5.2×10^4 stars with measured UBV and a much smaller quantity measured with more sophisticated systems.

From all these figures one is forced to conclude that spectral classification is still the quickest and most handy way of "classifying" stars and of sorting out objects for later study. It is clear that once an object has been summarily described, its detailed study should be taken up by other means, be it photometry or high-dispersion analysis. In this sense spectral classification is a first - but vital - step.

If we turn now to the limitations of spectral classification, we find essentially three:

a) Since the scheme is discontinuous, whereas nature produces a continuous variation of parameters, there exists a certain amount of built-in random error. To this error one adds other personal errors, coming from the fact that the classification is done by estimation and not by measurement.

b) It has become evident that a two-parameter classification system is insufficient and that at least a third parameter, related to chemical composition, has to be used.

c) Since the observations upon which the system is based are obtained in a certain wavelength range and with a certain dispersion and/or resolution, there exists no a priori reason why the system should be equally applicable to a different dispersion and a different wavelength region.

Let us examine next each of the three limitations just mentioned.

a) PRECISION

Probably the best studied limitation is the one related to precision. This has been the subject of several published and unpublished analyses and I quote simply the gross result. A spectral type said to be given in the MK system and taken from the literature is accurate within one tenth of a spectral class and 0.6 luminosity classes. If the classifications of only a single author working under the best conditions are considered, these precisions can be increased by a factor of less than 2.

Since obviously for general purposes one is practically always obliged to rely upon the work of different individuals, it seems that the precision quoted here is the one to be used in practice. Let us note in passing that although the MKK and MK system was defined with spectrograms giving 120 Å/mm at Hγ, the new MK system, as specified by the dagger types (Morgan and Keenan, 1973) is based upon 84-Å/mm (range O5-B9) and 125-Å/mm (A3-G2) spectrograms, both taken with grating spectrograms (range λλ 3600 - 4800). For these reasons not all refinements introduced recently can be seen on 120-Å/mm plates.

Coming now to the comparison of the errors of spectral classification with other classifications, for instance photometric ones, let me stress one essential point, namely that as a general rule it is much more difficult to get luminosities photometrically than spectroscopically, so that for most stars of the galactic field the spectroscopic luminosity is the only luminosity easily obtainable. This said, it is understandable that photometry will be superior to visual classification when it comes to compare order on the main sequence. In early-type stars this can be clearly seen, for instance in the region B6-B9 where the classifier working in the conventional range has very few spectral lines to hang on, and where consequently almost any photometry will be of equal or superior precision. In regions where there are several lines available, spectral classification is comparable to photometry. (See for instance Jaschek and Jaschek, 1973).

b) THE THIRD PARAMETER

Although the question of a third parameter, related to stellar composition, is mentioned in the introduction to the MKK Atlas, the idea was accepted only much later. The main opposition came apparently from the stellar atmospherists, who maintained the uniform composition of all stellar atmospheres. A real change came only when Baade's ideas concerning stellar populations made their full impact.

The first place where a third parameter was needed was in those late-type stars whose metallic lines can be either weak or strong. Starting with late-type stars (later than F), an increasing number of peculiarities was then found in all types of objects. Practically all of them are related to abnormal line strength. These peculiarity groups have been discussed many times in the literature; recently I gave a sumpary in IAU Symp. 72 (Jaschek, 1976). Table 1 summarizes the main peculiarity groups of the early-type objects (O - F). I would like to add that not all classes are recognizable at 120 Å/mm; for instance some Ap stars require 40 Å/mm or less and the CNO stars 60 Å/mm.

It is likely but not obligatory that the anomalies are linked to abundance anomalies. Other explanations can be put forward for at least some groups, based upon radiative pressure, magnetic field effects, gravity diffusion, surface nuclear reaction, etc. Let me insist however that our knowledge of the peculiarity groups is still very frag-

TABLE 1

Peculiarity types

Spectral type	Peculiarity type implying a third parameter
O	O-type subdwarfs WR
B	CNO stars He-strong stars He-weak stars B-type subdwarfs
A	Ap Am λ Boo δ Scu
F	Fm stars Stars with weak metal lines

mentary and that in consequence interpretations are even more problematic. Let me show this with just one example.

We have come to accept the idea that in late-type stars it is possible to define a metal abundance - the "z" - which summarizes the abundance of all metals with respect to some conveniently chosen standard. Such a gross characteristic is useful in, for instance, subdwarfs, where the metals are all rather weak. The generalization of a single "z" value, applying equally to all elements in a given stellar atmosphere, runs into difficulties in Am and Ap stars, where very striking exceptions are known. For instance all elements before atomic number A_0 are underabundant and thereafter, up to A_1, they are enhanced by a factor z, whereas after A_1, z adopts the value z_2, etc. Now in B stars, except for a few analyzed at very high dispersion, the situation is still worse. The main reason for this is the lack of metallic lines in the conventional wavelength region. In Table 2, I have specified the only atomic species observable in early-type stars.

The availability of a small number of rather intense lines in early-type spectra has made possible a process of systematization not yet feasible in other spectral types, which consists in posing the problem of how many criteria one should use to define the classification scheme at each place. Since a criterion is specified usually by a line ratio - thus implying two lines of different elements or of the same element in different ionization stages - one could choose in principle any two elements. In O-type stars we have to choose H and He, because nothing else is well visible, but in B-type stars we do have several elements. Walborn (1972) has suggested using He and Si both, because there do exist stars where carbon, nitrogen and oxygen behave abnormally. But He also can be abnormal, as illustrated for example in Table 1 by the He-strong and He-weak stars. This shows clearly that in this case, the idea of the "uniform z" seems to be an oversimplification. But because of the lack of lines of metals in the classical wavelength region, we do not know what happens with the metals, so that it is possible, but not very likely, that the "z factor" is identical for all metals in B-type stars. To solve this problem, we need the ultraviolet region, with its enormous number of metallic lines - provided that one can observe at such a dispersion to overcome the blending problem at least partially.

c) THE OBSERVATIONS

Because of the general acceptance of the MK system, one very important question has been left aside, namely to examine if when looking at other wavelength ranges one can group the stars the same way as in the MK system. The question is generally answered by saying that since the two basic parameters of the MK system are equivalent to temperature and gravity, and these two parameters are completely independent of the region of the spectrum from which they are derived, the analysis of other regions cannot be in contradiction with the MK classification. Such reasoning would be entirely convincing if one could be sure that one knows all the physics of the stellar atmospheres including the numerical values of all parameters. This clearly is not the case - remember for instance the discrepancies between the observed and the predicted fluxes below 3000 Å some years ago.

TABLE 2

Atomic species present in early-type stars regardless of luminosity, visible at 40-120 Å/mm in the region $\lambda\lambda$ 3700 - 4900 Å

Range	Atomic species
O5 - O9	H, He II, (He I)
O9 - B1	H, He II, C III, N III, Si III, Si IV
B1 - B5	H, He I, C II, N II, O II, Mg II, Si II
B5 - B9	H, He I, Mg II, Si II

One could also conceive of a situation where the behavior of one spectral region is dominated by one process which becomes unobservable (or difficult to observe) in other regions. For instance rapid rotation could have such an effect.

In view of such difficulties, it seems best to set up a classification scheme in the ultraviolet region and to examine then if it coincides with the system derived from another piece of the spectrum. If the results agree, one could be a little bit more sure that the basic physics is known; if not, one can start looking for the explanation of the discrepancies.

Probably some of you will feel that in view of all we know about stellar atmospheres, such a view is hardly justified. Let me thus reformulate the problem in other words. In Table 1, I have listed eleven groups of peculiar objects, to which for instance the Be stars could be added, as an example of another well defined group of peculiar stars. If we knew the physics of all the stars very well - this is just an assumption! - we should be able to explain each group of peculiarities in terms of an appropriate set of physical parameters. These could then be used to predict what the star should look like in another wavelength region - for instance in the UV. But this task has not been attempted by anybody yet, and so we are obliged to establish a system of classification in the UV in order to single out the peculiar objects by some "peculiar feature". By the same reasoning nobody can exclude a priori the possibility that new peculiarities might appear in the UV, which differentiate stars that look closely similar in the conventional range. I think thus that one cannot circumvent the necessity of establishing a new classification system in the UV - based upon ultraviolet criteria - and I think that this constitutes one of the most interesting and rewarding future developments in stellar spectroscopy.

BIBLIOGRAPHY

Morel et al. (1976) IAU Symp. no. 72, in press.
Jaschek C. (1976) IAU Symp. no. 72, in press.
Morel et al, (1976) IAU Symp. no. 72, in press.
Morgan W.W. and Keenan P. (1973) Ann. Rev. of Astr. and Astroph. 11, 29.
Walborn N.R. (1972) Ap. J. Suppl. 23, 257.

SPECTRAL CLASSIFICATION OF EARLY TYPE STARS FROM THE LOW DISPERSION ULTRAVIOLET SPECTRA

K. Nandy
Royal Observatory, Edinburgh

SUMMARY

The methods of spectral classification from the low dispersion ultraviolet spectra obtained with the S2/68 experiment in the TD1 satellite have been described. The bright stars, the spectra of which are photometrically accurate, can be divided into natural groups according to the spectral appearance of the features. These features vary in strength with spectral type and luminosity, and enable separation between main sequence and luminous stars. The limits for these stars are $V = 6^m.0$ at B0 to $5^m.0$ at A0. For fainter stars the spectral data have been combined to obtain narrow band magnitudes at several wavelengths. These photometric bands have an effective width of 100 A. An ultraviolet photometric system which enables determinations of spectral type and luminosity of early type stars is described and the results for about 3000 stars is presented. The photometric system considered here consists of the ultraviolet colour indices $(m_{2740}-V,)$ $(m_{2190}-V)$ and $(m_{1490}-V)$.

1. Introduction

The skysurvey telescope (S2/68) in the TD1 satellite (for details see Boksenberg et al 1973) provides stellar energy distributions on an absolute scale in the wavelength range from 1350A to 2350A with a resolution of 30A. In addition the experiment gives a broad band measurement at 2740A ($\Delta \lambda \sim 340A$). The survey has provided a homogenous set of spectral data of a large number of stars distributed all over the sky. The limits are $V = 9^m.0$. at B0 to $8^m.0$ at A0. Each object has been observed at least three times, more if the objects are nearer to the ecliptic pole.

Existing methods of stellar classification are based on line intensities and the colours in the optical wavelength range. In the ultraviolet wavelength region closely spaced atomic transitions of ionized metals (line blocking) occur and also the energy distributions of early type stars reach their Planck maximum. Therefore, the stellar classifications based on the ultraviolet data and their comparison with the classifications from the visible spectra would provide additional information on stellar physical parameters e.g. effective temperature and surface gravity etc. Since the S2/68 spectra have been taken with the same instrument a quantitative classification method will provide a homogenous classification of a large number of stars complete up to the limiting magnitude $V = 9^m.0$. In this paper we shall discuss the classification criteria which can be derived from the spectral features observed in the low dispersion spectra as well as from the observed flux distributions.

2. Spectral features observed in the low dispersion spectra

The absolute stellar fluxes derived from measurements obtained with the S2/68 experiment are based on the results of independent laboratory calibrations performed in Edinburgh and Liege (Humphries et al 1976). For each star a mean spectrum has been constructed from the repeated observations, and the accuracy of the mean spectrum is better than 3% for the bright stars ($V < 6^m.0$, and $E_{B-V} < 0.2$). A number of distinct features appear in the spectra, the strength of which vary with the spectral type and luminosity. Due to the low resolution ($\Delta\lambda = 30A$) the observed features correspond to the blends of a large number of appropriately situated atomic lines.

As a result of the chance accumulations of many faint lines the main contributors of the features depend on the temperature. The strong features which are important for spectral classifications and their main contributors are given in Table 1.

3. Classification criteria based on spectral features.

From a systematic examination of the spectra of about 400 stars brighter than $V = 6^m.0$ in the spectral type range from O to A2 the spectral appearances of the features, their approximate wavelength positions, and their

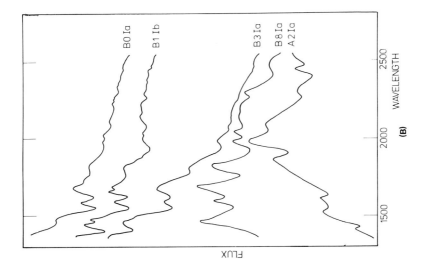
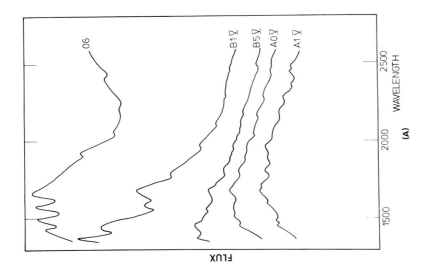

Fig. 1a-b. Typical S2/68 UV spectra of Class V and Class I stars.

dependance on the spectral types and luminosity classes have been studied and are summarised in column 3 of Table 1. For example, Fig 1 shows typical spectra of class V and class I stars. It is to be noted that some features are present only in luminous stars and are suitable for luminosity classification. According to the presence of the features the stars can be divided into natural groups. The range of MK spectral types of these groups and their characteristic features are given in Table 2. These observations yield an empirical two dimensional classification from a visual examination of the ultraviolet spectra; the classification scheme is described in Table 3. For an accurate classification, it is, however necessary to measure the relative strengths of the distinguishing features. For example, the strength of the 1450A feature, which is present only in O stars increases with earlier types.

Houziaux (1976) has found tentatively that the pseudo-equivalent width of this feature is 1A for O9.5V increasing to 5A for O4V. Cucchiaro et al (1976) have used the intensity ratio of the features occurring at 1410A and 1550A, and of the 1620A and 1550A features for a two dimensional classification of the stars in the spectral range from B0 to B8. For later types the strengths of the 2250A and 2450A are useful classification parameters.

The accuracy which is required to measure the relative strengths of the features observed in the low dispersion spectra restricts the classification to relatively bright stars whose spectra are photometrically accurate. The limits for these stars are $V = 6.^m0$ at B0 to $5.^m0$ at A0. For fainter stars the spectral data have to be combined into wider bands to give statistically useful results, and the classification parameters can be obtained from the study of the ultraviolet colours as described in the following section.

4. Ultraviolet spectrophotometry of early type stars.

4.1 Classification from the spectral energy distribution

The basis of classification is that the stars can be grouped according to their intrinsic flux distributions which are determined primarily by their effective temperatures, since the spectral range considered here contains most of the energy of the early-type stars. We have shown earlier that the luminous stars have a flux deficiency increasing with $1/\lambda$ as compared to the main sequence stars of similar spectral types, this probably being due to the luminous stars having lower effective temperatures than the corresponding main sequence stars (Humphries et al 1975, Nandy and Schmidt 1975). Therefore, by proper choice of colours the stars can be separated according to their temperature and surface gravity.

Photometric bands of effective widths of 100A have been constructed centred at 2500A, 2400A, 2300A, 2190A, 2100A, 2000A, 1900A, 1800A, 1700A, 1600A, and 1490A. The fluxes obtained at these wavelengths were converted to magnitudes m_λ, where $m_\lambda = -2.5 \log I_\lambda - 21.1$ (Oke and Schild, 1970) and I_λ is the mean flux at λ in erg s^{-1} A^{-1} cm^{-2}. The photometric error of the ultraviolet magnitude is $\pm 0^m.03$ for stars brighter than $V = 6^m.0$ rising to $\pm 0^m.12$ for fainter stars at wavelengths other than 2190A. Due to large interstellar extinction at 2190A, the uncertainty of m_{2190} can be as high as $\pm 0^m.3$ for moderately reddened stars ($E_{B-V} > 0.4$).

The problem of a two dimensional classification from the stellar flux distributions lies in choosing three colour-indices which fulfil the following requirements:-

(1) The interstellar reddening path and the thermal reddening path in the colour-colour diagram should be well separated,

(2) The separation between the main sequence stars and the supergiants of similar spectral types in the colour-colour diagram, and the change of the colour-indices per unit spectral class should be larger than the photometric errors.

Since the flux deficiency of the luminous stars increases with $1/\lambda$, the effect of luminosity on the ultraviolet colour $(m_{1490} - m_{2740})$ is very large as compared to the colours at longer wavelengths e.g. $(m_{2740} - V)$ (see Humphries et al 1975). We have chosen to use the colour-indices $(m_{2740} - V)$ and $(m_{1490} - m_{2740})$ as earlier results indicated that the first primarily determines the colour temperature (spectral type) while the second is sensitive to both temperature and luminosity. These colours have been corrected for interstellar extinction, using the colour-index $(m_{2190} - V)$ as a reddening indicator. In the intrinsic colour-colour diagram $(m_{1490} - m_{2740})_o$ vs $(m_{2740} - V)_o$ the main sequence stars are well separated from the luminous stars. (Nandy et al 1976). The change of colour index $(m_{2740} - V)_o$ and $(m_{1490} - m_{2740})_o$ per unit spectral class is $0^m.2$ for B-stars.

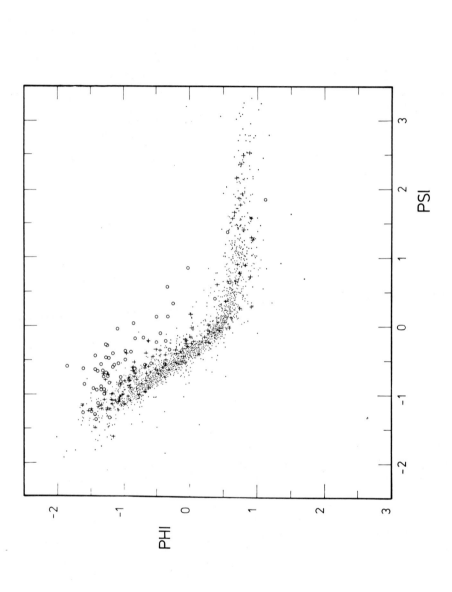

Fig. 2. The plot of PHI vs PSI. Luminosity Class I, III and V are denoted by circles, crosses and dots, respectively.

4.2 Classification parameters

For a two-dimensional classification, two extinction free parameters are defined as follows:-

$$\text{PHI} = (m_{2740} - V) - \frac{E_{2740 - V}}{E_{2190 - 2740}} (m_{2190} - m_{2740})$$

$$\text{PSI} = (m_{1490} - V) - \frac{E_{1490 - V}}{E_{2190 - V}} (m_{2190} - V)$$

The colour-excess ratios have been determined from the mean extinction law derived from the sample of several hundred reddened stars distributed in different galactic regions; for these samples no significant variation from the mean extinction law in the ultraviolet range concerned has been detected (Nandy et al 1976). The mean values of the colour excess ratios are:-

$$\frac{E_{2740 - V}}{E_{2190 - 2740}} \sim 1.1$$

$$\frac{E_{1490 - V}}{E_{2190 - V}} \sim 0.8$$

It is to be noted that the parameter PHI is nearly the colour difference, $(m_{2740} - V) - (m_{2190} - m_{2740})$. The relation between PSI and PHI for about 3000 stars is shown in Fig 2. The sample includes a large number of stars which show considerable amounts of reddening as indicated by the colour index $(m_{2190} - m_{2740})$. The points plotted fall naturally into two groups: most of the points lie in a fairly narrow region on the lower part of the diagram (as denoted by dots) and all of these are class V; also many points lie significantly above (open circles) and all of these belong to luminosity class I and II. Class III stars (denoted by crosses) tend to lie between the two sequences. A change of slope for the lower sequence occurs near (PSI $\sim$ -1, PHI $\sim$ 0.5); this is caused by the Planck maximum moving longward of 1500A for cooler stars.

The mean values of PHI and PSI as a function of spectral type and luminosity classes have been determined from those stars for which MK spectral types are known; these are taken from Blanco et al (1968). There are very few class I and II stars which are later than B8. The mean PHI-PSI relation for the class V stars is shown in Fig 3; the error box indicates the r.m.s. scatter of the mean

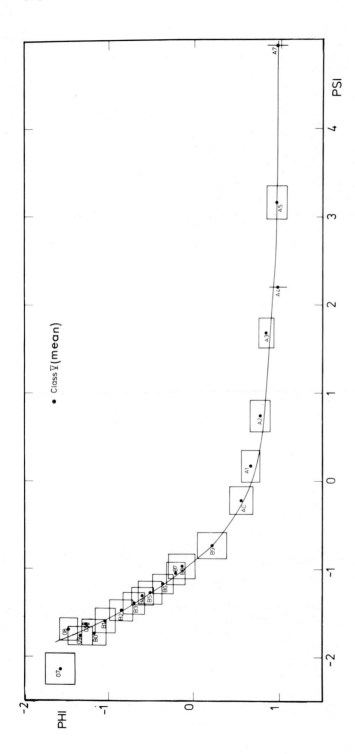

Fig. 3. The relation between PHI and PSI for Class V stars.

values of PHI and PSI for a given spectral class. The sample contains a few class II stars; the sequence for these stars is not well established, but in general they lie close to the supergiant sequence.

In establishing the correspondence between the classification parameters and the MK spectral types, we have excluded the known emission line and peculiar A stars. The Be stars are being studied by Houziaux (Private communication), while Jamar et al (1976) are investigating a large sample of peculiar A stars from the S2/68 data. The latter authors have found that the silicon stars are strongly deficient at 1400A as compared to the normal A stars, while the Hg,Mn stars are only weakly or not at all deficient.

Table 1

Spectral features in S2/68 UV spectra.

$\lambda(A)$	Main contributor	Appearance in sp. type
1400	Si IV Si II in late B's	O - B7 wavelength position of the minimum is $\sim$ 1380A for O-stars, $\sim$ 1410A for B-stars.
1450	FeV,CIII,CvIII,TiIII	O- stars
1550	CIV, SiII in late B's	Class V: O - B8 strong spectral type effect decreasing with later type. Wavelength (minimum) position appears to shift towards shorter wavelength for later types. Class I & II: seen till A2. Sharp and strong in comparison with class V stars.
1620	Fe II	Similar to 1550A feature.
1640	He II (emission)	WC
1720	Fe II, Fe III	BO - AO Present only in class I & II.
1909	He II (emission)	WC, WN

1920	Fe III	B0 – A2 Present only in class I & II stars. For later than B8 this feature seems to have double structure.
2050	Fe III, Cv II	Same as 1920A feature.
2250	Cv II, FeII, FeIII	Class V: Appears at A0 and increases with later type. Class I & II: Appears in B8 and strength does not appear to vary.
2340	Fe II, Cv II	Strong in late B and A class I & II stars; weak or absent in class V stars.
2420	Fe II and lines from singly ionized metals	Appears at B5 and increases with later type in Class V stars.

Table 2

Grouping of stars according to the spectral appearances of the features.

Spectral type range		Characteristic features
O-stars	class V:	1450A feature; wavelength position (minimum) of 1400A feature is 1380A±10. 1550A and 1620A features are sharp. No features in the spectral region between 1700A and 2550A. The strength of 1450A feature increases in the earlier type.
	Of:	1450A feature present but weak; 1720A feature strong. Other features same as in class V.

SPECTRAL CLASSIFICATION OF EARLY TYPE STARS 299

B0 - B4   class V: 1450A feature absent. 1400A,
                               1550A and 1620A features
                               strong. No 1720A feature.
                               1920A and 2050A features
                               absent or very weak. The
                               spectral region between
                               2050A and 2550A relatively
                               clear.

                class I & II: All the features which are
                               seen in class V are relativ-
                               ely strong. In addition
                               1720A, 1920A and 2050A
                               features are strong.

B5 - B7   class V: No 1450A feature. 1400A,
                               1550A and 1620A features
                               weakly present. No 1720A,
                               1920A and 2050A features
                               -2420A features appears at
                               B5 and increases with
                               later types
                               No 2250A and 2340A features.

                class I & II: No 1450A feature as in
                               class V. 1400A, 1550A,
                               1620A features are stronger
                               than in class V. 1720A,
                               1920A and 2050A features are
                               strong. 2340A feature
                               present, but no 2250A and
                               2420A features.

B8 - B9   class V: No 1450A feature. 1400A,
                               1550A, 1620A features absent
                               or very weak, 2420A feature
                               is moderate and other feat-
                               ures are absent.

                class I & II: No 1450A feature. 1400A,
                               1550A and 1620A, 1720A,
                               1920A and 2050A features are
                               present. 2340A feature is
                               strong and 2250A feature is
                               present. 2420A feature is
                               absent or weak.
                               1920A feature has double
                               structure.

A0 - A2   class V: No 1450A feature. 1400A,
                               1550A and 1620A are absent

or very weak. No 1720A, 1920A and 2050A features. 2250A feature appears at A0 and increases with later type. No 2340A feature. 2420A feature moderately strong.

class I & II: No 1450A feature. 1400A, 1550A and 1620A features present. 1920A, 2050A, and 2250A and 2340A features strong. 2420A feature weakly present. 1720A feature present in A0, becoming weak in A2.

Table 3

Spectral classification from S2/68 UV spectra

λ(A)	O	Of	B0-B4		B5-B7		B8-B9		A0-A2	
		class	V	I	V	I	V	I	V	I
1400	(1380A)	(1380)	S	S	P	S	W/A	S	A/W	P
1450	S	P(weak)	A	A	A	A	A	A	A	A
1550	S	S	S	S	P	S	W/A	S	A/W	P
1620	S	S	S	S	P	S	W/A	P	A/W	P
1720	A	S	A	S	A	S	A	P	A	P
1920	A	A	A	S	A	S	A	S	A	S
2050	A	A	A	S	A	S	A	S	A	S
2250	A	A	A	A	A	-	A	P	P	S
2340	A	A	A	A	A	-	W/A	S	W/A	S
2420	A	A	A	A	P	W	P	W	S	W

S = strong, P = present, W/A = weak in most cases and absent in some cases, A/W = absent in most cases and weakly present in some cases, A = absent.

References

Blanco V.M., Demers, S., Douglass, G.G., Fitzgerald, M.P. 1968, Pub. U.S. Naval Obs. Vol. 21.

Boksenberg, A., Evans, R.G., Fowler, R.G., Gardner, I.S.K., Houziaux, L., Humphries, C.M., Jamar, C., Macau, D., Macau, J.P., Malaise, D., Monfils, A., Nandy, K., Thompson, G.I., Wilson, R., Wroe, H. 1973, Mon. Not. R. astr. Soc. 163, 291.

Cucchiaro, A., Jaschek, M., and Jaschek, C. 1976, Astron. and Astrophys. (in press).

Houziaux, L. 1976, (in preparation)

Humphries, C.M., Nandy, K., Kontizas, E. 1975. Astrophys, J., 195, 111.

Humphries, C.M. Jamar, C., Malaise, D., and Wroe, H. 1976, Astron. and Astrophys, 49, 389.

Jamar, C., Praderie, F and Macau-Hercot, D. 1976, (in preparation)

Nandy, K., Humphries, C.M. and Thompson, G.I. 1975, I.A.U. Symposium No. 72 (Lausanne).

Nandy, K., and Schmidt, E.G. 1975, Astrophys. J., 198, 119.

Nandy, K., Jamar, C., Monfils, A., Thompson, G.I., Wilson, R. 1976, Astron and Astrophys., 51, 63.

Oke, J.B., and Schild, R.E. 1970, Astrophys. J., 161, 1015.

SPECTRAL CLASSIFICATION OF B AND A STARS FROM THE LINE FEATURES OF S2/68 SPECTRA

A. Cucchiaro - Institut d'Astrophysique, Liège, Belgium
M. Jaschek   Observatoire de Strasbourg, France.
C. Jaschek   "   "   "   "

The experiment S2/68 has supplied a large number of early type spectra in the wavelength region 1350-2730 A. The resolution is 37 A, which is about equivalent to a reciprocal dispersion of 1850 A/mm. We had at our disposal about one thousand spectra of stars brighter than 6^m5.

The availability of such material has lead us to establish a spectral classification system based only upon features visible in the spectrum. A second step was to compare this system to the MK classifications.

It turned out to be easy to establish a temperature sequence based upon convenient intensity ratios of line features and somewhat more difficult to establish luminosity criteria. Among the B type stars it is possible to distinguish main sequence objects, supergiants and intermediate objects.

The scheme has been applied to all spectra available, except those where considerable reddening is present. The results show a satisfactory coincidence with MK classifications.

The possibility of detecting peculiar spectra was also examined and the following results seem well established: a) Be stars cannot be segregated from stars of intermediate luminosity ; b) Ap stars (except those of the Mn type) can be segregated ; c) Am stars can be detected.

The complete results will be published in Astronomy and Astrophysics.

BEHAVIORS OF B STAR CONTINUA AND ABSORPTION FEATURES
DETERMINED FROM THE TD1 S2/68 "ULTRAVIOLET BRIGHT STAR
SPECTROPHOTOMETRIC CATALOGUE" AND FROM COPERNICUS SPECTRA

J.M. Vreux and J.P. Swings
Institut d'Astrophysique,
B-2400 Cointe-Ougrée (Belgium)

Ultraviolet low resolution spectra of most of the B stars observed with the S2/68 experiment onboard the TD1 satellite in the spectral range 1350 A - 2500 A are computer-investigated in order to define a grid of standard behaviors (J.M. Vreux and J.P. Swings, 1976 Astron. Astrophys., in press). The behavior of the relative intensities of the Balmer and Paschen continua is studied as a function of the spectral type and the visible photometric index Q : both relations are shown to be luminosity dependent. A comparison to theoretical predictions and to previous studies is also presented. A measure of the slope of the pseudo-continuum drawn between $\sim \lambda$ 1660 A and $\sim \lambda$ 2550 A is studied in the same way: the effect of luminosity is discussed in connection with recently proposed temperature scales for supergiants. The behavior of the depths of the absorption features at $\lambda\lambda$ 1550 A, 1940 A, 2000 A, 2055 A, 2105 A, 2340 A and 2395 A with respect to the spectral type, the Q index and the luminosity class is also briefly presented. Special emphasis is given to the absorption features at λ 1550 A in connection to the discussion of the FeIII lines in the ultraviolet spectra of early B stars by J.P. Swings, M. Klutz, J.M. Vreux and E. Peytremann (1976, Astron. Astrophys. Suppl. 25, 193). Preliminary results of a study of high resolution Copernicus spectra in the λ 1550 A region are also presented: it is shown that the role of FeIII remains predominant for stars of spectral type as early as about B1II - B1III (J.P. Swings and J.M. Vreux, to be published).

DISCUSSION

PECKER (paper by C. JASCHEK). I would like to complement the paper by Dr. Jaschek in noting that between the space ultraviolet spectrum, and the "conventional" part of the spectrum ($\lambda > 3700$ Å) there is a *large* part of UV (down to 3100 Å) easily reached from the ground and incorporated (for at least 30 years) in the IAP classification (Barbier, Chalonge, Divan). The IAP (BCD) classification is based only on the H - spectrum (Balmer jump, UV and visible gradients) and gives, in addition, UV (ground based) peculiarities of a high physical significance. Interstellar absorption is also easily corrected in the IAP classification.

CAYREL (paper by C. JASCHEK). *Comments on comparison between spectral classification and photometric systems.*
1) One fundamental advantage of spectral classification over photometric systems, which has not been mentioned, I believe, is the fact that spectral classification is not affected by interstellar reddening whereas photometry is.
2) There is no doubt that spectral classification at about 2 Å spectral resolution contains more information on the star than broad or intermediate band photometry. Nevertheless, most of this information is lost when the spectrum description is summarized into to integers, spectral type and luminosity class.

CODE (paper by C. JASCHEK). In one sense subdwarf O stars might be regarded as the normal stars, since unlike O main sequence stars they do not have extended atmospheres, mass loss etc...

PECKER (paper by K. NANDY, A. CUCCHIARO, C.M. JASCHEK, J.P. SWINGS and J.M. VREUX). 1. Congratulations to the Edinburgh - Liège groups for a magnificent work, for the good measurements, for making available to many groups their fine data. Congratulations and thanks !
2. However, I do not believe that UV measurements should not be used as "complements" to spectral classification. The UV spectrum is formed in a broad region of the star, it does not measure the same things as the MK spectral classification system (or any other classical system).
3. In particular, reddening has two components - an interstellar, - a circumstellar : now are the two indices being used to distinguish supergiants from main sequence stars could really be used that way ? They might show the existence, or not of a circumstellar cloud ! The latter may vary from star to star ! I do believe that the departures from correlations laws, the *dispersions* of the relations between UV indices and visible indices is *the* interesting thing,- not their rough agreement !

HOUZIAUX (papers by A. CUCCHIARO, C. and M. JASCHEK, and K. NANDY).

Nandy has proposed to classify the stars according to the values of the photometrically derived Φ and Ψ parameters. Cucchiaro and the Jascheks propose a system based on depth ratios of spectrometric features. Is there any correlation between the results obtained by these two approaches?

CUCCHIARO. Answer: Nandy has plotted the ratios r_1/r_2 and r_3/r_2 versus Φ. The results shown and the two figures below indicate a good correlation between our ratios and Nandy's quantity (dots denote main sequence stars, crosses class III objects, while circle refer to class I and II stars).

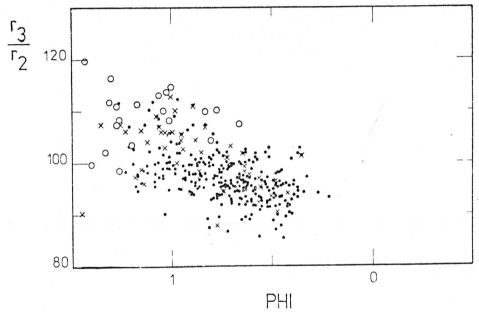

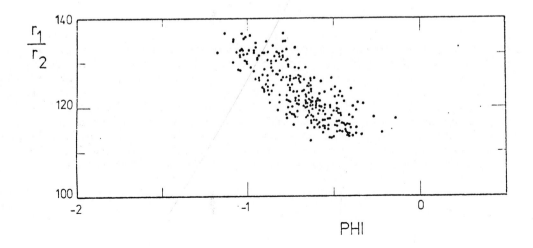

KODAIRA (paper by CUCCHIARO, C. and M. JASCHEK).
I suspect that you may have difficulty to detect Am stars in UV region which are classified as Am according to the weakness of CaII/ScII lines in the photographic region, just because of the same reason as far Mn-type Ap stars. Have you noticed this point in your analyses ?

CUCCHIARO. Answer. In our investigation, we have found that Am stars can be segregated fairly easily from usual stars on the basis of the strength of the features at $\lambda\lambda$ 1850, 1920, and 2400 Å. But stars in which the differences between the hydrogen line types and the metallic line types in the visual are very pronounced are more easy to detect.

SPECTRAL CLASSIFICATION USING ANS PHOTOMETRIC DATA

R.J. van Duinen and P.R. Wesselius
University of Groningen, Space Research Department

INTRODUCTION

In this contribution we will discuss the extension of visual photoelectrical photometry into the ultraviolet and its potential impact on stellar classification. As is the case in the visual, classification by photoelectrical photometry in the ultraviolet has some important advantages over classification by inspection of individual spectra. Studies that require global characteristics of stellar properties - such as studies of stellar distribution, extinction properties of interstellar dust, galactic structure - often require large numbers of observations to derive statistical properties. In photoelectric photometry such a large number of observations can usually be obtained in a reasonable amount of time and relatively conveniently processed without subjective criteria being applied to the data. Comparatively, the classical way to obtain stellar classification is a complex process that requires many steps.

In ultraviolet astronomy, because of the advent of small automated spacecraft with acurate pointing capabilities, the advantages of photoelectric photometry can be fully realised. A large number of observations can be made on relatively distant or highly obscured objects with good photometric accuracy.

This paper will report on status of a program carried out by the ANS satellite on stellar classification by photoelectric photometry. We will first briefly describe some characteristics of the photometric system. The observing program will be presented and some preliminary results discussed.

THE ANS ULTRAVIOLET PHOTOMETRIC SYSTEM

The ANS spacecraft was launched in August 1974 in a polar sunsynchronous orbit and has functioned from that date to April 1976 with an interruption of three months the ultraviolet experiment onboard - provided by the Space Research Department at the University of Groningen -

consists of a 22 cm diameter modified Ritchy-Chretien telescope followed by a five channel intermediate band photometer. Spectral dispersion is by means of a grating, the passbands are defined by masks in the image plane of the spectrometer, which results in sharp cut-offs in the spectral sensitivity curves. The entrance aperture is reimaged onto the five photomultiplier detectors, one for each channel. The field of view on the sky is 2.5 x 2.5 arcminutes (see also Aalders et al. 1975 and van Duinen et al. 1975). Observations are carried out by instructing the spacecraft to find specified star patterns in the field of an image dissector tube camera which uses the Ultraviolet experiment telescope. Once found, the object is centered on the entrance slot of the photometer, while the satellite tracks on a bright (up to 8.5th visual magnitude) star in the field of the camera (1.5 x 1.5 degree). Pointing is maintained for a predetermined duration up to 1000 seconds. Readout of samples from the counting registers in the onboard computer memory is controlled by the processor as are the experiment shutter position.

For weak sources the spacecraft is periodically pointing to the object and to blank sky with the register readout synchronized with the spacecraft pointing direction. Typically 5 such observations (on different objects) are performed per orbit in every third orbit (other orbits are devoted to X-ray experiments onboard ANS).

Observing programs are loaded in the onboard computer memory once per twelve hours during a pass over the groundstation. During this pass the experiment data are read-out. First printouts of results are available within three hours after the pass. Final data are in a few months after the pass.

The quality of the data obtained outside the radiation zones (mainly the South Atlantic Anomaly) is very good. Avoidance of radiation zones was quickly mastered by the people performing the experiment observation scheduling. Reproducibility of data obtained on the same object in different pointings is better than 0.5%. The sensitivity changes were obtained from repeated observations on standard stars at high ecliptic latitude. Corrections were derived by linear least square fits to the observed countrates for each visibility period.

SPECTRAL CLASSIFICATION

One of the aims of the observing program executed with the ANS is to set up a refined classification scheme for early type stars. We try to find parameters using linear combinations of ultraviolet colour indices that are correlated with intrinsic physical properties of stars. As an example we have constructed the parameter

$$\alpha = C_{2500-1800} - 0.257 \, C_{2500-2200}$$

This alpha has been calibrated against a sequence of some seventy early type main sequence stars and is correlated with spectral type, hence with effective temperature. At least another parameter independent of

alpha can be defined, viz. the difference between the narrow and wide band measurement at 1550 Å. This parameter is roughly correlated with the MK luminosity classification. Its physical interpretation is complicated because
1. for O type stars a strong CIV line is present showing a P Cygni profile.
2. for B stars up to B5 the depression at 1550 Å is a blend of mainly Fe III lines ; this depression seems luminosity dependent (Swings et al., 1976).

Luminosity effects, however, show up very clearly in ultraviolet colour-magnitude diagrams that we have obtained on galactic open clusters and associations. These observations may provide a basis for the choice of a luminosity parameter and at the same time provide an excellent luminosity calibration. In surface photometry of the LMC classification criteria have been developed on the basis of a multi-variant statistical analysis (Koornneef, 1976). These criteria give information on stellar population and reddening.

The results so far are based on the reduction of a small subset of the data. The bulk of the data will be reduced and analysed in the coming year using a set of computer programs that have just become available.

REFERENCES

- Aalders, J.W.G., v. Duinen, R.J., Luinge, W., Wildeman, K.J. 1975, Space Science Instrumentation $\underline{1}$, 343.
- v. Duinen, R.J., Aalders, J.W.G., Wesselius, P.R., Wildeman, K.J., Wu, C.C., Luinge, W., Snel, D. 1975, Astronomy & Astrophysics $\underline{39}$, 159.
- Koornneef, J. 1976, thesis.
- Swings, J.P., Klutz, M., Vreux, J.M., Peytremann, E. 1976, Astronomy & Astrophysics Suppl. $\underline{25}$, 193.

SPECTRAL CLASSIFICATION WITH OBJECTIVE-PRISM SPECTRA FROM SKYLAB

K. G. Henize, S. B. Parsons, J. D. Wray and G. F. Benedict
Department of Astronomy, University of Texas at Austin

I. INTRODUCTION

During the Skylab I, II and III missions, ultraviolet spectra were obtained in 188 fields with a 15 cm aperture objective-prism spectrograph. The instrument has been described by Henize, Wray, Parsons, Benedict, Bruhweiler, Rybski and O'Callaghan (1975; hereafter referred to as Paper I).

The spectra cover the wavelength region from 1300 to 5000 Å and have a resolution of 2 Å at 1400 Å and 12 Å at 2000 Å. Absorption and/or emission lines of C II λ1335, Si IV $\lambda\lambda$1394, 1403 and C IV λ1549 are visible in more than one hundred stars. The lines of Si IV and C IV are found to be particularly sensitive to stellar temperature and luminosity. Since these lines are visible in spectra of moderate to low resolution it is clear that they should be of special interest in any UV classification system for faint stars. This paper investigates the correlation of the intensities of the C IV and Si IV lines with MK spectral type, and presents a preliminary classification scheme for O4 to B2 stars based on these lines.

II. CORRELATION OF LINE INTENSITIES WITH SPECTRAL TYPE AND LUMINOSITY

The general behavior of the Si IV and C IV lines is illustrated in figure 1. In luminosity classes III and V the rapid changes in both Si IV and C IV intensities between O9 and B3 make the Si IV/C IV ratio an excellent indicator of temperature class. This is particularly true in the interval from B0 to B1 where the ratio changes dramatically from about one-fourth to about 4.

On the other hand, the increase in both C IV and Si IV intensities with luminosity suggests that total intensity of the lines should be a useful luminosity criterion. This is particularly true of Si IV in the O stars where the intensity ranges from zero in the main sequence stars to very strong in the supergiants. C IV is useful at all spectral

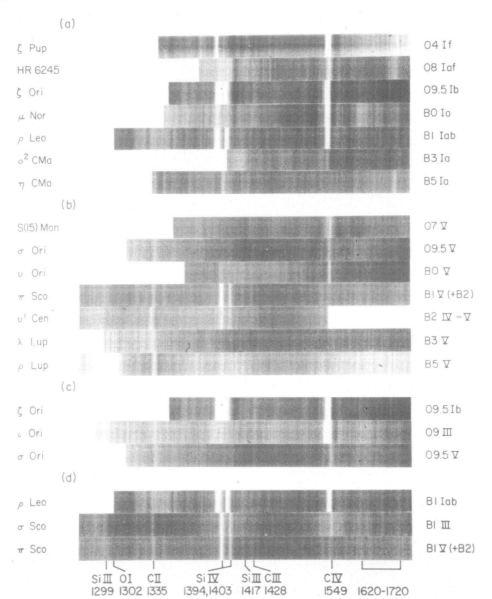

Figure 1. Computer-processed spectra in the region 1300-1800 Å. (a) Temperature effects among supergiants, (b) temperature effects along the main sequence, (c) luminosity effects at O9-O9.5, (d) luminosity effects at B1.

classes, especially at B1. Even though Si IV intensity appears to be a more powerful discriminant of luminosity in the O stars, we will give emphasis to CIV intensity largely because in our prism-dispersed spectra CIV is generally better exposed.

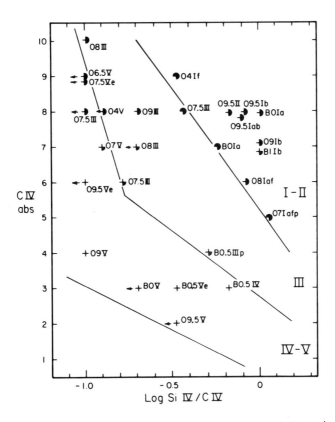

Figure 2. The C IV absorption vs. log Si IV/C IV diagram based on eye estimates of line intensities.

Another useful luminosity criterion is the appearance of a P Cygni emission-absorption profile in the higher luminosity stars. We find that this phenomenon correlates with bolometric luminosity more directly than with MK luminosity class since it varies diagonally across the two-dimensional regime of the MK system (see figure 3).

To further confirm and quantify the above noted trends, a special study has been made of all (a total of 20) O stars plus several B0 and B0.5 stars showing a well-exposed C IV line and reasonably well-exposed Si IV lines. In many cases the exposure in the Si IV region was weak but was sufficient to determine the presence or absence of a strong line. In these stars, eye estimates of the intensities of the C IV and Si IV lines and of the Si IV/C IV ratio were made by Henize from original plates. The results are plotted in figure 2.

In spite of estimated probable errors of ± 1 and ± 0.2 in C IV absorption intensity and log (Si IV/C IV) respectively, it is evident that luminosity classes are well separated in this diagram. A clear separation in temperature class is evident between the B0 stars and the O stars of luminosity classes III and V, but among the O stars the general weakness of Si IV makes the differentiation of temperature class difficult. However, in the class V stars the presence of P Cyg emission in CIV (this is noted by the filled sectors of the crosses in figure 2) clearly separates stars of class O7.5 and earlier from later spectral types. Although the supergiants are confined to a unique region in the upper right corner of the diagram, there is no clear separation of temperature classes.

The spectral types shown are taken from Walborn (1972, 1973), from Morgan and Keenan (1973) and from Hiltner, Garrison and Schild (1969) or Lesh (1968) in that order of priority. It is found that those stars which violate the luminosity regions indicated in figure 2 generally have alternate spectral classes which give better agreement with the UV data. For example, the O9.5 II star in the supergiant region is δ Ori. The class given is from Hiltner et al., but Conti and Leep (1974) give a class of O9.5 I in better agreement with the UV data. Another example is the O7.5 III star, Xi Per, which lies near the supergiant regime. The given class is from Walborn but the Conti class is O7.5 I. The absorption intensities as well as the strength of CIV emission give overwhelming evidence in the UV that this star is a supergiant.

The O7.5 III star in the class IV-V regime (BS 5680) appears to be a peculiar star in which the UV and visible data are truly discrepant. The given class is Walborn's but an alternate class by Hiltner et al., places it as a supergiant. However, there is no doubt on well-exposed UV plates that Si IV is exceedingly weak or absent, thus implying a luminosity class fainter than III.

The emission-line data in figure 2 confirm the conclusion reached in Paper I that the occurrence of the P Cyg profiles is a function of absolute bolometric magnitude (M_{bol}). This is examined in greater detail in figure 3 in which the eye estimates of CIV emission intensity are plotted against absolute bolometric magnitudes derived from the data referenced in Paper I. Three stars, not in figure 2, which show good data at CIV but not at Si IV have been added. This figure clearly illustrates the conclusion drawn in Paper I, that all stars at $M_{bol} = -8.4$ or brighter show emission at CIV while all those fainter do not. The single exception is the peculiar star V819 Cyg, which also shows a somewhat peculiar spectrum in the UV.

For those stars showing emission, the correlation between intensity and M_{bol} is evident but rough. The scatter may be partly cosmic but the majority of it almost certainly arises from errors in the data. The estimated probable error in the emission intensity is at least ± 1 and considerable uncertainty also pertains to the MK spectral class on which the M_{bol} is based. For example, the very low point at intensity 6 is Xi Per which would have $M_{bol} = -10.0$ if the Conti class of O7.5 I were used. The only extreme point for which a favorable revision can not be found is the high point at intensity 2 (λ Cep, O6 Ifp).

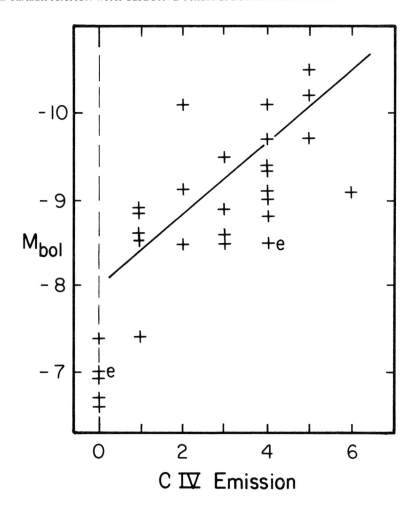

Figure 3. Correlation between CIV emission intensity and bolometric absolute magnitude.

In spite of the evident scatter, these data suggest that the CIV emission intensity in moderate dispersion spectra may be used as a direct indicator of M_{bol}. Such a simple luminosity criterion for the highly luminous stars would be invaluable in guaging the distances of faint O stars within our galaxy (where they serve as spiral arm tracers) or of external galaxies where this measurement might improve estimates of the luminosities of the brightest blue stars which are important distance indicators.

It may be debated whether the CIV emission, or the P Cyg profile in general, should be used as a classification criterion since it probably arises from outward mass flow in an unstable star whereas classical luminosity criteria generally pertain to physical processes within a stable atmosphere. However, the data indicate clearly that the instability itself is directly linked to M_{bol} in at least 90 percent of the stars in this sample. Therefore, it is difficult to deny the empirical usefulness of this criterion as an indicator of M_{bol}. More refined measures of emission-line strengths in a greater number of stars would be desirable to confirm this conclusion. Observations of Magellanic Cloud blue supergiants would be especially useful to both confirm and calibrate the relationship.

Further data on the use of the CIV absorption strength vs. log (Si IV/CIV) to indicate spectral type are given in figure 4. These central depth data have been derived by Parsons from tracings of the UV spectra. The eye estimates and central depth data for 14 stars common to the two sets are in good agreement.

Figure 4 confirms the findings of figure 2 and extends them to spectral class B3. Several instances where stars stray from their expected regimes are explainable in terms of available alternate spectral class. In several cases alternate spectral classes are given in parentheses. Three stars remain discrepant. Two of these, τ Sco (B0 V) and ϕ' Ori (B0.5 IV-V), which fall in the region of the supergiants, appear to have unusually strong Si IV. In both cases, comparison with eye estimates suggests that the Si IV central depths may have been exaggerated by uncertainties in the height of the continuum in underexposed regions of the spectrum. The star ς Cas shows unusually strong CIV for a B2 IV star as was previously noted by Henize, Wray, Parsons and Benedict (1976).

Although the scatter is large, a clear relationship between the Si/C ratio and temperature class for class III-V stars is evident in figure 4. As was previously noted in figure 1, the discrimination at classes earlier than O9 is poor while discrimination between O9 and B2 is much better. Similar data in the B0-B2 spectral range based on eye estimates is shown in figure 1 of Henize et al. (1976). Here we find the Si IV/CIV ratio ranging from 1/4 at B0 to 8 at B2.

Figure 4 also suggests that at spectral classes B1 and B2, the CIV strength is significantly greater in giants than in class V stars. However, the discrimination is small. The strength of CIV is also rapidly increasing with earlier temperature class and the validity of a CIV-based luminosity class depends critically on the accuracy of the temperature class.

III. A PRELIMINARY CLASSIFICATION SYSTEM

It may be concluded that the data presented in figures 2, 3, and 4 generally confirm, for a larger group of stars, the trends which are visually evident in figure 1. It thus appears that the strong lines of Si IV and CIV are useful in the spectral range from O4 to B2 to provide a two-dimensional spectral classification system which correlates reasonably well with the MK system. We propose the following as a preliminary classification system based on the dispersion of the Skylab

SPECTRAL CLASSIFICATION WITH OBJECTIVE-PRISM SPECTRA FROM SKYLAB

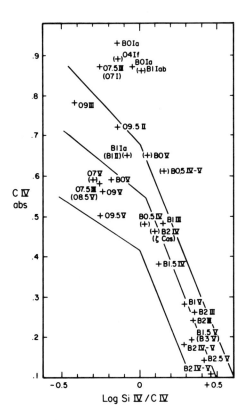

Figure 4. The C IV absorption vs. log Si IV/C IV diagram based on central depth measures.

spectra. Such a system, with appropriate modifications, should also be applicable to other spectra having resolutions between 1 and 6 A in the 1400 to 1600 A range.

A. If P Cyg emission is present in the Si IV or C IV lines, use the Si/C absorption line ratio to discriminate luminosity. If Si/C is very small, the spectral class is O4-O7.5 V. If Si/C is about 1, the spectral class is O4-B1 I. If Si/C is between the above, the spectral class is O4-O9.5 III. In any case, the emission intensity may be used to define a value of M_{bol} for the star.

B. If P Cyg emission is not present, use the Si/C ratio to define temperature class and C IV strength to discriminate luminosity. If the Si/C ratio is one-tenth or less, the spectral class is O8-O9.5 V. If Si/C is about one-fourth the spectral class is B0 III-V. If Si/C is about 1, the spectral class is B0.5 III-V. If Si/C is about 4, the spectral class is B1 III-V. And, if Si/C is about 8, the spectral class is B2 III-V.

This classification system should have significant practical applications inasmuch as it is based on strong lines that can be detected in spectra of moderate to low resolution. Indeed, in spectra of similar quality (most MK classification of early-type stars is currently done with slit spectra) it is to be expected that the UV system may be applied to fainter stars than the MK system since the total strength of the UV lines is significantly greater than the strength of the lines used for classification of visible wavelength spectra. In some areas (e.g., temperature class in the O9-B2 class III-V stars) the UV system has greater discrimination than the MK system, while in other areas (e.g., the temperature class of the O4-B1 supergiants) it has less discrimination. Thus it might be expected that, when UV spectra become more generally available, a hybrid system using the most sensitive data in both the UV and optical regions will evolve.

REFERENCES

Conti, P. S. and Leep, E. M.: 1974, Astrophys. J. 193, 113.
Henize, K. G., Wray, J. D., Parsons, S. B., Benedict, G. F., Bruhweiler, F. C., Rybski, P. M. and O'Callaghan, F. G.: 1975, Astrophys. J. Letters 199, L119.
Henize, K. G., Wray, J. D., Parsons, S. B., and Benedict, G. F.: 1976, in A. Slettebak (ed.), IAU Symposium 70, Be and Shell Stars, Reidel, Dordrecht.
Hiltner, W. A., Garrison, R. F., and Schild, R. E.: 1969, Astrophys. J. 157, 313.
Lesh, J. R.: 1968, Astrophys. J. Suppl. 17, 371.
Morgan, W. W. and Keenan, P. C.: 1973, Ann. Rev. Astron. Astrophys. 11, 29.
Walborn, N. R.: 1972, Astron. J. 77, 312.
―――――: 1973, Astron. J. 78, 1067.

DISCUSSION
(paper by K. HENIZE)

ROUNTREE-LESH. In his talk, Dr. Henize indicated that the ultraviolet spectrum of ζ Cas is peculiar, and that this peculiarity is also apparent in the OAO-2 data. Now the usual spectrum of ζ Cas is entirely normal, as is the ultraviolet spectrum from Mariner 9 data. This star is an MK standard as B2 IV, as well as a photometric and line-profile standard for comparison with variable stars. Could Dr. Henize describe in what way this star is peculiar ?

HENIZE. On 5019 plates, we find that C IV absorption is significantly stronger than is expected in a B2 IV star. Its strength is roughly equivalent to that of a B0.5 star. I have noticed that this strength is confirmed by spectra of OAO-2.

DWORETSKY. Would your overall correlations with spectral types determined from ground based spectra be improved by using the types of Conti and co-workers, which are based on higher dispersions and quantitative line ratios, and not visual inspection only ?

HENIZE. I have a somewhat subjective feeling that there is a tendency in the case of several stars which stray from the expected UV regime for the Conti spectral classes to be in better agreement with the UV data. However, I have not made a detailed study to see whether one classification system or the other really fits the UV data better.

BUSCOMBE. The spectra are beautiful, and I wish more of the hot stars were so well observed.
1. Are the visuel estimates from computer enhauced reproductions lacking identifications, or from original plates on which you inevitably know the region in the sky ?
2. Are you confident of the spectrophotometric calibration, which is more critical for lines as deep as Si IV & C IV than in the usual photographic spectral region ?
3. If τ Sco is deviant, for which some unusually well-resolved lines imitate high luminosity, do other slow rotators show similar effects ?

HENIZE. The visual estimates are from the original flight film. Concerning spectrophotometric calibration, yes we have a good calibration but since these data are based mainly on eye estimates, I am not sure that it much matters. And about τ Sco, I am sorry I haven't yet correlated our line strengths with rotation rates.

UNDERHILL. The Be and shell stars, which appear to have UV spectra corresponding to earlier types, when examined at high resolution from the ground do have earlier types than the MK type. For instance, in

1952 I showed ζ Tau was about B1 rather than B4 as given by the MK type. The UV spectrum confirms the type B1, as shown by S. R. Heap from Copernicus and sounding-rocket high-resolution spectra.

ROSENDAHL. It should be pointed out that the behaviour of the appearance of emission in the UV resonance lines at a certain limiting bolometric magnitude is also reflected in the optical region of the spectrum by the appearance of emission at Hα at approximately the same limiting magnitude. The behaviour of the optical emission is correlated with MK type. If it is assumed that the UV resonance lines and Hα formed in different regions of the atmosphere, then this behaviour implies that there is a strong coupling between various regions of the atmosphere and that it does make sense to attempt to order the UV behaviour by MK type.

AN ATLAS OF ULTRAVIOLET STELLAR SPECTRA

A. D. Code
University of Wisconsin

Ultraviolet stellar fluxes have been presented for 140 bright stars in the spectral region from 1200 Å to 3600 Å in a graphical and tabular form (A. D. Code and M. R. Meade, 1976). The spectra represent a subset of OAO-2 spectrometer data on file at the National Space Science Data Center. The monochromatic flux is given in units of ergs cm^{-2} sec^{-1} Å^{-1} with a spectral resolution of about 22 Angstroms in the region from 3600 Å to 1850 Å and approximately 12 Angstroms from 1850 Å to 1160 Å.

The monochromatic flux, F_λ, is based upon the absolute calibration described by Bless, Code and Fairchild (1976).

Figure 1a and 1b shows the digital output for a single scan, with 8 second integration time at each step plotted against step number for spectrometer 2 and spectrometer 1 respectively. These counts have been corrected for digital data overflows, based upon the analog output. A background count has been subtracted. The background count determination for spectrometer 2 is relatively straight forward since shortward of the LiF cutoff at 1050 Å the measured counts are just the background as shown on Figure 1a. The spectrometer 1 background is based on the data from late-type stars and the correlation of the background with total counts at longer wavelengths. Typically background counts correspond to about 5 counts.

Figure 1c and 1d show the superposition of 8 scans and 3 scans respectively where the step position for Lyman α or MgII 2800 respectively, have been determined for each scan before superposition. This assignment is usually good to ±1/2 step.

Figure 2a and 2b show the mean curve fitted to the data of Fig. 1. The wavelength assignment is based on the determination of the step positions for Lyman α and MgII along with the empirical relation between angstroms and steps. The data for each scan has been corrected for the small time dependent system degradation by an algorithm constructed to reduce all scans to the sensitivity at orbit zero. The

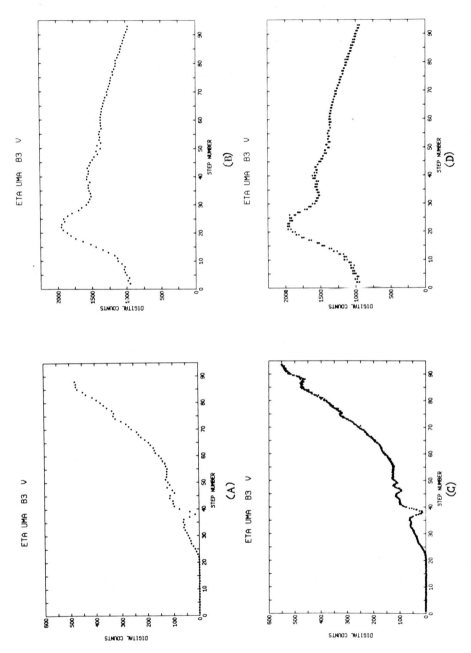

Fig. 1. (a) Digital output for a single spectrometer 2 scan, (b) spectrometer 1, (c) Superposition of 8 spectrometer 2 scans, (d) Superposition of 3 spectrometer 1 scans.

AN ATLAS OF ULTRAVIOLET STELLAR SPECTRA

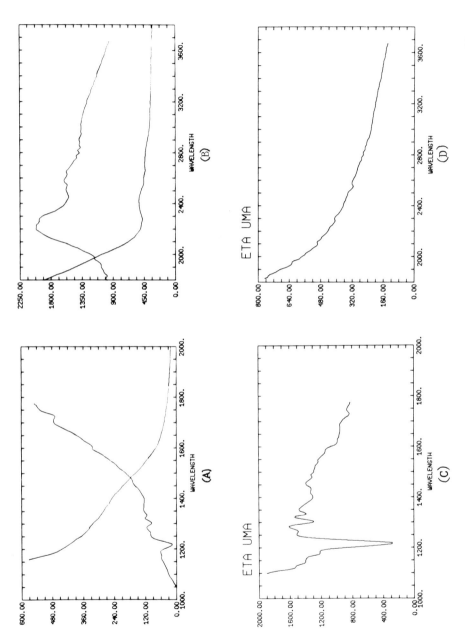

Fig. 2. (a) Mean digital output vs wavelength and relative sensitivity for spectrometer 2, (b) Mean digital output vs wavelength and relative sensitivity for spectrometer 1, (c) & (d) Stellar energy distribution.

investigation of the change of system response over the life time of the satellite for spectrometer 1 is given by C. Navach and M. R. Meade (1976) and for spectrometer 2 by A. Holm and M. R. Meade (1976).

Also plotted on Fig. 2a and 2b are the relative sensitivities of each spectrometer. The product of these two curves yields the stellar energy distribution shown in Figures 2c and 2d. These sensitivity curves are in units of flux per spectrometer count. In the figure the spectrometer 1 data has been normalized to ground based data, on the system of Hayes and Latham (1975), in the neighborhood of 3600 Å. The spectrometer 2 data remains on the OAO-2 calibration system of Bless, Code, and Fairchild.

The general behavior of the ultraviolet stellar spectra at the resolution obtained here is illustrated in Figures 3-8. Most spectral features are a blend of several lines. Underhill **et al.** (1972) have listed possible contributors to twenty absorption features present in early-type stars on OAO-2 spectra. Panek and Savage (1976) have identified the principal contributors to the strongest features present in these spectra in O and B stars. Table 1 reproduces their identifications and comments on the sensitivity of these line features to spectral type and luminosity.

Figure 3 shows the variations in the appearance of the ultraviolet stellar spectra for main sequence stars from O9 through B1. The line at 1215 Å is primarily interstellar Lyman α and shows significant variations from star to star. At O9 the strongest stellar feature is λ1550 C IV which decreases through the sequence as the λ1400 Si IV line grows in strength. The Si III lines at 1300 Å become prominent by B1. The interstellar reddening varies among these early-type stars and is evident in differences in the continuum gradient. Figure 4 shows the luminosity effects at O9.5 and Figure 5 for B0.5. The Si IV 1400 and C IV 1550 resonance lines increase in strength with luminosity. The broad blends at λ1600-1640 and λ1720 are sensitive to luminosity through the B stars and early A stars. Figure 6 shows spectral type variations from B3 V to B7 V. The λ1300 Si III, Si II, blend increases and the ratio of λ1300 to λ1340 is a good spectral type indicator. The luminosity effects at B5 are illustrated in Figure 7. Main sequence A stars at this resolution are characterized by broad blends primarily of Fe II lines that depress the continuum in the 2200-2500 Angstrom region. Figure 9 shows the spectra of later type main sequence stars longward of 2000 Å. The dominant feature at 20 Angstroms resolution is the λ2800 Mg II doublet. The discontinuity at about 2610 Å is present in giant and supergiant stars as well. Inspection of these figures and the Atlas plots show many other systematics with spectral type and luminosity class.

A variety of programs have been carried out or are underway utilizing the data presented here. For 32 early type stars for which angular diameters were available from the Narrabri intensity interferometer Code **et al.** (1976) have determined effective temperature

TABLE 1

STRONGEST FEATURES OBSERVED IN OAO-2 FAR ULTRAVIOLET SPECTRA OF O AND B STARS

λ[Å]	Principal Contributors	Comments
1175	C III (1175-1176, UV4)	Maximum strength near B1, insensitive to luminosity.
1215	H I (1216, Lα) Si III (1207, UV2) N V (1239-1243, UV)	For stars hotter than B2 feature is mostly due to interstellar absorption. For cooler stars the stellar line dominates (see Savage and Panek 1974).
1300	Si III (1295-1303, UV4) Si II (1304-3109, UV3)	Si III dominates in early B stars, Si II dominates in late B stars. The feature increases in strength toward cooler spectral types and increases slightly in strength with increasing luminosity.
1400	Si IV (1394-1403, UV1)	Maximum strength near B1, very sensitive to luminosity. P Cygni profiles are apparent for the very luminous O stars.
1550	C IV (1548-1551, UV1)	Maximum strength occurs for stars earlier than O9, very sensitive to luminosity. P Cygni profiles are apparent for very luminous O stars.
1600-1640	Fe III (1601-1611, UV118) Aℓ III (1600-1612) N II (1627-1630) He II (1640, UV12) and unidentified lines at 1621 and 1632	The strength of this broad blend of lines is relatively insensitive to spectral type but is sensitive to luminosity.
1720	N IV (1719, UV7) C II (1720-1722, UV14.02) Aℓ II (1719-1725, UV6)	The strength of this feature is relatively insensitive to spectral type but is sensitive to luminosity (see Underhill et al. 1972).

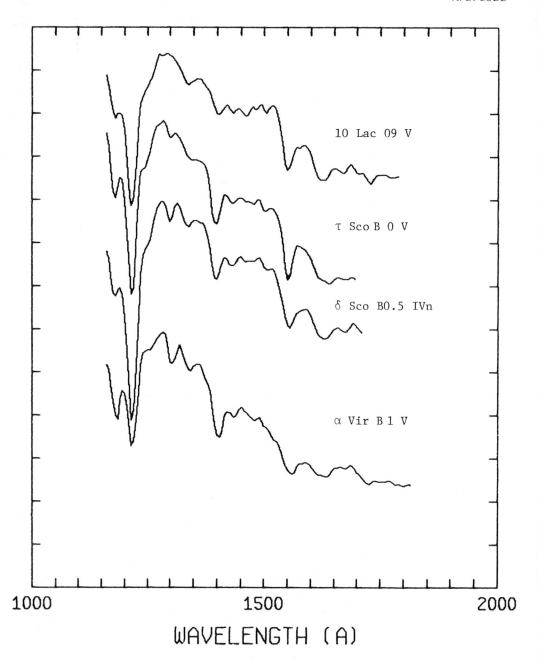

Fig. 3. Spectra of main sequence stars from O9 through B1.

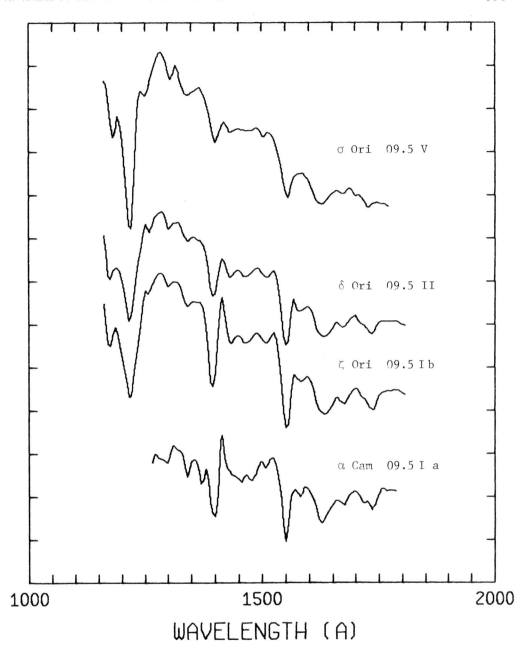

Fig. 4. Luminosity effects at O9.5.

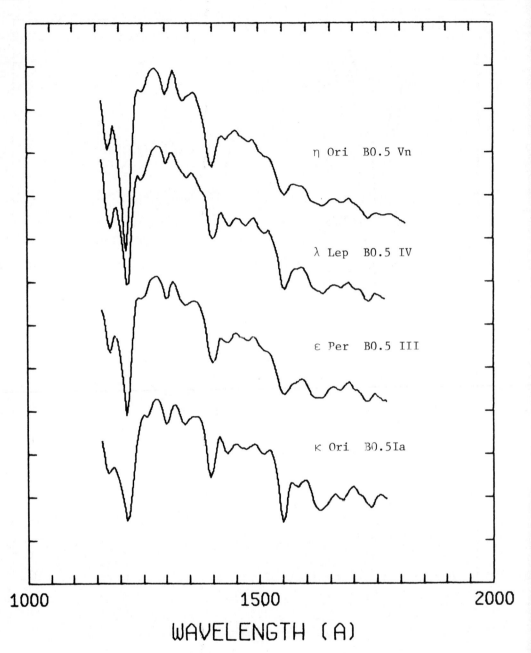

Fig. 5. Luminosity effects at B0.5.

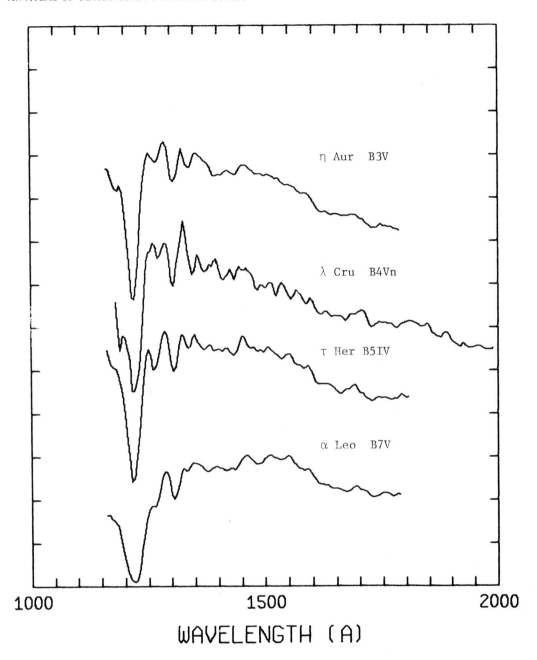

Fig 6. Spectra of main sequence stars B3 through B7.

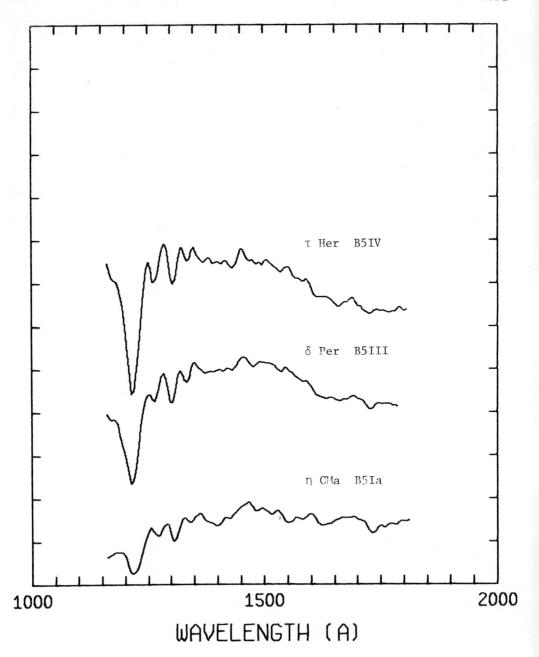

Fig. 7. Luminosity effects at B5.

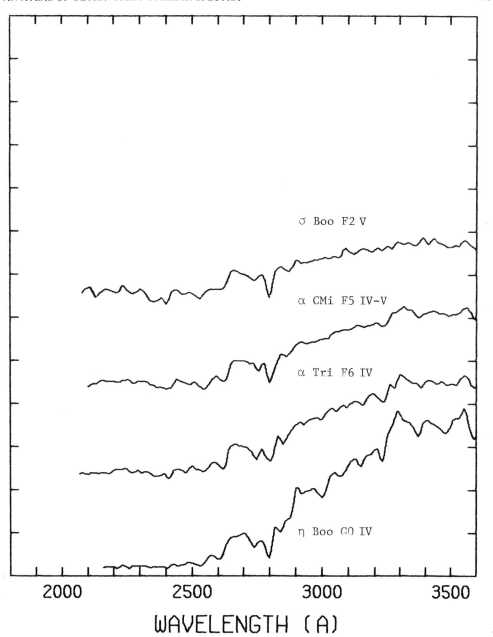

Fig. 8. Spectra of main sequence stars F2 through G0.

and bolometric corrections. Code and Tobin (1976) have recently
determined bolometric corrections for the remainder of the stars
appearing in this atlas in order to improve the statistical relation
between bolometric correction and spectral type. Among the other
investigations currently being carried out are a comparison of these
spectral energy distributions with model atmosphere calculations (see
Code (1975)), a determination of line blanketing coefficients in the
ultraviolet for early-type stars and discussions of the effects of
rotation and spectral peculiarities on the observed ultraviolet flux.

Leckrone (1973) has shown that the blue Ap stars are underluminous
and cool for their observed UBV colors. Members of these class of stars
had already been known to be bluer in B-V than implied by their
published MK spectral types. Both the ultraviolet flux and spectral
features are more consistent with the MK type than with colors. It
is possible that enhanced ultraviolet opacities are responsible for
modifying the UBV colors.

Examination of the ultraviolet spectra of Be stars indicates a
somewhat earlier spectral type than found in the visual. This may
be due to the variation of effective temperature from the pole to the
equator in these rapidly rotating B stars.

The effects of interstellar extinction on many of the early type
spectra is obvious from inspection, especially the strong extinction
bump near 2200 Å. The 2200 Å bump shows a good correlation with B-V
color excess and is detectable in stars if the color excess exceeds
$0^m.02$. For peculiar stars for which there is no basis for an a priori
knowledge of the normal color or spectral distribution, a correction
for interstellar extinction can be made based upon the appearance of
this 2200 Å feature.

An extension of observational material such as presented in the
atlas described in this report would provide a powerful tool both
for diagnostics of stellar atmospheres and for studies of galactic
structure.

References

Bless, R. C., Code, A. D., and Fairchild, E.T.: 1976, Astrophys. J.
203, 410.
Code, A. D.: 1975, in A. G. Davis Philip and D. S. Hayes (eds.),
Multicolor Photometry and the Theoretical HR Diagram, Dudley
Observatory Report, Albany, No. 9, p. 221.
Code, A. D., Davis, J., Bless, R. C. and Brown, R. H.: 1976,
Astrophys. J. 203, 417.
Code, A. D. and Meade, M. R.: 1976, Wisconsin Astrop. No. 30.
Code, A. D. and Tobin, W.: 1976, in preparation.
Hayes, D. S. and Latham, D. W.: 1975, Astrophys. J. 197, 593.
Holm, A. V. and Meade, M. R.: 1976, Wisconsin Astrop. No. 29.

Leckrone, D. S.: 1973, Astrophys. J. 185, 577.
Navach, C. and Meade, M. R.: 1976, Wisconsin Astrop. No. 28.
Panek, R. and Savage, B. D.: 1976, Astrophys. J. 206, 167.
Underhill, A. B., Leckrone, D. S., and West, D. K.: 1972, Astrophys. J. 171, 63.

A NEW TEMPERATURE SCALE FOR B STARS BASED ON OAO-2 DATA

Janet Rountree Lesh
Department of Physics and Astronomy, University of Denver

The empirical temperature scale of Code, Davis, Bless, and Hanbury Brown has been recalibrated in terms of an ultraviolet color derived from OAO-2 photometry, and mean temperatures have been computed for MK spectral types from O9 to A0. The dereddened color $(1910-V)_o$ was chosen as a temperature indicator because it is not strongly affected by lines or continuum edges. This parameter was computed for 16 stars with spectral types earlier than A2, luminosity classes III-V, and normal spectra, whose temperatures had been measured by Code et al. Plotting θ_{eff} against $(1910-V)_o$, we obtained the relation $\theta_{eff} = 0.111(1910-V)_o + 0.565$. Next, the quantity $(1910-V)_o$ was computed for over 150 stars in the same spectral-type range which had been observed by OAO-2, and a mean value was obtained for each spectral type. (Class III stars were excluded for types B5 and later because they were found to be systematically redder than classes IV and V - possibly an evolutionary effect.) Finally, the mean values of $(1910-V)_o$ were converted to effective temperatures using the formula given above. The resulting temperature scale is in good agreement with the scales of Morton and Adams, Schild, Peterson and Oke, and Code et al. for types B3 and later, but between types B0.5 and B2 the new scale and the Code scale are significantly hotter than the others. A particular application of the new calibration is the determination of small temperature changes occurring in early-type variable stars, since in this case the large and uncertain reddening correction drops out. For the β Cephei variables δ Cet and γ Peg, we obtained $\Delta T = 550°K$ and $185°K$, respectively. This work was supported by the National Aeronautics and Space Administration under grants NSG 5004 and NSG 5069.

EXTREME ULTRAVIOLET OBSERVATIONS OF WHITE DWARFS

Michael Lampton, Bruce Margon, and Stuart Bowyer
Space Sciences Laboratory, University of California, Berkeley

ABSTRACT

Observations shortward of the hydrogen Lyman limit provide sensitive determinations of stellar temperatures and interstellar absorption. Such data are of particular value in studies of hot white dwarfs, for which a large fraction of the emission occurs in the extreme ultraviolet band (100-1000 Å). Observations of HZ 43 and Feige 24 have been obtained with the Apollo-Soyuz extreme ultraviolet telescope; both stars are copious EUV emitters, with 4×10^{-9} and 3×10^{-9} erg/cm^2 sec in the 170-620 Å band respectively. The EUV data combined with optical spectrophotometry, allow their temperatures to be estimated as 80,000 and 60,000 K respectively. The corresponding interstellar neutral hydrogen column densities are $\sim 4 \times 10^{18}$ cm^{-2}.

I. INTRODUCTION

A variety of astrophysical questions motivate the observation of nearby stars in the 100-1000 Å extreme ultraviolet band. These questions concern stellar evolution and the space density of hot, evolved stars, the composition of the atmospheres of hot white dwarfs, and the density and ionization state of the interstellar medium (ISM). A major historical objection to attempts at stellar EUV observations has been the presumed opacity of the interstellar gas. However, recent spectroscopic studies of the ISM towards nearby stars indicate that in many directions neutral hydrogen concentrations are as low as 0.01 to 0.1 atom/cm^3 (Rogerson et al. 1973; Bohlin 1975; Dupree 1975). At these low densities, EUV photometry of hot stars should be possible to distances as great as 20-100 pc, according to the photoelectric cross-sections of Cruddace et al. (1974).

The nine-day Apollo-Soyuz mission offered the first opportunity to make a systematic sensitive survey of a number of candidate classes of EUV sources, and to study the distribution of the diffuse foreground and background radiation. Of the thirty targets examined, by far the strongest EUV emitters were sources in Coma Berenices and Cetus, which we have identified as the hot DA stars HZ 43 and Feige 24 respectively. These are the subjects of this report.

II. INSTRUMENTATION

The <u>Apollo-Soyuz</u> extreme ultraviolet telescope (Margon and Bowyer 1975; Lampton et al. 1976) consisted of a nested set of parabolic grazing-incidence reflectors having an aperture of 37cm, a six position filter wheel, and two channel electron multiplier photon counters. The filter wheel rotated continually at 10 rpm to give nearly continuous coverage of stellar fluxes in five wavelength bands; the sixth opaque position permitted the detector dark count rate to be monitored. The instrument's field of view was circular, with selectable diameters of 2.5 or 4.3 according to which of the two detectors was commanded into the axial focus position. Count rates from both detectors were telemetered each 0.1 sec, along with auxiliary information. Preflight laboratory calibration was conducted using a variety of wavelengths between 44 Å and 2650 Å. Absolute photon fluxes were determined with NBS vacuum-photodiode standards longward of 200 Å, and with primary standard propane counters at shorter wavelengths. The results of these calibrations are summarized in Table 1, where we list the filters employed (1), the system bandpasses at 10% of peak sensitivity (2), and the energy-integrated effective area or "grasp" $G = \int A(E)dE$, (3); column (4) gives the effective central energy of each band, $E_e = \int EA(E)dE/G$.

Filter material (1)	Bandpass (2)	Grasp cm^2 eV (3)	E_e eV (4)
Parylene	73-225 eV (55-170 Å)	590	142
Beryllium	83-109 eV (114-150 Å)	60	100
Aluminum	20-73 eV (170-620 Å)	270	46
Tin	16-25 eV (500-780 Å)	108	21
BaF$_2$	8.0-9.2 eV (1350-1540 Å)	0.47	9

Table I. Characteristics of the EUV Telescope

The aspect determination for each observation was conducted using the <u>Apollo</u> spacecraft inertial guidance system. Prior to the first EUV targets, the telescope alignment was verified to 0.3 accuracy by a raster scan of the stars ι and κ Aquilae, which as planned gave strong signals in the UV barium fluoride filter band. Throughout the mission, frequent sightings of UV stellar fluxes verified both the telescope aspect angles and the constancy of its sensitivity.

III. HZ 43

On 22 July 1976, observations were conducted on the ultrasoft X-ray object in Coma Berenices (Hayakawa et al. 1975; Hearn and Richardson 1975; Margon et al. 1976a; Hearn et al. 1976) identified by Hearn and Richardson as the DA white dwarf HZ 43 (Humason and Zwicky 1947; =EG 98, Eggen and Greenstein 1965; =L1409-4, Luyten 1949; =FB127, Greenstein and Sargent 1974; =29550, 33767, and 33965, Turner 1906). The star has coordinates $\alpha(1950) = 13^h 14^m0$, $\delta(1950) = +29° 22'$. As recently reported (Lampton et al. 1976) the Apollo-Soyuz observations revealed intense fluxes in the 170-620 Å, 114-150 Å, and 55-170 Å bands at a position compatible with the HZ 43 identification. The count rates and inferred fluxes are listed in Table II; a plot of the inferred spectral energy distribution has been published (Lampton et al. 1976).

Filter material	Count rate c/s	Derived Fluxes		
		raw; ph/cm^{2}s eV	corrected for atmos. ph/cm^{2}s eV	mfu
Parylene	22±1	0.037	0.039	3.7
Beryllium	8±0.5	0.13	0.15	9.9
Aluminum	160±3	0.59	1.0	30
Tin	<50	<.46	<1.2	<17
BaF$_2$	<25	<53	<53	<325

Table II. Observations of EUV Source in Coma

A variety of simple spectral energy distribution functions were found to be compatible with the EUV and soft X-ray data provided that the models' parameters were appropriately chosen. However, when we additionally constrain the models to fit the measured optical flux of U = 11.44 (Eggen and Greenstein 1965; corrected by Graham 1970) we find that power law and optically-thin bremsstrahlung functions are ruled out, and that blackbody models are tightly constrained to have temperatures near 110,000 K. This result suggests that the EUV and soft X-ray flux is thermal radiation from the star's photosphere. At this temperature, HZ 43 would be the hottest known white dwarf.

The customary optical technique for estimating DA temperatures is based on the colors, taking into account the size of the Balmer discontinuity (Greenstein and Sargent 1974). However, at temperatures above 50000 K the technique becomes much less sensitive because the

UBV wavelengths lie on the Rayleigh-Jeans portion of the Planck function. Thus we do not regard Shipman's (1972) estimate of Teff = 50000 K as definitive.

Recently, other studies of HZ 43 have been conducted. Image tube spectrophotometry of the white dwarf shows only broad shallow Balmer lines (Margon et al. 1976b) confirming its DA classification. That group has also obtained a parallax and, for the star's dwarf M companion, a spectroscopic distance modulus which combined place the system at a distance of 65±15 pc. At this distance, the luminosity of the DA star is 7 L☉, making HZ 43 the most luminous white dwarf known. The stellar radius derived from the brightness temperature is 5000 km. And, at 65 pc, the interstellar hydrogen density required to fit the EUV data is 0.02 cm^{-3}.

Durisen et al. (1976) have compared the EUV data to a set of high gravity solar-abundance model atmospheres, and find compatibility for effective temperatures in the vicinity of 125,000 K. Auer and Shipman (1976) have examined helium- and metal-deficient models, and find compatibility with effective temperatures between 60000 and 90000 K depending on composition. Further refinement of our knowledge of the evolutionary state of this star will be possible when its composition (which controls the opacity, particularly in the EUV) is better known.

IV. FEIGE 24

As part of the Apollo-Soyuz extreme ultraviolet observing program, observations were conducted on the very blue white dwarf Feige 24 (Feige 1958; =EG20, Eggen and Greenstein 1965; =FB24, Greenstein and Sargent 1974) at α(1950) = 02^h 32^{m}5, δ(1950) = +03° 31' in the constellation Cetus. The data reveal a strong stellar flux of about 200 counts/sec in the 170-620 Å band, but no discernible flux in the other filter bands: the count rates in the 55-170, 114-150, and 1350-1540 Å bands remained at their background values of 3.4, 1.5, and 600 counts/sec respectively. Thus the EUV source in Cetus has a spectral energy distribution which is radically different from the Coma source, being comparably strong at ∼300 Å but no more than 10% as intense at ∼100 Å. The inferred fluxes and upper limits have been plotted in Figure 1, from a forthcoming detailed discussion of these data (Margon et al. 1976c). In contrast to the Coma source, the soft X-ray flux is too weak to be detectible by current experiments.

The EUV data can be used to constrain simple spectral energy distribution functions. We find that these functions must be substantially softer, i.e. steeper, than for HZ 43 in order not to violate the upper limits shortward of 170 Å. In particular, blackbody models fit provided that the temperature T < 90,000 K and the neutral hydrogen column density N_H > 2 x 10^{18} cm^{-2}.

In attempting to identify this EUV source, we have combined the characteristics of the telescope field of view with the detailed count rate modulation and spacecraft attitude variation to define a position box. This box contains Feige 24, and also the hot sdO star Feige 26; however if the subdwarf classification is correct, Feige 26 is almost certainly too distant to be the EUV source due to the strong interstellar absorption in the 300 Å band. (For example, the sdO star BD+28°4211 is 2 mag brighter than Feige 26, yet our Apollo-Soyuz EUV upper limit on the former star is 6×10^{-10} erg/cm^2 sec in the 170-620 Å band, i.e. 20% of the flux of the Cetus source.) We thus adopt the Feige 24 identification. In Figure 1, we have included the U-band photometry of Oke (1974) and OAO-2 UV photometry of Holm (1976). Key features of these data are the exceptionally blue color of the optical and UV continua, and the steep EUV spectrum.

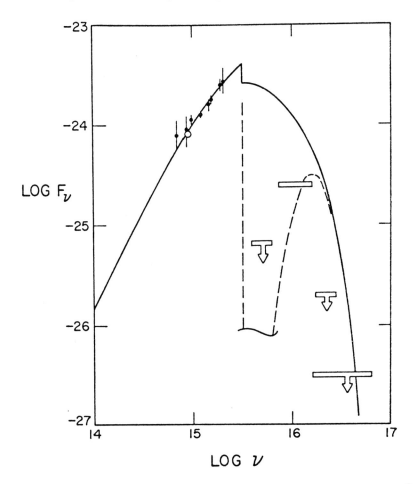

Figure 1. Spectral energy distribution of Feige 24, in erg cm^{-2} sec^{-1} Hz^{-1}. Boxes: EUV data, present paper; dots: UV data, Holm 1976; circle, Oke 1974. Also shown is one model having Teff = 60,000 K and log g = 8, without (solid line) and with (dashed line) interstellar attenuation.

It is of interest to determine whether these data can be explained by thermal emission from a white dwarf atmosphere, as is the case for HZ 43. Feige 24 is known to be an unresolved binary. Recent spectrophotometry by Liebert et al. (1976) confirms the DA classification of the blue component of the system. Thus the appropriate model atmosphere grid with which to compare the data would be one for high gravity and a nearly pure hydrogen composition.

Because no such published models exist, we have constructed an LTE code to obtain continuum fluxes at visible, UV, and EUV wavelengths, for pure hydrogen atmospheres at high gravity. In this code, 1000 layers uniformly span the log-Rosseland-mean interval -6 to +4, and the emergent flux is computed at 35 wavelengths. The opacity contributions considered were free-free, bound-free from n=1, 2, and 3 levels, and electron scattering. Pressure ionization effects were included by lowering the ionization potential. Convection was ignored, as is appropriate for hot, pure hydrogen atmospheres, and LTE was assumed throughout, justified for continuum processes by the high collision frequencies associated with high gravity. The code was tested and found to be satisfactory, both with regard to internal consistency (i.e. flux constancy) and agreement with existing general-purpose codes.

High surface gravity ($\log g = 8$) models were computed for a variety of temperatures and normalized to the $\lambda 3340$ point of Oke (1974). We find that none of these pure hydrogen models fit the EUV observations, even when allowance for an arbitrary amount of interstellar absorption is made: models cooler than 40-50000 K have insufficient 300 Å flux, yet models hotter than 30000 K have more 100 Å flux than is observed. An example of one of these models (60000 K) is shown in Figure 1.

One possible resolution of this discrepancy is to suppose that the atmosphere of Feige 24 contains a small amount of helium. It might be possible to introduce sufficient opacity at $\lambda < 228$ Å to allow a fit to the 55-170 Å data, yet not violate the apparent lack of He II lines in the optical spectrum. Further model atmosphere work is called for in this regard. A second possibility could be the presence of an appreciable amount of interstellar ionized helium; a He II column density of 3×10^{19} cm^{-2} would permit the pure hydrogen models to be reconciled with the data. In either case, stellar effective temperatures of about 60000 K are called for. To satisfy the 600 Å upper limit an interstellar hydrogen column density of 4×10^{18} cm^{-2} is required; i.e. the line-of-sight average $n_H = 0.027$ (50 pc/d), where d is the distance to the star. The stellar radius at a brightness temperature of 60000 K is 8400 (d/50 pc) km.

V. CONCLUSIONS

The brightest EUV objects thus far observed are hot white dwarfs, whose EUV fluxes are explainable as thermal radiation from their photospheres. EUV data appear capable of extending white dwarf temperature determinations into the relatively inaccessible region above 30,000 K. In addition, useful constraints can be placed on the interstellar medium's column density towards these stars. Forthcoming spectroscopic EUV measurements will play an important role in measuring the compositions and opacity sources of these stars, opening the way to a better understanding of late stages of stellar evolution.

We wish to acknowledge the support of NASA contract NAS9-13799.

REFERENCES

Auer, L., and Shipman, H. L.: 1976, Ap. J., submitted.
Bohlin, R.: 1975, Ap. J. 200, 402.
Cruddace, R., Paresce, F., Bowyer, S., and Lampton, M.: 1974, Ap. J. 187, 497.
Dupree, A. K.: 1975, Ap. J. (Letters) 200, L27.
Durisen, R. H., Savedoff, M. P., and Van Horn, H. M.: 1976, Ap. J., in press.
Eggen, O. J., and Greenstein, J. L.: 1965, Ap. J. 141, 83.
Feige, J.: 1958, Ap. J. 128, 267.
Graham, J. A.: 1970, Contrib. Kitt Peak Natl. Obs. No. 376.
Greenstein, J. L., and Sargent, A. I.: 1974, Ap. J. Suppl. 28, No. 259, 157.
Hayakawa, S., Murakami, T., Nagase, F., Tanaka, Y., and Yamashita, K.: 1975, presented at IAU/COSPAR Symposium on Fast Transients in X- and γ-rays, Varna, Bulgaria, May 1975; Ap. and Sp. Sci., in press.
Hearn, D. R., and Richardson, J. A.: 1975, IAU Circular 2890, June 17.
Hearn, D. R., Richardson, J. A., Bradt, H. V. D., Clark, G. W., Lewin, W. H. G., Mayer, W. F., McClintock, J. E., Primini, F. A., and Rappaport, S. A.: 1976, Ap. J. (Letters) 203, L21.
Holm, A. V.: 1976, Ap. J. (Letters), submitted.
Humason, M. L., and Zwicky, F.: 1947, Ap. J. 105, 85.
Lampton, M., Margon, B., Paresce, F., Stern, R., and Bowyer, S.: 1976, Ap. J. (Letters) 203, L71.
Liebert, J., Margon, B., and Kuhi, L.: 1976, in preparation.
Luyten, W. J.: 1949, Ap. J. 109, 528.
Margon, B., and Bowyer, S.: 1975, Sky and Telescope 50, 4.
Margon, B., Lampton, M., Bowyer, S., Stern, R., and Paresce, F.: 1976c, Ap. J. (Letters), submitted.
Margon, B., Liebert, J., Gatewood, G., Lampton, M., Spinrad, H., and Bowyer, S.: 1976b, Ap. J., in press.
Margon, B., Malina, R., Bowyer, S., Cruddace, R., and Lampton, M.: 1976a, Ap. J. (Letters) 203, L25.
Oke, J. B.: 1974, Ap. J. Suppl. 27, No. 236, 21.
Rogerson, J. B., York, D. G., Drake, J. F., Jenkins, E. B., Morton, D. C., and Spitzer, L.: 1973, Ap. J. (Letters) 181, L110.
Shipman, H. L.: 1972, Ap. J. 177, 723.
Turner, H. H.: 1906, Astrographic Catalogue, Oxford Section (Oxford Observatory).

THE NEAR ULTRAVIOLET SPECTRUM OF FE III AS A CLASSIFICATION CRITERION

R. Faraggiana
Astronomical Observatory, Trieste, Italy
H.J.G.L.M. Lamers
The Astronomical Institute, Utrecht, the Netherlands
M. Burger
Astrophysical Institute, Vrije Universiteit Brussel, Belgium

The Utrecht Orbiting Ultraviolet Stellar Spectrophotometer S59 on board the ESRO TD-1A satellite has observed ultraviolet spectra of about 200 stars in the wavelength regions 2060-2160 Å, 2490-2590 Å and 2770-2870 Å with a resolution of 1.8 Å (cf. de Jager et al., 1974). The spectra are analyzed in order to find UV criteria for stellar classification. Particular attention has been given to Fe III lines since no strong lines of this ion occur in the visible part of the spectrum. As one of the most striking results, it is found that the feature at 2078 Å, which is mainly due to Fe III, is very sensitive to luminosity.
We calculated the ratio R between the residual flux in the central part of the feature (2077-2080 Å) and the mean residual flux in the two adjacent wavelength bands (the local continuum : 2074-2076 Å and 2081.5-2082.5 Å) in order to avoid inaccuracies introduced by drawing the continuum. Stars with a high signal to noise ratio have been selected and by preference stars of which more than one observation is available. Figure 1 shows the relation between R and Q, where Q is defined as $Q = (U-B) - s(B-V)$ (cf. Heintze, 1973). The values of s proposed by Heintze for main sequence stars have been adopted also for giants and supergiants. The standard deviation for stars observed more than once is indicated, except for class III, IV and V, where only stars with $\sigma \leq .015$ were taken.
The separation between the supergiants and the other stars is clear; therefore we are able to confirm the classification given by Hiltner et al. (1969) of γ Ara (B1 Ib) and υ Sco (B2 IV), which is in contradiction with the classification given by Hoffleit (1964) (B1 III and B3 Ib respectively).
The Be stars have Fe III at least partly in emission ($R \geq 1$). The shell stars ζ Tau and o And show a very strong Fe III absorption feature at 2070 Å, the strength of which is comparable with that of a B5 Ia or B8 Ia star respectively. This suggests that Fe III is mainly formed in the shell surrounding the star.
A comparison is made with theoretical values derived from line spectra, which have been computed with classical methods (plane-parallel atmosphere, hydrostatic and radiative equilibrium, LTE) (cf. Burger and van der Hucht, 1976). The resulting spectra were convoluted with a profile

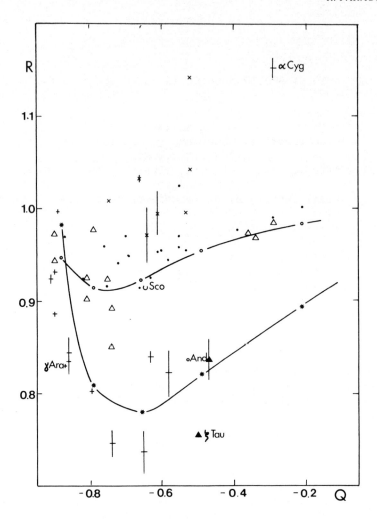

Figure 1. The relation between R and Q (see text for definition).
- Luminosity class IV, V; △ class III; + class I, II;
× Be stars; ▲ shell stars
o——o theoretical values for log g = 4 (ξ_t = 0 km/s)
—— theoretical values for log g = 2.5 or 3 (ξ_t = 10 km/s).

of 1.8 Å halfwidth, after which the ratio R was derived in exactly the same way as was done for the observed spectra. The difference between the minimum value of R for stars of class IV or V and the supergiants, which can be seen both in the observations and in the theoretical curves, reflects the shift of the ionization equilibrium as a function of wavelength.

More details will be published elsewhere.

REFERENCES

Burger, M. and Hucht, K.A. van der : 1976, Astron. Astrophys. 48, 173
Heintze, J.R.W. : 1973, IAU Symposium N° 54, p. 231
Hiltner, W.A., Garrison, R.F. and Schild, R.E. : 1969, Astrophys. J. 157, 313
Hoffleit, D. : 1964, Catalogue of Bright Stars, Yale Univ. Obs., New Haven
Jager, C. de, Hoekstra, R., Hucht, K.A. van der, Kamperman, T.M., Lamers, H.J., Hammerschlag, A., Werner, W. and Emming, J.G. : 1974, Astrophys. Space Sci. 26, 207

A CATALOGUE OF 0.2 Å RESOLUTION FAR-ULTRAVIOLET STELLAR SPECTRA MEASURED WITH COPERNICUS

Theodore P. Snow, Jr. and Edward B. Jenkins
Princeton University Observatory

Of the nearly 300 stars which have been observed in the ultraviolet with Copernicus, 60 have been chosen for publication of their complete intermediate-resolution spectra, which consist of scans made with photomultiplier U2 at a nominal resolution and step length of 0.2 Å. The spectra cover the wavelength range 1000 to 1450 Å and are expressed in the form of direct numerical tabulations, compressed-scale plots, and synthetic photographic spectrograms. These three modes of presentation are expected to satisfy the needs of various types of research on stellar spectra, ranging from detailed and rather specialized analyses of line profiles, which require accurate numerical results, to broad, comparative studies of various qualitative features over different spectral types, for which the plots or photographs are best suited. From the qualitative comparisons one might expect to synthesize more exacting criteria for spectral classifications of normal stars, using ultraviolet instead of visible spectra. The catalogue has been submitted for publication in the Astrophysical Journal Supplements.

The 60 stars included in the catalogue were chosen to give maximum coverage in the H R diagram, and they represent a good distribution of luminosity classes among spectral types O4 to A1. Wolf-Rayet stars, Be and shell stars, and other known peculiar objects were avoided, since the primary purpose of the catalogue is to present data on normal stars for a variety of temperatures and luminosities.

All of the data have been corrected for background and stray light contamination, and are generally of good quality, although minor flaws may still be present.

The numerical tables include an indication of the number of repeated scans averaged together at each wavelength, so that the noise amplitude due to photon statistics may be estimated directly. In the synthetic photographic spectrograms a granularity proportional to the relative noise amplitude is superposed. This granularity mimics the noise seen in ordinary photographic spectra of faint objects without degrading the actual spectral information. The eye can easily judge relative qualities of spectra from different stars and at a glance one can differentiate between noise fluctuations and authentic stellar and interstellar lines.

Included with the catalogue are figures showing the estimated absolute sensitivity of the U2 detector, including information on the change of response with time. The determinations of instrument sensitivity are based on observations of stars whose absolute fluxes have been measured (at $\lambda \gtrsim 1150$ Å) or computed (for $\lambda \lesssim 1150$ Å). From these sensitivity curves and the observation dates, which are also given, it is possible to reconstruct approximate far-UV absolute flux distributions of the stars in the tables.

The compressed plots, which are presented at the same linear wavelength scale as the synthetic photographic spectrograms, can be used to make a qualitative examination of luminosity and temperature effects. P-Cygni profiles indicative of mass loss, and line-blanketing (especially in the B supergiants) are particularly prominent features of these plots. Much the same sort of information can be derived from the synthetic photographic spectrograms.

It is important that our Princeton colleagues be acknowledged for their part in planning and carrying out the observations included in the catalogue, for their role in the development of the data-handling and correction procedures which were utilized, and for permitting us to use data originally acquired for other purposes. Several Copernicus Guest Investigators were similarly generous. Dr. W. Bidelman provided some of the initial encouragement to undertake this effort.

SPECTRAL CLASSIFICATION FROM COPERNICUS DATA

William P. Bidelman*
Warner & Swasey Observatory-CASE WESTERN RESERVE UNIV.
Cleveland, Ohio, U.S.A.

ABSTRACT. The suitability of Copernicus U2 spectrometer scans (resolution 0.2 Å, spectral range roughly 1000-1450 Å) for purposes of spectral classification is discussed. The main conclusion is that while many features in the spectra complicate the situation (numerous interstellar features and those related to the phenomenon of mass loss), the general behavior of the lines in the spectra of most of the 22 stars studied is entirely consistent with the presently accepted classifications of these objects. Whether a significant increase in the accuracy of determination of all of the relevant classification parameters can be obtained from this material is not yet certain. It is emphasized that study of the high-resolution Copernicus scans should be of great value for forthcoming classification programs necessarily utilizing spectral data of considerably lower resolution.

You have just seen what Copernicus spectra in the range $\lambda\lambda$ 1000-1450 look like, and have heard how these simulated photographic spectra were obtained. As has already been said, the homogeneity of the data is best in this spectral region, so when I started to think about the spectral classification problem a year or so ago it was precisely a portion of the material that you have heard described that I started to utilize. It is however true that I had, up to a few weeks ago, access only to the spectral scans and not to the simulated spectra. The spectra are certainly easier to intercompare than the scans, and more thought-provoking, as well. For a comparative qualitative study of the Copernicus spectra, I was fortunate enough to obtain from the Princeton investigators and others involved in the project spectral scans of twenty-two presumably normal stars, all of which had supposedly good spectral types. The stars involved ranged from 9 Sgr to α Cyg, and there was a good representation of both the most luminous supergiants and dwarfs. For the intermediate-luminosity stars the situation was not so favorable, as nothing was available between σ Sco at B1 and ζ Dra at B6; so while I have worked on the giants somewhat I do not propose to discuss them at this time.

*Guest Investigator with the Princeton University Telescope on the Copernicus satellite, which is sponsored and operated by the National Aeronautics and Space Administration.

*Edith A. Müller (ed.), Highlights of Astronomy, Vol. 4, Part II, 355-359. All Rights Reserved.
Copyright © 1977 by the IAU.*

Useful work in spectral classification necessitates, as is well known, comparable material covering a wide range of wave-lengths. I do have fairly comparable material, though there are substantial differences in the count rates among the various stars, which, of course, affect the appearance of the scans. The most serious deficiency, however, is the limited spectral range. Since most conventional classification work on the early type stars draws heavily on the lines of hydrogen and helium, it is unfortunate indeed that these lines are essentially unavailable in the material that I have used. The central portions of both Ly α and Ly β are of course completely dominated by interstellar absorption, and only the profiles of the far wings can occasionally give any useful information. No neutral helium lines are present at all, and the only available He II line, at $\lambda 1085$, is hopelessly blended with N II.

When you ask a spectral classifier to classify one of your spectrograms, he is quite apt to excuse himself from giving a definite answer on the grounds that he "is not used to that dispersion". And further, if the spectral region is not just the one that he has specialized in, he will probably refuse to even hazard a guess. In discussing the Copernicus data both of these excuses come in handy! The resolution of the material is comparable to that of earth-based coudé spectrograms, and the spectral region is certainly novel. Since it is unlikely that we will be obtaining similar high-resolution scans for many fainter or previously unclassified stars, at least in the foreseeable future, I have not attempted to set up a classification scheme as such, in the way that Miss Maury and Miss Cannon did at the turn of the century, E.G. Williams did in the 30's, and Morgan, Kopylov, Conti, and Walborn have done more recently. What I have rather attempted to do up to now is to simply inquire whether the Copernicus spectra are qualitatively consistent with the spectral types presently assigned. Even this has not been as trivial a question as one might have thought.

When I started this work only one of the stars that I had scans of had been identified, namely the B5 supergiant η CMa (Underhill, 1974). There are by now several other line-lists available, but most of my effort still necessarily has had to go into identifying the spectra. This is nothing new: if you read the old Harvard work you will find that along with their attempts at spectral classification they also measured and attempted identifications of the spectral features that they were using. Miss Maury (1897) writes: "It was necessary to conduct the investigation systematically, by successively comparing each typical star with those most nearly resembling it in the nature of their spectra, in order to avoid errors in the identification of the lines."

Now for a word about the general nature of the spectra. First, they are predominently absorption, rather than emission, spectra, in this spectral range. And second, at least in the earliest types there is a reasonably well-defined continuous spectrum. However, there is a dramatic change in the middle B's, when the lines become so numerous and strong that the continuum disappears and you are left with an excessively choppy spectrum that makes quantitative analysis difficult if not impossible. This phenomenon happens earlier and is much more prominent in the more luminous stars, and generally resembles the situation in the conventional spectral region when one goes from A-type stars toward those of later types. The amount of line-blocking in the $\lambda\lambda1000$–1500 spectral region is surprising, though anticipated (see Gaustad and Spitzer, 1961).

Now for some details. When one takes a casual glance at those simulated spectra one is struck by a considerably greater-than-expected diversity in their appearance. For this diversity differences in atmosperic temperature and pressure are only partly to blame: there are two additional factors that one does not have to contend with in the usual spectral range that are extremely important in the ultraviolet: (1) interstellar features, and (2) spectral features associated with mass loss. Since others are working on both of these things, I hesitate to say much about them, but in view of the important role that they play in the appearance of the spectra I feel bound to say something.

As previous work by the Princeton investigators has shown, there are a very large number of interstellar atomic and molecular hydrogen features in this spectral range and they can be of very substantial strength. In the earliest stars the majority of the most noticeable features in the spectra are apt to be interstellar. Shortward of $\lambda1100$ the H_2 absorption can drastically change the whole character of the spectrum. Once recognized it can be allowed for, but it is an unfortunate and important nuisance nevertheless. And the interstellar atomic lines are even worse, as they occur over the entire spectral range and are in many cases exactly the same lines that can be strong in a star as well. For example, carbon, nitrogen or silicon lines can be of interstellar origin in the earlier types and stellar in the later B's. In very broad-lined stars the interstellar features can be distinguished by their sharpness, but in the stars that I have been studying this doesn't work as most of my objects are fairly sharplined. Thus, one must rely on the fact that only the lowest-level lines will show up in interstellar space, as well as on good judgment as to what the physical conditions in the star are. The three N II lines near $\lambda1085$ are very helpful in this connection as their intensities reverse between interstellar and stellar conditions. I would warn anyone using these spectra that even little-reddened but moderately distant stars like 10 Lac, 15 Mon, or λ Ori can exhibit very prominent interstellar lines. A further cautionary word is that there is little correlation between the strenghts of the atomic and moelcular interstellar features. Luckily, in nearby dwarfs like γ Peg the interstellar lines appear to be essentially absent, except, of course, for hydrogen. I should add that in my material I have seen no evidence for sharp circumstellar--as distinct from interstellar--features.

The second thing that strongly affects the appearance of the spectra is the occurrence of broad longwarded-displaced emission and shortward-displaced absorption that is associated with very strong, mainly low-level, lines in the earlier and more luminous stars. This phenomenon is very marked in the supergiants, and can be seen nicely in S IV at $\lambda1070$, C III at $\lambda1175$, N V at $\lambda1240$, and Si IV at $\lambda1400$ in α Cam. However, it exists for other ions and in stars of somewhat lower luminosity as well. I will say nothing further about this, except to stress that one must be on the lookout for features of this sort. The strenghts of these emissions and displaced absorptions can, I think, be considered a luminosity criterion to the same extent as the somewhat similar phenomenon observed at Hα in the more luminous early type stars, i.e. as suggestive but not absolutely conclusive evidence of high luminosity.

Let me now turn to the problem of identifying the stellar lines. Happily, the resolution of the Copernicus U2 spectrometer is sufficiently high that blending is not a severe problem except in the later B stars; the wavelenghts can be read off the scans with an accuracy of .2 A or better, permitting line identifications to be done with fair ease. One finds quickly, however, that one needs all of the atomic data presently available, and then some. After plausibly identifying as many of the lines as one can, one still finds that perhaps 10% or so of the stronger lines are still unidentified. One can usually, of course, find a line in Kelly and Palumbo at the right wavelength, but it probably isn't the right line. I was at first very excited to find, for example, that the strongest line of Li II in this region, at $\lambda1198$, seemed to be in several of the hotter stars, but this line, which is the analogue of the well-known $\lambda3889$ line of neutral helium, arises from 59 e.v. and one may suspect that the identification is, to say the least, doubtful. In general I would say that the identifications, at least in the stars that I have studied, can be done with some conficence without the necessity of going to any exotic elements. The higher stages of ionization of elements like C and Si, N and P, and O and S account for many strong lines. Si III is much in evidence in the somewhat cooler stars. In the O-type stars we find many strong lines of Fe V and some of Ni IV (Fe IV is apparently inadequately studied in the laboratory). The great complication of the spectra of the middle and later B's is primarily due to the appearance of strong lines of Cr III and Fe III, along with a few of Mn III and Ti III. I have not attempted identifications as late as B8 but it may be safely assumed that the singly-ionized metals would account also for many lines there.

Finally, at last, to the problem of spectral classification. As yet I have studied carefully only spectra in the range O9-B5. I have carefully intercompared the supergiant sequence, the dwarf sequence, and the stars of differing luminosity with each other. My over-all conclusion is that the general behavior of the lines in most of the stars is entirely consistent with the presently accepted classifications of the stars. P V, C IV, Si IV, C III and N III weaken as one goes to later types, C II, N II, Si II, C I and N I strengthen. Si III goes through a maximum as one would expect. The lines of the metals behave just as anticipated. As far as luminosity is concerned, the wings of Ly α are much wider in the stars of low luminosity, at least in the later types. There is a general tendency for most of the lines to be substantially stronger in the supergiants than in the dwarfs, the same phenomenon that we see in the usual spectral region but at later spectral types than here. But at the same spectral type the level of ionization is evidently somewhat higher in the supergiants, as the lines of lower ionization potential tend to be stronger in the lower-luminosity stars.

The question of abnormal abundances in the CNO group is a matter of considerable current interest, and the ultraviolet spectra might be expected to shed some light on this matter. According to Walborn (1976) ρ Leo is somewhat nitrogen-enhanced and ε Ori somewhat nitrogen-deficient. Neither of these things is entirely obvious in my material, though such effects could easily be masked by slight differences in temperature and pressure among the stars being compared. As E. G. Williams (1934) well said: "There seems to be no line in type B suitable for classification purposes which is independent of luminosity." More marked elemental abnormalities should be more noticeable. This is not to say that everything is entirely clear. There are a few features in the spectra that defy easy explanation. I do not really understand the rather marked differences in the appearance of Ly α seen in some of the simulated photographic spectra. From a preliminary inspection the spectrum of ζ Dra seems rather peculiar, as the neutral C and N lines seem too strong for its spectral type. Originally noted as peculiar in the Yerkes Spectral Atlas, this star has been recently classified B6 III. ρ Leo may also be a bit odd in some ways. But I prefer to emphasize the normalities in the data rather than the abnormalities.

I would like to close with a word about work at lower resolutions. It is evident that the very-high-resolution data that I have been discussing will be mainly utilized for studies of the individual stars involved, and it seems very likely that substantial programs of spectral classification in the ultraviolet will in fact be done with data of considerably lower resolution. Consequently the chief application of the present work to that problem may well be in the assistance that the Copernicus data can give in determining the spectral features involved in the necessarily badly-blended spectra observed at lower resolution. We must thus, whether stellar astrophysicisists or workers in spectral classification, be enormously grateful for the splendid spectroscopic material that all those involved with Copernicus have made available to today's and tomorrow's astronomers.

Acknowledgements

The writer is happy to acknowledge that this investigation was suggested, encouraged, and greatly assisted by Drs. Donald G. York and Theodore P. Snow, Jr. of the Princeton Copernicus project. He is further indebted to the University of British Columbia's Department of Geophysics and Astronomy, where part of the work was carried out, and to Barbara Lysakowski, in Vancouver, and Nancy Smith, in Cleveland, for substantial help in the analysis of the data.

References

Gaustad, J. E. and Spitzer, L.: 1961, Astrophys. J. **134**, 771.
Maury, A. C.: 1897, Harvard Ann. **28**, 3.
Underhill, A. B.: 1974, Astrophys. J. Suppl. **27**, 359 (No. 249).
Walborn, N. R.: 1976, Astrophys. J. **205**, 419.
Williams, E. G.: 1934, Publ. Astron. Soc. Pacific **46**, 292.

THE ULTRAVIOLET SPECTRUM OF BETA LYRAE

M.Hack[1], J.B.Hutchings[2], Y.Kondo[3], G.E.McCluskey[4]
[1]Osservatorio Astronomico,Trieste
[2]Dominion Astrophysical Observatory,Victoria B.C.
[3]Johnson Space Center,Houston
[4]Lehigh University,Bethlehem

ABSTRACT. Selected regions of the ultraviolet spectrum of Beta Lyrae were observed in 1973 and 1974 at the epochs of the two eclipses and of the two quadratures with the <u>Copernicus</u> Princeton University spectrometer. The results were published in the Astrophysical Journal (Hack et al. 1975, 1976). Here we summarize the results of a third series of observations with the same instrument covering the whole period of 12.93 days. The following spectral regions were observed in June 1975 during 13 consecutive days:
  λ 1036-1060; 1300-1326; 1398-1416; 2050-2098; 2580-2632; 2777-2812 in the low resolution mode; λ 1172-1177 and 2795-2799 in the high resolution mode. The main results are the following :
1) The radial velocity curve for λ 1175 C III is almost 180° out of phase with the orbital velocity curve of the primary star and the mean velocity is - 240 km s^{-1} ; K = 70 km s^{-1} . This suggests that λ 1175 C III is formed in an expanding envelope associated with the secondary.The mass ratio can be determined : m_2/m_1 = 2.7 $^{+0.2}_{-1.1}$. 2) The continuum variation confirms the results obtained with OAO-2. At λ 2100 we observe the light of a continuum indicating that the temperature of the body eclipsing the primary star is lower than that of the primary and of about 9000 K. The light curves at λ 1900, 2100 can be explained with the contribution of Fe III emission in the circumbinary plasma. On the contrary, the depth of the secondary minimum at λ 1311-1326 and at
 λ 1410 cannot be explained with the Si III and Si IV emissions. A possible explanation is the presence of a hot spot in the region of the eclipsing body (probably a disk surrounding the companion) where the stream from the primary impinges on the disk. The hot spot can also explain the behavior of the infrared light curves observed by Jameson and Longmore (1976).
   The full paper will be submitted to the Astrophysical Journal.

REFERENCES

Hack,M.,Hutchings,J.B.,Kondo,Y.,McCluskey,G.E.,Plavec,M.,and

Polidan,R.S. 1975, Astroph.J. 198,453
Hack,M.,Hutchings,J.B.,Kondo,Y.,McCluskey,G.E., and Tulloch,M.K. 1976, Astroph.J. 206,777
Jameson,R.F., and Longmore,A.J. 1976,Monthly Notices Roy.Astron. Soc. 174,217

THE Mg II FEATURES NEAR 2800Å AND SPECTRAL CLASSIFICATION

Yoji Kondo
NASA Johnson Space Center

The Mg II resonance lines at 2795.523 and 2802.698Å and their respective subordinate lines at 2797.989 and 2790.768Å are probably among the most prominent and interesting spectral features in the ultraviolet; they are perhaps *the* most significant in the mid-ultraviolet. They are also observable in one form or the other in stars of practically all spectral types. We shall discuss relatively high (about 0.4Å) resolution observations of these features.

I. GENERAL BEHAVIOR OF THE Mg II FEATURES

a. Main-sequence Stars

In the early spectral types both the resonance and subordinate lines are seen primarily as absorption lines. An interstellar absorption of varying strengths is superimposed on the photospheric absorption of the resonance lines. The strengths of the photospheric resonance and subordinate lines increase from O to B, e.g., Lamers et al. (1973) and Kondo et al. (1975). The subordinate lines begin to merge with the resonance lines in late-B stars. In mid-A type stars, the resonance and subordinate absorption strengths become maximum. In F-type stars, the photospheric absorption strenghts continue to decrease. Chromospheric emissions become definitely detectable in F-type stars (Kondo et al. 1972). In a G2 V star, the sun, the chromospheric emission is fairly prominent at the core of relatively weak photospheric absorption. In K and M-type stars, this region is presumably dominated by the chromospheric emissions of Mg II resonance lines with the photospheric absorption becoming negligible in late-K stars; the only extant observation in this region is that of ε Eri (K2 V) (McClintock et al. 1975).

b. Giants

The general behavior of the Mg II lines in giants is fairly analogous to that for main-sequence stars. The emission features in G and K giants often show asymmetry (Kondo et al. 1976a) as a result of the chromospheric turbulence or the mass flow.

c. Supergiants

The Mg II features in supergiants differ significantly from those in the main-sequence stars as follows. In stars of mid-B to mid-A the resonance lines are affected by shortward shifted additional absorption arising in the outward moving shell, e.g., Kondo et al. (1976b). In K and M spectral types, in which the Mg II features are effectively entirely in resonance emissions, the 2795Å component is markedly asymmetric due to the selective absorption near 2795Å by neutral metals in the circumstellar shell whereas the 2802Å component is symmetric (Kondo et al. 1972, 1976b; Bernat and Lambert 1976).

II. ABSOLUTE MAGNITUDES VERSUS Mg II EMISSION WIDTHS

The relationship between the absolute magnitudes of stars and the Ca II K line emission widths, W (expressed in terms of the Doppler velocity km sec^{-1}), was first discovered by Wilson and Bappu (1957). There are several reasons why one might expect the Mg II emissions to be more prominent and easier to detect than Ca II K emission as discussed by Kondo et al. (1972); the possible existence of a relationship similar to the one found by Wilson and Bappu was first suggested in that paper. This relationship was subsequently explored by several workers. According to the latest study by Kondo et al. (1976b), the relationship may be expressed as

$$M_V = -12.45 (\log W) + 28.78.$$

III. ABSORPTION WIDTHS AND SPECTRAL TYPES

The absorption widths increase from O to late-B; in late-B stars, the widths of the subordinate lines increase from luminosity class V stars to luminosity class I stars. In the main-sequence and giant stars, the absorption widths continue to increase to late-A, where maximum values are attained; the widths then decrease into F and G types. In mid-B to mid-A type supergiants, the widths are also increased by the excess absorption taking place in the outward moving shell.

REFERENCES

Bernat, A. P. and Lambert, D. L.: 1976, Ap. J., 204, 830.
Kondo, Y., Giuli, R. T., Modisette, J. L. and Rydgren, R. E.: 1972, Ap. J., 176, 153.
Kondo, Y., Modisette, J. L. and Wolf, G. W.: 1975, Ap. J., 199, 110.
Kondo, Y., Morgan, T. H. and Modisette, J. L.: 1976a, Ap. J., 207, 167.
Kondo, Y., Morgan, T. H. and Modisette, J. L.: 1976b, Ap. J., in press.
Lamers, H. J., van der Hucht, K. A., Snijders, M. A. J. and Sakhibullin, N.: 1973, Astron. & Astrophys., 25, 105.
McClintock, W., Henry, R. C., Moos, H. W. and Linsky, J. L.: 1975, Ap. J. 202, 733.
Wilson, O. C. and Bappu, M. K. V.: 1957, Ap. J., 125, 661.

DISCUSSION

WEIDEMANN (paper by M. LAMPTON). Congratulations for your observations. In your paper (Ap. J. Oct. 1976, in press) you state a temperature of 11,000°K, whereas now you have 60,000 - 90,000°K. This is now more convincing, since a hydrogen Balmer line of W_λ =9 Å as observed for HZ 43 cannot be formed in an atmosphere much hotter than 50,000°K. What you have detected is, however, important from the point of view of white dwarfs : the absence of the He II discontinuity at 229 Å shows clearly that hot DA white dwarfs are already helium deficient, this gives you the right to use a pure H atmosphere.

FRIEDJUNG (paper by M. LAMPTON). Your two white dwarfs had red companions. How far are they from their companions ? Is there a possibility of their companion being close enough for mass transfer and the formation of an accretion disc.

LAMPTON. It is true that both HZ 43 and Feige 24 have dwarf M companions. However the mass loss rates are very small for M dwarfs, and binary separations are large. At present, I discount mass transfer as an energy source.

DWORETSKY (paper by M. LAMPTON). Is there any possibility of observing the nearest and hottest O subdwarfs in the EUV ? O subdwarf stars have helium and should be observable at wavelengths longward of 228 Å. However, we presently have upper limits throughout the EUV on BD +28°4211.

UNDERHILL (paper by R. FARRAGIANA et al.). Your Fe III criterion isolates only an extended atmosphere. It does not tell whether the underlying star is a supergiant or a shell star. You use *other* criteria to determine that ζ Tau and o And are shell stars whereas β Ori, η CMa and ε Ori are supergiants. In the ultraviolet of B5 stars the strength of the absorption from resonance lines (C II, P III, Si IV) is a good luminosity criterion.

BUSCOMBE (paper by R. FARRAGIANA et al.). The star ν Sco is certainly a member of the Scorpio-Centaurus association and lies close to the main sequence. The wrings of the hydrogen lines are broad, and the Bright Star Catalogue erroneously quotes a luminosity based on a defective spectrogram. On the other hand, the star θ Arae to which J.P. Swings referred, is far beyond Sco-Cen and has high luminosity well shown on beautiful high dispersion plates obtained by T. Dunham.

BUSCOMBE (paper by Y. KONDO). How do you resolve the interstellar components, and were comparisons made with Hobbs'work on those of other ions ?

KONDO. Answer. We have developed a technique to compute the photospheric absorption profiles for stars of this type (Kondo, Modisette and Wolf, Ap. J., 199, 110, 1975). We evaluate the interstellar absorption as the difference between the computed photospheric absorption and the observed absorption if no circumstellar absorption is present. In the case just discussed, our results are in reasonable agreement with those obtained from other ultraviolet interstellar lines.

BIDELMAN (paper by Y. Kondo). α UMa is a close visual binary, the companion being a few magnitudes fainter but of considerably earlier spectral type than the primary. Is it possible that you are seeing, at the Mg II lines, some effect of the continuous spectrum of the companion ?

KONDO. Answer. According to the information given in the Catalogue of Bright Stars, the close visual companion to α UMa should not have noticeable effects on the continuum of the observed spectrum. However, I shall look into it.

KHARADZE (paper by W. Bidelman). I would like to ask Dr. Bidelman what is his opinion. How much strongly the results obtained from orbital experiments are able to devaluate that whole multitude of determined spectral types and luminosities which are done, published and applied for researches as the ground based observations.

BIDELMAN. Answer. I am sorry that I cannot give a quick answer to that question. Certainly much more study will be required to fully investigate the proper role of the ultraviolet in spectral classification. For the stars that I have studied the phenomena of mass loss represent new and very important factors, and for all of the stars of later type I would anticipate much new information on chromospheric phenomena in general to become available. Whether this new information will radically change our approach to spectral classification is at present an unaswered question.

CONCLUDING REMARKS

L. Houziaux

We have had today for the first time a confrontation of the views of many astronomers about the use of ultraviolet spectroscopic and photometric data in the field of spectral classification.

Dr. Jaschek recalled us that much work has still to be done from the ground since only 14% of the stars up to the tenth visual magnitude have M.K. spectral types. The current accuracy is around one tenth of a spectral type and 0.6 of a luminosity class. Jaschek notes that the luminosity classification from spectral inspection is superior to what is obtained from photometric data, a fact which is not surprising if we recall that lines are much more sensitive to pressure effects than the continuum. However, the introduction of a third parameter seems necessary, although it is not clear what this parameter should be.

Can the ultraviolet observations improve the situation as far as the nature and the accuracy of the classification are concerned ? It is perhaps too early to answer this question, but we may try to summarize the various trends of thought over this question.

Low resolution (30 Å - 50 Å) observations, as summarized by K. Nandy indicate the gross characteristics of the spectral features. It is not obvious that the study of such features will lead to a better classification than the visual data, especially if the photometric accuracy is not very high. Nandy has proposed a new photometric system based on the reddening-free Φ and Ψ parameters, somewhat similar to the Q parameter. The luminosity effect is apparent when a Φ, Ψ diagram is used, while Φ is essentially a spectral type indicator. Van Duinen and Wesselius reported on a similar exercise using more accurate photometric data from the ANS satellite. They propose to use an index α combining colour indices at 2500 Å, 2200 Å, and 1800 Å in order to determine the star's effective temperature, while a quantity similar to β (in the uvby β system), and based on wide and narrow filter measurements centered on 1550 Å would be luminosity dependent.

A classical but very attractive approach has been followed by K. Henize and his coworkers, who center their efforts on two important spectral features around 1400 and 1550 Å, mainly due to Si IV and C IV respectively in the O and early B types. These features are easily measurable on 2 to 12 Å resolution objective prism spectrograms. With such a material, it is easy to distinguish emission components in the C IV line. Henize proposes a classification scheme based on the strength of the C IV feature (for luminosity) and on the Si IV / C IV ratio (for the spectral type). Because of the wide range of variation of such a ratio, this scheme leads to an appreciable refinement in the spectral type, at least in the O9-B2 range, while emission components may be linked to absolute luminosity. The quality of the spectra makes it is possible to locate the areas where UV observations are likely to lead to some progress.

While the first systems mentioned above tried to provide classification systems based almost exclusively on UV measurements, Henize's work represents an important step in integrating ultraviolet criteria in a more general scheme, using both visible and UV spectra, that may ultimately result in a new, more adequate classification system.

A useful luminosity criterion, based on the Fe III lines around 2078 Å has been mentioned by Farragiana, Lamers and Burger. It requires spectra with a resolution of about 2 Å and may lead to improvements in the luminosity classification of B0 to B5 stars. On the other hand, the classification of very hot objects will certainly benefit from the EUV observations as those obtained by Dr. Lampton.

Dr. Code has presented to us a very complete atlas of the OAO-2 spectra and there is no doubt that this work represents a great wealth of data not only for spectral classification but also for a more direct determination of atmospheric parameters, as we have seen from the work of Dr. Rountree-Lesh.

We have just seen what detailed spectra look like in the 1000 - 1450 Å range thanks to the high resolution synthetic photographic spectrograms and tracings.

We will have to wait for their publication for examining the richness of such material. According to W.P. Bidelman, the general behavior of the lines is entirely consistent with the adopted classification of these objects. It is not clear that improvements in the classification may be easily reached, and this is not surprising because of the enormous amount of lines seen in this wavelength range ; it means that most of the features we observe are blends, to say nothing about the many interstellar lines spread all over the spectrum. Using a comparable resolving power, Dr. Kondo has been able to show that the study of the strength, shape and width of the Mg II resonance lines (at 2795.5 and 2802.7 Å) and of the subordinate lines (at 2798 and 2790.8 Å) are very useful indicators of spectral type, luminosity chromospheric features, over a wide variety of stars (from B to M), despite the presence of interstellar components. Width measurements of the emission features will be particularly useful for absolute magnitude determination.

In conclusion, my impression is that we still do not know exactly how to use the ultraviolet data in order to integrate them in the classification schemes. In order to be useful the photometric data will certainly need to reach a good accuracy, comparable to what is obtained in the visible region. Otherwise, although colour indices may vary in a wider range than in the visible, this gain might be lost, especially if we consider the important effects of interstellar and circumstellar continuous absorption. As far as the use of line features is concerned, it seems that a resolution of 2-5 Å is needed in order to derive sufficiently accurate criteria. At high resolution, care must be taken to discriminate carefully the stellar, circumstellar and interstellar components. As far as the additional parameters to our traditional two dimensional classification scheme is concerned, no clear indication emerges so far. However, it is already certain that ultraviolet spectra will be good indicators of phenomena like abundance anomalies, mass loss and chromospheres.

JOINT MEETINGS

JOINT MEETING OF COMMISSIONS
25, 27, 29, 35, 36, and 42

OBSERVATIONAL EVIDENCE OF THE HETEROGENEITIES OF THE
STELLAR SURFACES

(Edited by M. Hack and J.P. Swings)

CONTENTS

OBSERVATIONAL EVIDENCE OF THE HETEROGENEITIES OF THE STELLAR SURFACES

(Edited by M. Hack and J.P. Swings)

D.J. MULLAN / Heterogeneity of the Solar Atmosphere	377
M. HACK / The Heterogeneity of Surfaces of Magnetic AP Stars	389
D.S. EVANS / Starspots on BY DRA-Type Stars	395
D.M. POPPER / Starspots on AR Lac Type Stars	397
J.W. HARVEY, C.R. LYNDS, and S.P. WORDEN / Direct Observations of the Heterogeneity of Supergiant Disks	405
R.E. GERSHBERG / On the Spottedness and Magnetic Field of T TAU-Type Stars	407

HETEROGENEITY OF THE SOLAR ATMOSPHERE

D.J. MULLAN

Bartol Research Foundation of the Franklin Institute
Swarthmore, Pennsylvania, USA

1. Introduction

Heterogeneities in the solar atmosphere exist on many different length scales ranging from values as large as the solar radius ($\sim 10^6$ km) down to features which are identifiable only by interferometry ($\sim 10^2$ km). Rather than simply cataloguing the observed parameters of each and every known type of heterogeneity, I would like to concentrate on a few types of heterogeneities, with a view to identifying the information which is currently available concerning the physical mechanisms responsible for creating the inhomogeneities. It is only if we can first identify the physics of each type of heterogeneity that we can hope to take even the first step towards predicting how each particular heterogeneity should scale to other stars. Since the present session is a joint discussion among mainly stellar astronomers, I feel that this approach is probably the most favorable method to present some of the large amount of information now available on solar features. Of course we expect that our solar information will be of most use to stellar astronomers in interpreting observations of stars which have similar spectral types to the sun. Nevertheless, we hope that nature will be kind enough to allow us to scale at least some of our information over a non-negligible area in the H.R. diagram.

2. Classification of Heterogeneities

There are two broad categories of heterogeneities which are unfortunately not mutually exclusive, but which can at least serve as an initial classification scheme. There is one category associated with hydrodynamic effects: these heterogeneities would probably exist even in the absence of a magnetic field. The second category is associated with effects in a magnetized plasma.

3. Heterogeneities due to Hydrodynamic Effects

3.1. DIFFERENTIAL ROTATION

The largest scale heterogeneity on the white-light sun is differential rotation. The equator rotates 10-20% faster than the polar regions. Several models have been proposed to explain this observation, including detailed numerical solutions of convection in a deep rotating spherical shell (Gilman, 1974). This type of model is sufficiently complicated that it is not yet obvious which physical parameter one may use to predict the degree of differential rotation on other stars. Some observational evidence for differential rotation on the surfaces of red dwarfs exists, but is not yet conclusive (Vogt, 1975).

3.2. CONVECTION

Heat transfer near the surface of cool stars occurs by convection, and the molecular viscosity of the gas is so small in most cases that it is almost inevitable that the convective flow is turbulent. It is generally believed that granulation on the solar surface is the physical manifestation of turbulent convection cells which happen to lie nearest to the top of the solar convection zone. The cells in quiet regions are in general polygonal during their lifetime, with hot gas rising at the center and cool gas sinking at the edges. The optical contrast between hot and cold gas on white light photographs at disk center is about 15%. Thus these granules are relatively small heterogeneities which, when viewed on a large scale, give the impression of an almost homogeneous solar surface. The velocities involved in the cells are several km/sec near $\tau \simeq 1$.

Convective cell sizes cover a range of values from several hundred km up to 2-3 thousand km, but there is a preference for a mean value of 1-1.5 thousand km. The existence of a preferred cell size has until recently been difficult to understand theoretically. Early attempts to derive growth rates for convective instabilities using linear perturbation analysis showed no preferred cell size. Recently, however, Deupree (1976) has published results of numerical work on non-linear convection, in which he finds that the vertical depth H of the small cells within a convective layer does have a preferred value of about one pressure scale height, H_p, although the depth determination is imprecise. Near the solar surface, the cell depth should therefore be 300-400 km. It has been an _assumption_ of the earlier models of stellar convection that the cell depth should indeed be about one pressure scale height: Deupree's results provide some much-needed support for this assumption.

However, visible granule sizes are _horizontal_ scale sizes, and very little information is currently available on what value to choose for the ratio of cell diameter to cell depth, D/H. Deupree's value is 0.8, but he admits that this may be an artifact of his numerical scheme. The classical value at marginal stability is D/H ≃ 3, but it is not known whether or not this carries over to conditions at large Rayleigh numbers. If (and this is very uncertain) indeed D/H ≃ 3 is valid in stars, then Deupree's results allow us to estimate granulation sizes on other stars: the pressure scale height is the relevant parameter. Another relevant parameter might be the vertical extent of the subphotosphere region with high superadiabatic temperature gradient. In either case, granules on dwarf stars will be small features, as in the sun. They will not be detectable photometrically, although their contribution to spectral line profiles may limit the accuracy with which the radial velocity of a star may be measured, unless one chooses a spectral line which is formed sufficiently high in the atmosphere to be above the convection zone (Dravins, 1975). On the other hand, on giants and supergiants, Schwarzschild (1975) has shown that individual granules might indeed be detectable. However, it is worth reiterating that the assumption of D/H=3 which enters this result is quite uncertain, and even $H=H_p$ is by no means a result of complete certainty at the present time.

Whatever can be said about the <u>mean</u> sizes of granules, there appears to be a theoretical <u>lower limit</u> on the expected horizontal size of convection cells, set by the existence of molecular viscosity. This limit is $\lambda_m \sim 200$ km at $\tau \leqslant 1$ in the sun (Spiegel, 1966). This lower limit coincides with the smallest scales observed, and unless this is an effect of limited observational resolution, this result provides a physical basis for estimating minimum granule sizes in other stars.

3.3. SUPERGRANULES

There is a preferred scale size of supergranules of order 30 thousand km on the sun, with a velocity pattern reminiscent of cellular convection. This may be a convective cell pattern associated with a vertical cell depth of order $(1-5) \times 10^4$ km. It is not at present clear why a depth of $(1-5) \times 10^4$ km should be preferred, although suggestions have been made that an opacity maximum, or helium ionization, or the bottom of the convection zone, are involved. A recent suggestion by Gough et al. (1976) is that just as molecular viscosity imposes a lower limit on the scale size of small-scale convection near the upper boundary of the convection zone, so eddy viscosity may impose a lower limit on the larger scales of convection. Gough et al.(1976) use the kinematic eddy viscosity in a solar model to estimate that large scale convection in the sun should be cut off at scales less than 20 or 30 thousand km. This indeed agrees with the lower limit of observed supergranule diameters. Thus this suggestion may provide a method of scaling supergranule minimum sizes to the surfaces of other stars.

Unfortunately, the assumption that supergranules are a direct convective process is not confirmed at the present time. Any existing temperature differential between rising gas at the center of the cell, and sinking gas at supergranule boundaries, is masked by magnetic heating effects. The general temperature structure over a supergranule cell is not apparently appropriate to convective energy transport (Worden, 1975). It seems permissible to conclude that almost no solar flux is carried by supergranules, and so the regions where supergranules carry the bulk of solar flux must lie far below the photosphere (if such regions do in fact exist at all). This uncertainty in understanding the observed physical properties of supergranules means that there is essentially no reliable way at the present time to scale supergranule sizes at the surface of the sun to supergranule sizes on other stars.

3.4. HEIGHT DEPENDENCE

Still considering hydrodynamic effects, we must realize that the sizes of the heterogeneities quoted above refer to a particular depth in the atmosphere, $\tau = 1$. Granulation cells are not visible higher in the atmosphere than $\tau \simeq 0.1$, which means that although convective overshoot must occur, the fraction of flux carried at $\tau \leqslant 0.1$ is quite small. Overshoot does, however, occur, and the question arises how does the scale size of the granulation pattern vary with height? We might try to get information on this by probing the atmosphere at different wavelengths λ to reach optical depth $\tau_\lambda = 1$, since penetration into the atmosphere varies with λ. At a wavelength of 350μ, we see above the temperature minimum, and at that level, some 500 km above the photosphere, studies of center to limb variations (Lindsey and Hudson, 1976) indicate that heterogeneities are indeed present in the solar atmosphere, with both vertical and horizontal scale sizes of order 1500 km. Thus the vertical scale size is several times larger than that of the photospheric granules. The vertical scale of roughness in the solar atmosphere must increase strongly with altitude: the chromosphere has a much rougher surface than the photosphere.

In the ultraviolet, we can also probe the high layers of the atmosphere, near the temperature minimum using continuum observations at about 1650 Å. An NRL rocket group has found (Brueckner and Bartoe, 1976) that the brightness temperature at this level of the atmosphere varies between $4200°K$ and $4800°K$ at different parts of the surface. There are differences in intensity by a factor of almost 2 between the brightest and the darkest features (excluding sunspots). The heterogeneities have sizes of 1"-2" which are comparable to the sizes deduced from infrared observations. However, these features are not associated with convective overshoot, for, contrary to the situation in the photospheric granulation, the brightest gas is moving downward towards the sun, rather than upwards. The downward flow is funnelled in such a way that heterogeneities which are only 1"-2" in diameter

HETEROGENEITY OF THE SOLAR ATMOSPHERE

at the temperature minimum, spread out to 2"-5" in the transition zone. This is due to magnetic field effects, and now we may turn our attention to heterogeneities caused by the presence of a magnetic field.

4. Heterogeneities due to Magnetic Effects

4.1. GENERAL

The occurrence of turbulent cyclonic convection in a differentially rotating star almost inevitably leads to the generation of magnetic fields. In theories of the solar dynamo, the solar field is usually calculated as a dipole or a quadrupole field (Stix, 1976), and there certainly is a general magnetic field of the sun which causes heterogeneities on a large scale (order of 1 solar radius) in the corona. This large scale general field has a period of 22 years, but scaling this period to other stars is currently a very uncertain art on account of the complexity of dynamo models.

Besides the general field, there are more intense fields, confined locally in features which are small compared with the solar radius. There are large areas of somewhat enhanced field, called active regions, with areas up to a few percent of a hemisphere, and with mean fields of order 100 gauss. Within these large areas, there are large numbers of small compact flux tubes, no more than 100-300 km across, where the field may exceed 2000 gauss. These tight bundles of strong field (which also exist outside active regions, but with smaller number densities) may be the sites of strongly convergent gas motions, and much theoretical work is currently being devoted to understanding how such compact flux tubes are created by the small-scale convective circulation. Viewing the sun as a star, there may be a chance of discovering features analogous to the active regions on other stars, but there is no chance of being able to discover the analog of the small compact flux tubes. In the present discussion therefore, which is directed mainly to stellar astronomy, I will confine my attention to the larger scale magnetic heterogeneities.

The horizontal scale sizes of active regions are probably determined by the diameter of a magnetic flux rope, but other effects may also be important. Supergranules may play a role. Perhaps the depth of the convection zone H_C is important. If the latter is true then stars with convection zones having depths of an appreciable fraction of the stellar radius would be prime candidates for looking for active regions. But we again stress that it is not clear why solar active regions have a certain size. Also we must ask what causes the mean field to be about 100 gauss? The answer is not clear, Parker (1975) suggests that it is an effect of magnetic buoyancy: if the field gets larger than about 100 gauss, then buoyancy carries the flux tube up through the convection zone so fast that there is not enough time to allow local amplification of the weaker general dipole field of the star. Moreover, Parker suggests that the place where the dynamo field is being generated is in the very lowest levels of the convection zone. Therefore, if Parker's suggestions are correct, in order to scale to other stars, we will need to know the structure of the deepest layers of the convection zones.

4.2. WHITE LIGHT FACULAE

The field strength now becomes one extra parameter which must be known before we can scale reliably from solar results to stellar conditions.

The interactions between magnetic flux and the gas and the radiation field are complex. Effects of a field depend on the altitude in the atmosphere at which one observes. At the photosphere the effect of a field is such that at small fluxes, the localized magnetic areas are brighter than normal (white light faculae), while at large fluxes, the localized magnetic areas are darker than normal (sunspots). White light faculae are local heatings in the upper photosphere, visible at optical wavelengths with greatest contrast (60% brighter than normal) near the limb and <10% contrast at disk center. They cover up to 0.5% of the solar surface at solar maximum, but it is not clear what types of magnetic structures they are associated with. Perhaps they are associated with lateral heat influx into small magnetic flux tubes (Spruit, 1976). Perhaps they are associated with closed field loops confined close to the surface, where hydromagnetic waves are trapped and dissipate. The non-thermal flux required to power the white light faculae may be very high, perhaps some 30% of the entire thermal flux passing outward through the surface (Wilson, 1971). Despite this large flux, the faculae are not spectacular heterogeneities because the energy is dumped too low down in the atmosphere. If there are white light faculae on other stars, then in order to predict brightness contrasts we may need to know the diameters of flux tubes, or the heights of closed magnetic loops on the surface of the stars (perhaps related to a granule scale size), and also the efficiency of conversion of thermal to mechanical flux. None of these quantities is known with any certainty at the present time.

4.3. SUNSPOTS

Sunspots are the most obvious heterogeneities on the white light photospheric surface. They can occupy areas up to 0.1% of the solar hemisphere, and can have effective temperatures of $4000°K$ and lower, with flux deficits of 80% of the normal flux. Fields at the spot surface are usually almost vertical and have strengths in excess of 1200-1400 gauss and very few (<5%) have fields greater than 3000 gauss, according to current investigations. Little evidence is currently available about depth-dependence of the field strengths. It is remarkable that field strengths in spots are on the whole confined to within such a narrow range (factor of 2). Why such field strengths are preferred is not certainly known, although the kinetic energy density in models of the deep solar convection zone is roughly equivalent to the magnetic energy density of a 2000 gauss field (Danielson and Savage, 1968). Thus if equipartition were a valid argument, then this might provide a method of estimating spot field strengths in other stars, if one had believable models of their deep convection zones. However, according to numerical dynamo work by Nagarajan (1971), there seems little or no reason to believe that equipartition will in fact be a valid concept in a turbulent dynamo. It may be that the turbulent dynamo has time to generate fields of order only a few hundred gauss, as seen in active regions, and then we must rely on subsequent instability to amplify the general active region field into locally strong fields with local energy density much larger than local equipartition would permit. If such a two-stage process is in fact at work, then it becomes doubly difficult to know how to scale magnetic field strengths to spots on other stars.

The darkness of sunspots (i.e. their effective temperatures) are determined physically by whatever mechanism carries away the missing flux. The flux which compensates for the missing flux of sunspots is an elusive quantity. One line of thought (cf. e.g. Meyer et al. 1974) is that by means of modified convection just around the spot, the missing flux is redistributed below the photosphere over an area much larger than the spot, so as to form and undetectable excess brightening over an area much larger than the spot. Alternatively, the missing flux may be in the form of Alfvén waves (Mullan, 1974) which are comparatively difficult to dissipate and can therefore propagate along field lines to distant regions of the sun. The propagation may occur either upwards along the open vertical field lines in the umbra into the corona, or downwards into the deep interior of the sun. Beckers (1976) claims to have discovered an appreciable flux of Alfvén waves in the surface layers of a sunspot: as much as 20-50% of the missing flux may be carried by the waves, according to Beckers. A decision between the two sunspot missing-flux mechanisms must be made before one knows how to scale the missing flux to starspots.

Sizes of sunspots are determined by physical processes which are as yet only partially known. In certain cases, sunspots have a tendency to have areas equal to supergranule areas (Dmitrieva et al. 1968), but since we do not know for sure what causes supergranules to have their characteristic sizes, we are uncertain about spot sizes. Lifetimes of spots are determined probably as a result of erosion by the surrounding turbulent gas motions. An eddy diffusivity can be

defined to describe the sunspot erosion, and reproduce the observed
decay of spots at certain phases of their evolution, during which
the area declines linearly with time (Meyer et al., 1974). With models
of convection zones for other stars, eddy diffusivity might be estimated and lifetimes predicted, but again, this requires knowledge of
supergranule sizes.

4.4. HEIGHT DEPENDENCE OF HETEROGENEITIES

It is a general rule that the degree of heterogeneity becomes
more pronounced with increasing height. The reason is that as we
increase height we must make a conceptual transition from what is
usually called an ionized gas to what must be called a plasma (cf.
Alfvén and Arrhenius, 1973). As long as we concentrate on material
near the photosphere, the gas pressure exceeds the magnetic pressure,
and the degree of ionization is not too large. In these conditions,
cosmic electrodynamics can be studied fairly well with the "first
approach", i.e. with assumptions of homogeneous models, with infinite
conductivity, zero parallel electric fields, frozen-in field lines,
and neglecting instabilities. But at greater altitudes, specifically
at $h \gtrsim 1500$ km, where the chromosphere-corona interface becomes extremely rough, the gas pressure need no longer be large compared to
the magnetic pressure, and the ionization need not be small. In these
conditions, the "first approach" to cosmic electrodynamics must be
abandoned for it may lead to conclusions and conjectures totally divorced from reality. Alfvén and Arrhenius argue that a second approach to cosmic electrodynamics becomes necessary, in which the plasma must be allowed to have a complicated heterogeneous structure,
with electrical conductivity depending on the current, sometimes with
zero conductivity, with electric fields parallel to the magnetic
field lines, and with current lines and the electric circuit just as
important to consider as the magnetic field lines. These currents
automatically produce filamentary structure and flow in this sheets.
Many plasma configurations are unrealistic because they are subject
to one of at least 32 known plasma instabilities. And the frozen-in
field line picture is often completely misleading. Thus at great
altitudes we expect to find a totally heterogeneous solar atmosphere
in which the heterogeneities, rather than being simply minor perturbations superposed on a fairly homogeneous background, now become
the dominant constituents of the atmosphere. Thus, whereas in the
photosphere, faculae are small perturbations on a generally homogeneous sun, with increasing altitude, the faculae show up as plages
of ever increasing contrast. Moreover, the functional dependence of
contrast on magnetic field strengths becomes fundamentally different
as the altitude increases: active regions remain bright and spots
remain dark up through the chromosphere, but then at altitudes in
the transition region, this behavior reverses, and spots become the
brightest features in active regions (Foukal et al., 1974). This is

due to a widening of the transition zone, but how to scale this widening to other stars is not known.

Pictures of the corona in optical lines (showing prominences) and in X-rays show examples of the extremely heterogeneous nature of the solar atmosphere at these altitudes. In general, the heterogeneities in the corona follow the active regions although not all active region loops are filled with emitting material. There are also bright X-ray points associated with emerging flux regions all over the sun. In trying to scale these heterogeneities to other stars, we can make almost no progress at all. The physical mechanism responsible for hot coronal regions above active regions may involve dissipation of trapped Alfvén waves, as Wentzel (1974) has suggested. If this is so, then in order to scale to other stars, we need to know at least the scale sizes of the trapping magnetic field arches, the field strengths, the densities, and the period of the Alfvén waves. None of these data are currently available.

The most interesting large scale type of heterogeneity in the solar corona is a coronal hole, where densities and temperatures are reduced below the quiet coronal values as a result of enhanced solar wind flux along locally open magnetic field lines. These holes may extend more than $90°$ in latitude, and some tens of degrees in longitude, and their most remarkable feature is that they rotate essentially rigidly: the amount of differential rotation of a hole boundary is an order of magnitude less than the differential rotation of the photosphere and the chromosphere (Timothy et al., 1975). This observational fact may contain information about the internal rotation of the sun, but such information has not so far been unravelled in a completely unambiguous fashion. Other stars will also presumably have open field lines at certain parts of their surface, and so, coronal holes should also be a feature of stellar coronae, but what their physical characteristics will be is not yet clear.

4.5. FLARES

Finally, I would like to turn briefly to the transient heterogeneities called flares, for these are the examples par excellence of heterogeneities caused by plasma instabilities in the upper atmosphere of the sun. Flares were of course historically one of the first types of solar features to be observed in other stars. A solar flare involves conversion of magnetic field energy to thermal energy in the upper chromosphere by a process which may involve the mediation of rapid electric current dissipation, or large fluxes of Alfvén waves. The flare process involves complex plasma-field interactions. The process is such that it is reasonable to expect that flare plasma should have a beta (= gas pressure/magnetic pressure) of order unity (Moore and Datlowe, 1975). This is an important physical constraint, for if we have some way to determine gas densities and temperatures in the flare plasma (e.g. from X-ray data), then we can estimate the order of magnitude of the field strength in the chromosphere near the flare. Time scales for flare decay are determined in the sun mainly by conductive energy losses. It seems to be true that the conductive time scale is also the relevant physical parameter which we must scale in order to predict decay times of flares on other main sequence stars (Mullan, 1976).

The relation between flares and the missing flux in sunspots is not yet clear, but it certainly is energetically favorable to tap the reservoir of missing spot energy (wherever it is) and energize large solar flares (De Jager, 1968). Applying this idea to stars, we note that spotted stars and flare stars should be related: in fact the two groups are essentially identical. There is even a quantitative reproduction of rates of occurrence of flares as a function of amplitude by this model (Mullan, 1975).

5. Conclusion

I have tried to summarize certain aspects of heterogeneities in the solar atmosphere, stressing our ignorance in identifying at the present time the physical parameters which control the sizes and lifetimes of many of the heterogeneities. Thus we have no yet gotten enough information to provide a starting point in attempting to scale these heterogeneities to other stars. There is such a wealth of observational data currently available on heterogeneities in the solar atmosphere that at first sight it appears to provide a goldmine of valuable information from the point of view of interpreting observations of other stars. Unfortunately, in many cases the relevant physical parameters which determine the characteristics of solar features are not well known, and so the scaling from solar to stellar atmospheres is not on firm ground at the present time. Hopefully, in the next few years improved knowledge of the internal structure of the solar convection zone, and improvements in our knowledge of how small-scale magnetic flux tubes are formed, may help to put this scaling on a firmer basis.

References

ALFVÉN H. and ARRHENIUS G.: 1973, Astrophys. Space Sci., 21, 117

BECKERS, J.: 1976, Astrophys. J. 203, 739

BRUECKNER, G., and BARTOE, J.D.F.: 1976, paper presented at COSPAR meeting, Philadelphia.

DANIELSON, R. E. and SAVAGE, B. D.: 1968, Kiepenheuer (ed.) Structure and Development of Active Regions, Reidel Publ. Co., Dordrecht, p. 112

DeJAGER, C.: 1968, in K.O. Kiepenheuer (ed.) Structure and Development of Active Regions, Reidel Publ. Co., Dordrecht, p. 480.

DEUPREE, R.G.: 1976, Astrophys. J. 205, 286.

DIMITRIEVA, M.G., KOPECKY, M., and KUKLIN, G.V.: 1968, in K.O. Kiepenheuer (ed.), Structure and Development of Active Regions, Reidel Publ. Co., Dordrecht, p.174.

DRAVINS, D.: 1975, Astron, Astrophys. 43, 45.

FOUKAL, P.V., HUBER, M.C.E., NOYES, R.W., REEVES, E.M., SCHMAHL, E.J., TIMOTHY, J.G., VERNAZZA, J.E., and WITHBROE, G.L.: 1974, Astrophys. J. Letters 193, L143.

GILMAN, P.: 1974, Ann. Rev. Astron. Astrophys. 12, 47.

GOUGH, D.O., MOORE, D.R., SPIEGEL, E.A., and WEISS, N.O.: 1976, Astrophys. J. 206, 536.

LINDSEY, C., and HUDSON, H.S.: 1976, Astrophys. J. 203, 753.

MEYER, F., SCHMIDT, H.U., WEISS, N.O., and WILSON, P.R.: 1974, Monthly Notices Roy. Astron. Soc. 169, 35.

MOORE, R.L. and DATLOWE, D.W. 1975, Solar Phys. 43, 189.

MULLAN, D.J.: 1974, Astrophys. J. 187, 621.

MULLAN, D.J.: 1975, Astrophys. J. 200, 641.

MULLAN, D.J.: 1976, Astrophys. J. 207, 289.

NAGARAJAN, S.: 1971, in R. Howard (ed.), Solar Magnetic Fields, Reidel Publ. Co., Dordrecht, p. 487.

PARKER, E. N.: 1975, Astrophys. J. 198, 205.

SCHWARZSCHILD, M. 1975, Astrophys. J. 195, 137.

SPIEGEL, E.A.: 1966, Trans. IAU 12B, 539.

SPRUIT, H. C.: 1976, Solar Phys. (in press).

STIX, M.: 1976, Astron. Astrophys. 47, 243.

TIMOTHY, A. F., KRIEGER, A.S., and VAIANA, G.S.: 1975, Solar Phys. 42, 135.

VOGT, S.S.: 1975, Astrophys. J. 199, 418.

WENTZEL, D.G.: 1974, Solar Phys. 39, 129..

WILSON, P. R.: 1971, Solar Phys. 21, 101.

WORDEN, S.P.: 1975, Solar Phys. 45, 521.

THE HETEROGENEITY OF SURFACES OF MAGNETIC AP STARS

M. HACK

Osservatorio Astronomico
Trieste, Italy

Abstract. The observations of spectrum-variability and light-variability of Ap stars are reviewed. It is shown that these variations are interpretable as due to the changing aspect of the spotted surface as the star rotates. It is stressed that we understand fairly well the geometry of the phenomenon but the physics is very far from being understood.

1. Introduction

Magnetic Ap stars are probably those where the presence of a spotted surface is very evident. Their spectrum-variability (profiles, line-intensity and radial velocity), light-variability and magnetic field variability, all occurring with the same period, are explained in a simple way if we assume that these variations are due to the changing aspect of the spotted surface as the star rotates. The oblique rotator model was proposed by Babcock in 1949 and by Stibbs in 1950 and was worked out in great detail by Deutsch (1954). This model allows us to explain the magnetic field variation from some + 1000 to some − 1000 gauss in a few days; it explains the crossover effect, the line-width versus period relations, the line - intensity and radial velocity variation, and in part also the light curves. The main objection against the oblique rotator hypothesis was the supposed existence of many irregularly variable magnetic stars. However, the large number of observations accumulated in the last twenty years indicates that probably all magnetic Ap spectrum-variables are regular variables with periods which are generally of a few days, but includes a small group of long period variables (100 days up to 23 years for HD 9996). The light variability, which is the quantity measurable with the highest precision, has often remained undetected, because the amplitude is always small, in many cases few hundreths of magnitude.

Strong support to the oblique rotator theory is given by the relation v sin i versus P (Preston, 1971), where P is the period of variability. This relation indicates that all Ap stars fall on or below the curve rotational velocity $\underline{v}$ versus rotational period P, where

$v = 2\pi R/P$ and P is assumed to be equal to the period of variability, and none falls above it. Thus it is possible to compute $\underline{i}$ from the relation

$$\sin i = (v \sin i)/(2\pi R/P)$$

where $v \sin i$ is measured from the line broadening, R is given by the relation $L = 4\pi R^2 \sigma T_e^4$ and P is the period of spectral or light variability. The knowledge of $\underline{i}$ allows us to construct models of the spotted surface which are able to reproduce the observations. It is very important to map the surface spots and derive their physical parameters in order to ascertain in which degree the determination of the surface chemical composition is affected. The Mn-stars are not included in these considerations; though they show several abundance peculiarities similar to other groups of Ap stars, none of them has been found variable with certainty.

2. Spectrum-Variability

The complex variable profile of several spectral lines indicates that each line is split into several components having different intensity and radial velocity, each of them representing a different spot. A general behavior is the phase relation between velocity and intensity variation, with Wλ reaching its maximum value when the RV is zero (relatively to the velocity of the stellar center of mass). This means that when the spot crosses the center of the disk its Wλ is maximum and the corresponding RV is zero; when the spot approaches the limb, projection effects and limb darkening produce a diminution of Wλ. This correlation between line - intensity and RV variation was not always evident in the old works, because the low spectral resolution did not permit the separation of the different components. The importance of using high resolution spectrograms in studying these problems must be stressed here.

A rough determination of the latitude of the spots can be made through the RV curve of each line component. The RV of a spot on the parallel of latitude β varies from $-v \sin i \cos \beta$ to $+v \sin i \cos \beta$. Hence the amplitude of the sinusoidal curve $v \sin i \cos \beta \sin \emptyset$ where $\emptyset$ is the phase ($\emptyset = 270°$ at the approaching limb, $\emptyset = 0$ at the center of the disk, $\emptyset = 90°$ at the receding limb) permits the determination of β. Moreover, the visibility of a spot will last longer if it is on the hemisphere where the pole is visible. An exemplary application is given by the observations of HD 133029 by Aslanov (1975).

Generally certain groups of elements are concentrated in the same spots, and other elements are diffused over the whole surface or concentrated in other spots, but we cannot recognize a general behavior in the separation of the elements. For instance in some cases iron peak elements present a different distribution than rare earths, in other cases they present the same behavior; sometimes members of the iron peak behave like some rare earths or heavy elements, and other members of the iron peak behave differently. This intriguing behavior was observed by Deutsch in 1947; more detailed observations during the past thirty years have not been able to clarify the situation.

Two methods for mapping the distribution of spots on the stellar surface have been employed: a mathematical one developed by Deutsch (1957, 1970), consisting of a harmonic analysis of the spectrum, and an empirical one develped by Khokhlova (1975). Applications of these two methods have been made by Pyper (1969), by Rice (1970), by Megessier (1974, 1975), and by several researchers of Shemakha Astrophysical Observatory, Azerbaidzhan, USSR, and of the Zentral Institut der Deutschen Akademie der Wissenschaften, DDR, who have started an extensive and systematic research program making use of spectroscopic and photoelectric observations (Aslanov et al.,1973; Ryabchkova, 1974; Glagolevskii et al., 1974).

3. Light Curves and Spectrum-Variability

The light variability always occurs with the same period as spectrum and magnetic field variability. Generally maximum light (in V or in UBV system) is in phase with maximum intensity of Eu and other rare earth lines. In hot Bp, generally, V is in antiphase with He line intensity and in phase with Si line intensity. We cannot say if this is a general behavior because no sufficient data are available. From the existing observations it follows that rare-earth spots are generally hotter than the photospheric background while He spots seem to be cooler than the photospheric background. This fact, at least in the case of rare earths, can be explained with the blanketing in the far ultraviolet due to the crowding of a large number of lines of rare earth ions. This blanketing produces a backwarming effect in the visible part of the spectrum, where the number of rare earth lines is less numerous, as shown by the OAO-2 observations of several Ap stars (Molnar, 1973, 1975, Leckrone, 1974).

As an example of the importance that mapping the surface of Ap stars can have on empirical tests of the theories proposed to explain their abundance peculiarities, let us consider the case of HD 193722 (Aslanov et al., 1973). The UBV light curves are in phase with the Eu II lines and this can be explained by blanketing in the ultraviolet and backwarming in the visible. Spots of Si II (Si-rich) are located on a normal background because silicon is strongly in excess. In the same spots there is also a He-concentration, (He-rich), located on a He-deficient background, because the total intensity of helium lines indicates strong He-deficiency. Hence, at least in this star, we have He and Si excess in the same spots. According to the diffusion theory (Michaud, 1970) we expect, instead, He-deficiency and Si-excess in the same spot. We must consider the possibility that the temperature may be different in the spot from that in the surrounding photosphere and can affect the determination of the abundance.

Several researchers have systematically observed light curves of Ap spectrum-variables (see, for instance, Provin, 1953a, 1953b; Rakos, 1962a, 1962b, 1963; Stepien, 1968a, 1968b, 1968c; Wolff and

Wolff, 1970, 1971). The majority of these observations have been made in the UBV system. A complete review of the photometric properties of the Ap stars was made by Preston in 1971. The results are confusing because no general behavior can be identified. After this review came the beautiful results from OAO-2 ultraviolet observations proving that blanketing and backwarming effects, already suspected by Peterson (1970) and by Wolff and Wolff (1971), are the main reason for the variation of amplitude and shape of the light curve with wavelength. Now the recent series of 10-color photometric observations (from 3400 Å to 7600 Å) of light curves made by Schöneich et al.(1976 a, 1976 b) and by Schöneich and Staude (1976) relative to 16 Ap stars, show that each star is a singular case: an enormous variety of light curves is observed. However, a general behavior can be recognized: the amplitude decreases suddenly at the Balmer limit and rises again in the red. According to these authors, backwarming is not sufficient to explain these curves. The presence of various spots with different concentrations of elements producing different effects of blanketing and backwarming, and possibly a different atmospheric structure in the spots, due to the presence of the magnetic field, might explain the observations. But different combinations of the many variables of the problem can reproduce the observations: it is difficult or impossible to find a unique solution.

Conclusion

The geometric explanation of spectrum-variability of magnetic stars is very satisfactory; that of the light curves is less satisfactory, although ultraviolet observations have given an important contribution for explaining the phenomenon. However, we do not understand the physics of the problem. Why some elements should concentrate in some spots and others in other spots? And, if this behavior is correlated with the magnetic properties of the elements, why do they not always follow the same behavior, and are they not always associated in the same group?

How stable are the spots? Observations repeated at a distance of months or years agree well generally, suggesting that the life of these spots is much longer than that of the solar spots. However, the existing observations made at a distance of several tens of years very often disagree. It is not clear if the disagreement is entirely real or partly due to the different spectral resolution, which in the old times was too low to allow us to resolve the various components of the spectral lines.

We do not know how the magnetic field affects the atmospheric structure within the spots. Our attempts to understand the phenomena characterizing the magnetic Ap stars is often based on the solar model, however we must recall that Ap stars have general fields of hundreds or thousands of gauss against the solar one, which is less than one gauss: and that an Ap star has little or no convection in the subphotospheric regions while the sun has a thick convective zone.

We still need high quality observations consisting of high resolution spectrograms, magnetic field measurements, light curves in several colors, from far ultraviolet to infrared, and to repeat these observations for several years at least for a few typical magnetic stars. On the other hand, we need theoretical work in order to understand the behavior of the various elements in the presence of the magnetic field and its effect on the atmospheric structure.

References

ASLANOV I.A., HILDEBRANDT G., KHOKHLOVA V. L., and SCHÖNEICH W:
    1973 Astrophys. Space Sci. 21,477
ASLANOV I.A.:
    1975 Pis'ma Astron.Zh. 1,39 (Sov. Astron.Lett.1,64)
BABCOCK H. W.:
    1949 Observatory 69,191
DEUTSCH A. J.:
    1947 Astroph.J. 105,283
    1954 Trans. IAU 8,801
    1957 IAU Symp. 6,209
    1970 Astroph.J. 159,985
GLAGOLEVSKII YU.V., KOZLOVA K.I., and POLOSUKHINA N.S.:
    1974 Astrofizika 10,517 (Astrophysics 10,327)
KHOKHLOVA V. L.:
    1975 Astron.Zh. 52,950 (Sov.Astron. 19,576)
LECKRONE D.S.:
    1974 Astroph.J. 190,319
MEGESSIER C.:
    1974 Astron, Astroph. 34,53
    1975 Astron, Astroph. 39,263
MICHAUD G.:
    1970 Astroph.J. 160,641
MOLNAR M.R.:
    1973 Astroph.J. 179,527
    1975 Astroph.J. 80,137
PETERSON D.M.:
    1970 Astroph.J. 161,685
PRESTON G. W.:
    1971 Publ.Astron.Soc. Pacific 83,571
PROVIN S.S.:
    1953 a) Astroph.J. 117,21
    1953 b) Astroph.J. 118,489
PYPER D. M.:
    1969 Astroph. J.Suppl. 18,347
RAKOSCH K.D.:
    1962 a) Lowell Bull. 5,227
    1962 b) Z. Astrophys. 56,153
    1963   Lowell Bull. 6,91

RICE J. B.:
    1970 Astron.Astroph. 9,189
RYABCHIKOVA T.A.:
    1974 Astron.Zh. 51,761 (Sov.Astron.18,451)
SCHÖNEICH W., HILDEBRANDT G., and FÜRTIG W.:
    1976 a) Astron.Nachr. 297,39
SCHÖNEICH W., KRIVOSHEINA A.A., KHOKHLOVA V.L., and ASLANOV I.A.
    1976 b) Astron.Nachr. 297,207
SCHÖNEICH W., and STAUDE J.:
    1976 in press
STEPIEN K.:
    1968 a)Astron.J. 73, S 36
    1968 b)Astroph.J. 153,165
    1968 c)Astroph.J. 154,945
STIBBS D.W.N.
    1950 Monthly Notices Roy. Astron. Soc. 110,395
WOLFF, S.C. and WOLFF R. J.:
    1970 Astroph. J. 160,1049
    1971 Astron.J. 76,422

STARSPOTS ON BY DRA-TYPE STARS

D.S. EVANS
University of Texas, Austin, Texas

The first detection of spots on dwarf M stars was by Kron on YY Geminorum in 1948. Since then phenomena attributable to starspots have been found by numerous authors beginning with Kraft and Krzeminski on the star now known as BY Draconis and on CC Eridani by Evans. There is a close correlation between spottedness, Balmer emission and flaring in dwarf M stars. Evans appears to have been the first to attempt to analyse the small range light variations on spotted stars in terms of modulation caused by axial rotation. In comparison with the solar case one must assume that the spots are very large and long lived. There is also an ambiguity in dealing with very large dark spots as between a star with a dark area or a cool star with a bright area (the zebra effect-- is a zebra a black animal with white stripes or a white animal with black stripes?). It is difficult to arrive at a solution for the spot parameters which is unique. In particular several analyses which have appeared have assumed that maximum light corresponds to the immaculate star whereas in fact the maximum light in the variation corresponds to the immaculate star dimmed by the light of any constant star abstraction.

In the case of BY Draconis for which the orbit of Bopp and Evans gives an inclination of 30^0 approximately it seems possible to arrive at a definitive solution. In a paper by Oskanyan, Evans, Lacy and McMillan a long series of photometric observations made at Byurakan Observatory in Soviet Armenia and at McDonald is analysed. These can all be represented by small range sinusoidal variations and the following plausible assumptions can be made. (a) That since the light curve shows no level portion the spot lies wholly in the polar area always exposed to view and that no part of it lies in the equatorial zone where it would sometimes be exposed and sometimes hidden at each revolution. (b) That a sinusoidal variation through a range of a few hundreths of a magnitude is most economically explained as due to a single active area. The data string is best represented by a series of different sinusoidal trains with parameter changes between trains. A simple mathematical treatment relates amplitude of variation to depression of the mean level as a function of latitude of the spot. The results depend on the assumption that all the spots observed have been on the brighter component (only a single periodicity has been found) but are not dependent on the assumed luminosity ratio between the components. We infer that spots occur in high latitudes, even in the extreme polar region. The phase jump of 2.1 days found by Krzeminski (approximately half the rotation period of 3.9 days) is then readily explained as due to migration of the spot over the polar region. The spot has migrated both in latitude and longitude and there is no evidence of a systematic change of period with latitude. Consideration of all data covering the last 20 years suggests that a large dark spot appeared in 1965, migrated over the pole in 1968, and was replaced by a bright area in 1970 (when the value of B-V showed a reduction). This bright area is not due to line emission which in recent years has been weak.

STAR SPOTS ON AR LAC TYPE STARS

D. M. POPPER

University of California
Los Angeles
USA

There is no generally accepted definition of AR Lac Stars, and the term RS CVn stars is used interchangeably or to refer to a particular subgroup. For the purposes of this discussion I use the term AR Lac stars to refer to detached close binaries showing Ca II emission in at least the cooler component outside eclipse, the hotter component being a main-sequence or subgiant star of spectral type F or G. Most of the systems show irregularities in their light curves as well as period changes. In order to determine whether a system is detached, one must know both the mass ratio and the relative radii. The determination of minimum masses is a fairly straightforward spectroscopic task, and provisional values are available for 22 of the systems, two or possibly three of them being non-eclipsing. All but 3 (AD Cap, RT Lac, RV Lib) have masses of the two components within 30% of each other. Because of appreciable irregularities in the light curves, the radii are subject to considerable uncertainty even when photometry of good precision is available. Nevertheless the 9 systems with very provisional radii all appear to be detached. These all have mass ratios near unity. We may assume, as a working hypothesis, that the other systems with mass ratios near unity are also detached and hence also belong in the AR Lac group. Most of the data referred to are to be found in IBVS 1083.

In a presentation before Commission 42 I will discuss the mass-radius and color-luminosity relations for these stars and suggest that their present condition may be ascribed to normal stellar evolutionary processes combined with a moderate amount of mass exchange, perhaps through the medium of a stellar wind enhanced in the binary system. Evidence for the flow of matter in these detached systems is provided by the appreciable period variations in a number of them and possibly by the radio flaring observed in the brightest systems. These systems are distinguished from non emission systems of similar mass by having more evolved components. It appears that the stars tend to misbehave as they move into the Hertzsprung gap after they have depleted their core hydrogen fuel, have developed convective envelopes, and have started to evolve more rapidly.

So much for what I mean by AR Lac systems. What about starspots ? I have never published a statement or otherwise taken the position that starspots exist on these stars, and I do not know whether they exist. I don't know either why I didn't have enough sense to turn down the invitation to participate in this discussion. Potential evidence for the existence or non-existence of starspots must come primarily from photometric observations, and I am principally a spectroscopic observer. But since I am here, it behooves me to make some pertinent comments.

The potential evidence referred to lies in irregularities in the light curves of a number of these systems. There is only one set of irregularities known to me that seems to provide a measure of direct evidence for the presence of starspots in systems of this type. They are Kron's original observations of AR Lac itself, during the partial phases of primary eclipse obtained many years ago. The individual observations have been published only in graphical form. Some roughness of the light curve during these phases was attributed to the covering or uncovering of spots. To my knowledge these observations have not been repeated, nor has the effect been found in other systems. Observations of exceptionally high quality are required.

There is much more evidence on irregularities in the light curves of systems in this group outside eclipse, and I suppose it is these observations that the organizers of this discussion had in mind. By far the best studied of the systems photometrically is RS CVn, principally by virtue of the work of Catalano, Rodonó, and others at Catania. Three slides are shown of their famous investigations. There have been a number of discussions of the cause of the travelling wave in the light curve of RS SVn which I do not intend to review here. There have been suggestions of such waves in a few other systems, but its presence in none of them has been as clearly established as for RS SVn. In that system the wave has been followed pretty closely for a full cycle. Will it suffer the fate of other apparently regular phenomena in variable-star astronomy and fail to perform according to prediction in the future ?

Next I would like to show a sampling of partial light curves obtained at Kitt Peak in 1972 and 1973. These are all systems, as is RS, CVn, having total eclipses. It should be emphasized that ellipticity and reflection effects should be negligible in all of them. LX Per appears to be a well behaved system at the observed epoch, but it is not distinguishable in its fundamental properties (mass, radius, state of evolution, etc.) from the other systems. Fluctuations in the light of RU Cnc, RW UMa and UX Com (Figures 1 and 2) are shown.

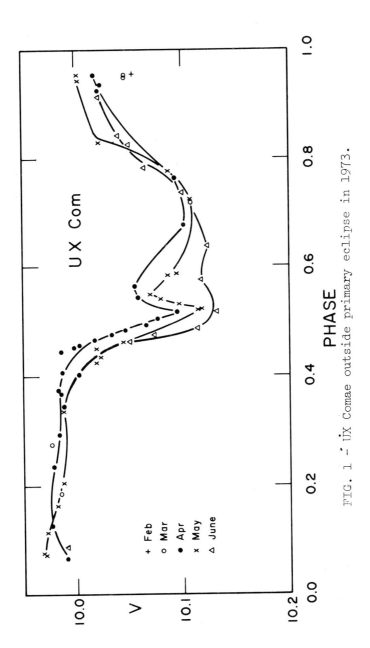

FIG. 1 - UX Comae outside primary eclipse in 1973.

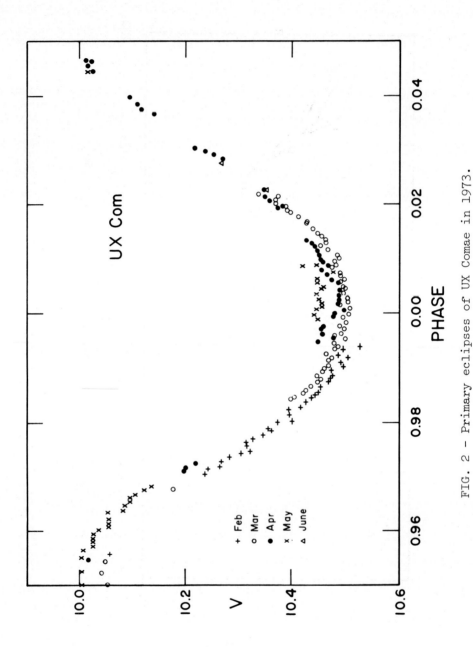

FIG. 2 – Primary eclipses of UX Comae in 1973.

The ranges of the fluctuations in units of the light of the larger, cooler star are listed.

RU Cnc	UX Com	RS CVn	RW UMa
15: %	15: %	30: %	30: %

The general form of these fluctuations in a given system persists for at least a few orbital periods, so that the fluctuations have roughly the orbital period and presumably the period of axial rotation as well. What evidence exists is in favor of equal orbital and axial periods.

To obtain some clue as to the origin of these variations we may examine color-magnitude plots for the light outside primary eclipse. In each of these four systems from my own observations (Figure 3) the system is redder when it is more luminous. This result points very strongly to the redder, larger star as the source of the variations in each system. In AR Lac itself the fluctuations are reported to be associated mainly with the hotter component. Each system needs to be observed carefully. Evidence in favor of the cooler components of the four systems as the sources of the fluctuations outside primary eclipses comes from the equality of the fluctuations in B and in V light when they are expressed as fractions of the light of the cooler star. Independent though less strong evidence (Figure 3) comes from the similarity in the color-luminosity relation for points outside eclipses and inside secondary eclipse. Since the two stars in a given system differ in color by only about 0.5 mag, since the light of the larger star is diluted by that of the hotter, and since the fluctuations are 15-30% of the star, moderate changes in the color of the fluctuating star itself would cause second-order effects, detectable only with observations of extremely high precision.

I can think of three potential ways of changing the contribution of the light from the cooler star in a reasonably neutral way: dark spots, obscuration by dust or gas, and pulsation. One can construct *ad hoc* models of obscuring clouds and of drifting spot groups, but the clouds would appear to be very difficult to account for dynamically, and spot groups giving up to 30% light variation may be just as difficult to support on a physical basis. With respect to pulsation, it has been usually ruled out because of the color variations expected. But large irregular light variations accompanied by only small color variations are found among variable stars (RV Tau, SR classes), perhaps caused by non-radial oscillations in more than one mode. Is it possible that we see counterparts on a smaller scale in our stars, with the oscillations excited in some way by aspects of the binary system ? I hasten to add that I do not support any particular explanation but rather

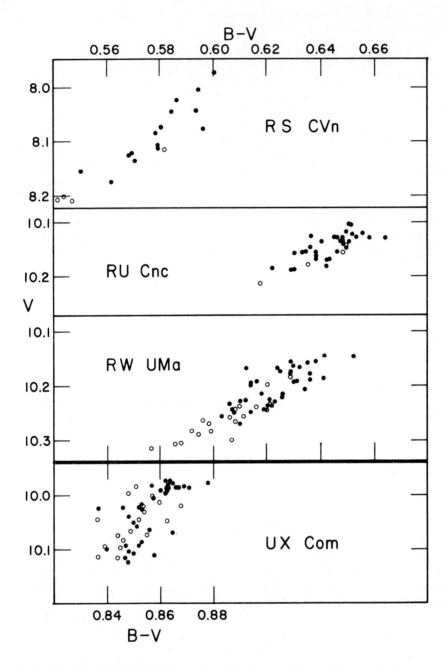

FIG. 3 – Correlation between magnitude and color index for AR Lac systems. Filled dots: outside eclipses; open circles: within secondary eclipses.

point out that oscillations provide a potential explanation perhaps worth investigating. I am indebted to R. K. Ulrich for discussions on this point.

Additional observations are, of course, essential. Several of these systems need to be followed photometrically over a wide range of wavelengths. Some work by D. S. Hall and E. F. Milone has indicated modest infrared excesses. These need to be confirmed and extended. Variations in light as large as 30 % should show in the relative intensities of spectrum lines. The kind of effect reported by Kron for AR Lac needs to be looked for.

In conclusion, I repeat that I do not know whether starspots are present on AR Lac stars.

DIRECT OBSERVATIONS OF THE HETEROGENEITY OF SUPERGIANT DISKS

J. W. HARVEY, C. R. LYNDS
Kitt Peak National Observatory
S. P. WORDEN
Sacramento Peak Observatory

Resolved images of the disks of the largest stars observed with the largest telescopes can be constructed using the class of techniques called speckle imaging. The observations must be made with narrow passbands ($\sim$ 10 nm), short exposures ($\sim$ 20 ms) compensation for atmospheric dispersion, high magnification and good signal-to-noise ratio. One specific technique applied to α Ori (Lynds et al.,1976) shows slight but apparently real differences in the images of the disk corresponding to low and high opacity in the stellar atmosphere which we interpret as due to temperature differences. There are also significant differences in the star's diameter and/or limb darkening at the two different opacity wavelengths.

The stellar disk images can be significantly sharpened using a technique developed by McDonnell and Bates (1976). An improved image of α Ori shows a dark feature on the disk which does not seem to be associated with a temperature difference. In addition, a diameter of $0\overset{''}{.}066 \pm 0\overset{''}{.}006$ with a limb darkening coefficient of 0.6 ± 0.3 could be deduced by McDonnell and Bates.

More general speckle imaging techniques have been proposed (e.g. Nisenson et al., 1976) and offer the promise of improved imaging of stellar disks over a wide range of wavelengths.

References

LYNDS, C. R., WORDEN, S. P., and HARVEY, J. W.: 1976, Astrophys. J. <u>207</u>, 174.
McDONNELL, M. J., and BATES, R. H. T.: 1976, Astrophys. J. <u>208</u>, 443.
NISENSON P., ELM, D. C., and STACHNIK, R. V.: 1976, Proc. SPIE <u>75</u>, 83.

ON THE SPOTTEDNESS AND MAGNETIC FIELD OF T TAU-TYPE STARS

R. E. GERSHBERG

Crimean Astrophysical Observatory

There are several observations that suggest the existence of surface heterogeneities on the T Tau-type stars. Firstly Dr. Ismailov from Shemakha has found strong variations of emission line profiles for 5 - 10 minutes: 2-component profiles become 3 or 1 component and vice-versa. The results were confirmed recently in the Sternberg Institute. Secondly, variations of polarization degree and relative intensities in the IR calcium triplet were found when the stellar brightness were being constant. Then, the relative intensities of the Ca triplet are not equal to what must be expected for both optically thin or thick but homogeneous emission regions. On the basis of the data Petrov and Sheberbakov from Crimean Observatory have proposed that T Tau stars have active regions with strong magnetic fields similar to stellar ones: they can give noticeable heterogeneities for calcium emission, for polarization effects in dust envelopes and local processes for profile variations. The existence of rather strong and local magnetic fields on T Tauri stars may give a key for explaining the mysteries of these variables. Small variations of strong magnetic fields on stars with a convective zone can give a new mechanism for stellar variability. As it is known, only a quarter of the energy flux that reaches the bottom of the convective zone, is emitted from a sunspot in form of light and the main part of the energy escapes in form of hydromagnetic waves. Then, the sunspot exists if the field is not less than 1 kilogauss. Therefore the field acts as a relay: if the field is low we have light, if the field is strong we have mainly hydromagnetic waves. The transition from low to high field may occupy only a small part of the whole field range. The field variations on T Tau stars can give strong brightness variations **without** additional energy sources but only by magnetic relay operation. I hope that this scheme gives a possibility to explain a set of features of T Tau stars (such as their high ratio of chromospheric to photospheric radiation, IR excesses, the nature of their flares) as well as main features of FU Ori-type flares.

The full paper by Mr. Petrov and myself will be published in October in Astron.Zh. Letters.